CONSTRUCTION MANUAL:

ROUGH CARPENTRY

By T.W. Love

Craftsman Book Company
6058 Corte del Cedro, Box 6500, Carlsbad, CA 92008

Library of Congress Cataloging in Publication Data

Love, T   W
      Construction manual, rough carpentry

      Includes index.
      1. Carpentry--Handbooks, manuals, etc.
I. Title.   II. Title: Rough carpentry.
TH5608.L69   694'.2   76-21704
ISBN 0-910460-18-3

© 1976 Craftsman Book Company
Thirteenth printing 1991

Contents

Chapter 1
Selecting Wood Products

As a carpenter you should have a reasonable understanding of the material you use every day. Under ordinary circumstances you need not have a comprehensive understanding of such phases of the wood industry as lumbering, milling and the scientific aspects of tree growth. However, it is quite necessary that you understand those phases that deal with the selection, use, and care of lumber.

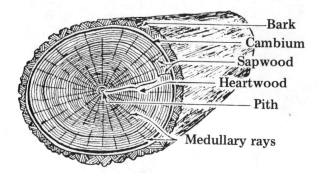

Cross section of a tree
Figure 1-1

Description Of Wood Structure

Trees are generally divided into the *hardwoods* or broad-leaved trees and *softwoods* or needle-leaved trees. Various species of the latter are used for rough carpentry.

The pith in the center of the tree (Figure 1-1) is the growth of the first year. A series of concentric annual rings surrounding the pith are formed by the growth of the tree, one ring being added on the outside each year. When these rings are large, due to fast growth, the grain is coarse. When the rings are narrow, as in hardwoods, the grain is said to be fine.

The cambium is a spongy layer just beneath the bark. It conveys nourishment for the tree through the trunk.

The medullary rays are fibers or cells that run from the pith to the bark of the tree and convey nourishment from the outer cambium layer to the inner part of the tree.

The heartwood is that part of the tree that is near the pith and consists of annual rings whose

cells have become hard and dark in color. This wood forms the bulk of the tree and merely serves as a support. It provides the best lumber. The sapwood is the section between the cambium layer and the heartwood. It is not as desirable as heartwood because it is more subject to shrinking and warping since it dries more because of its large cells.

The annual rings (Figure 1-2) are made of fibers or long tubes, running parallel to the trunk of the tree. These tubes are crossed by the medullary rays. Lumber is said to be green when the tubes or cells contain moisture, and dry when the cells have collapsed and have been drained of moisture.

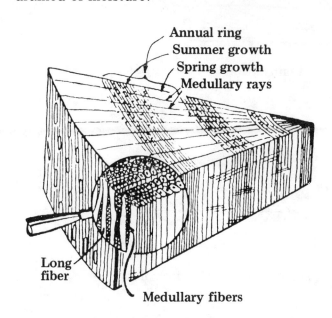

Enlarged section
Figure 1-2

Sawing Of Lumber

The method of sawing a log has a direct bearing on its durability and quality, and on its ability to resist wear and hold its shape.

Plain sawed lumber (sometimes called slash or flat sawed) is cut from the log as shown in Figure 1-3. The log is first squared by sawing boards off the outside, leaving a rectangular

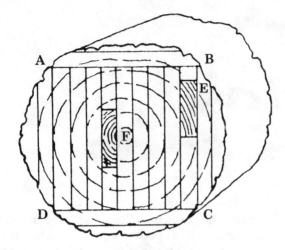

Plain sawed log
Figure 1-3

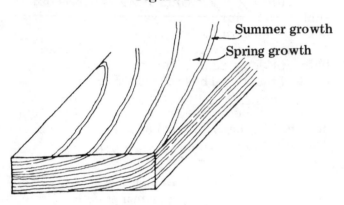

Summer growth
Spring growth

Plain sawed outer edge of log
Figure 1-4

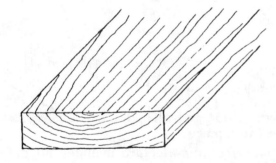

Plain sawed inner section of log
Figure 1-5

that will not warp as readily as that sawed by other methods. Figure 1-7 shows a board cut in this manner. Since the annual rings are perpendicular to the surface of the board in this type of sawing, the stresses caused by the drying of the lumber will be equal in all parts of the width of the board. Furthermore, since the most shrinkage takes place in the direction parallel to the annual rings, the shrinkage in the thickness of this board is proportionally greater than in the width. Sections taken from the log as shown at A, Figure 1-8 present very short and uniform segments of the annual rings. Sections taken from B, Figure 1-8 present long rings over the width of the board.

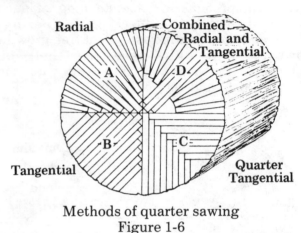

Radial   Combined Radial and Tangential

Tangential   Quarter Tangential

Methods of quarter sawing
Figure 1-6

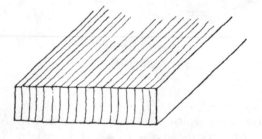

Radial sawed
Figure 1-7

section A, B, C, D, which is then cut up as shown by the parallel lines. This is a common method of cutting framing lumber. A board cut out of section E in this manner is shown in Figure 1-4. A plain sawed board from the inner section (F) of the log is shown in Figure 1-5.

The methods of sawing shown in Figure 1-6 produce lumber of higher quality in some respects, but they also cause more waste in sawing. The radial method shown at A is perhaps the best because it produces lumber

The tangential cut (B, Figure 1-6 and Figure 1-9) is used to accomplish approximately the same results as the radial cut but the tangential cut is a more economical method of sawing.

The quarter tangential cut (C, Figure 1-6 and Figure 1-10) is an even more economical method and, except for plain sawing, is the one most commonly used for framing lumber.

Because of the large amount of waste, the method of sawing shown at D, Figure 1-6 is used only for cutting large timbers. The pieces left over can then be sawed into smaller pieces.

The lumber shown in Figures 1-7, 1-9 and 1-10 is called quarter sawed, riff sawed or edge grain. These methods are most frequently used

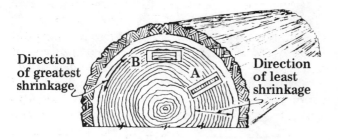

Direction of greatest shrinkage

Direction of least shrinkage

Shrinkage of wood
Figure 1-8

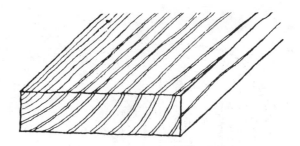

Tangential sawed
Figure 1-9

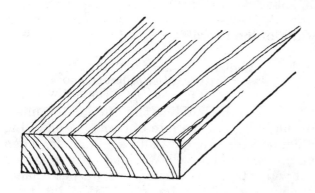

Quarter tangential sawed
Figure 1-10

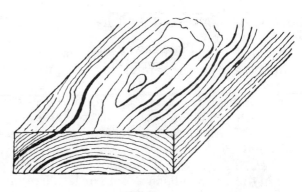

Shake
Figure 1-11

in hardwoods to bring out the beauty of the grain, but this factor need not be considered in softwoods used for framing.

Surfacing Of Framing Lumber

Manufactured framing lumber is classified as *rough* lumber (lumber as it comes from the saw) and *surfaced* lumber (lumber that has been dressed by running it through a planer). It may be surfaced on one side (S1S), two sides (S2S), one edge (S1E), two edges (S2E) or other combinations of sides and edges. Most framing lumber is S4S.

Defects In Lumber

In the grading of softwood yard lumber, it is necessary to classify and define the defects and blemishes which may occur. Some of the commonly recognized defects and blemishes as given by the American Lumber Standards are:

A *defect* is any irregularity occurring in or on wood that may lower some of its strength, durability or utility value.

A *blemish* is anything, not classified as a defect, marring the *appearance* of the wood.

A *bark pocket* is a patch of bark partially or wholly enclosed in the wood.

Bird's eye is a small central spot with the wood fibers arranged around it in the form of an ellipse, so as to give the appearance of an eye. A

bird's eye, unless unsound or hollow, is not considered a defect.

A *cross break* is a separation of the wood cells across the grain, such as may be due to tension resulting from unequal shrinkage or mechanical stresses.

Decay is a disintegration of the wood substance due to the action of wood-destroying fungi. The words *dote* and *rot* mean the same as decay.

A *gum spot* or *streak* is an accumulation of gumlike substance occurring as a small patch or streak in a piece.

Imperfect manufacture includes all defects or blemishes which are produced in manufacturing, such as chipped grain, loosened grain, raised grain, torn grain, skips in dressing, hit and miss, variation in sawing, miscut lumber, machine burn, machine gouge, mismatching, and insufficient tongue or groove.

A *knot* in lumber is caused by the growth of a branch, the inner end of which is embedded in the main stem of the tree. The location of a knot in a board may seriously affect the structural strength of the board. However, if the knot is solid and small, it may not do any particular harm.

A *shake* is the separation of the wood between the annual rings running lengthwise in

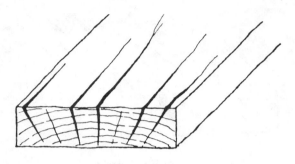

Checks
Figure 1-12

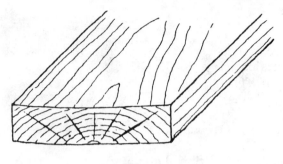

Warp
Figure 1-13

the board (Figure 1-11). This defect greatly weakens the board.

A *check* is a lengthwise separation along the grain and across the annual rings (Figure 1-12). Checks are commonly seen on the ends of lumber and are caused by too rapid and uneven drying. They not only weaken the lumber but make it difficult to nail, as the nailing may cause additional splitting.

Pitch is a poorly defined accumulation of resin in the wood cells in a more or less irregular patch.

A *pitch pocket* is a well defined opening between rings of annual growth, usually containing pitch, either solid or liquid. Bark also may be present in the pocket.

A *pitch seam* is a shake or check which is filled with pitch.

A *pitch streak* is a well defined accumulation of pitch in a more or less regular streak.

Warp is a bending of the lumber from a flat plane (Figure 1-13). As a board drys, it shrinks more along the long annual rings than along the short ones. Therefore. the board will tend to assume the shape shown in Figure 1-13. Note that the cupped face is on the same side as the longest annual rings.

Dry rot in lumber is caused by a fungus. The term dry rot is misleading as it only occurs in the presence of moisture where the free circulation of air is prevented. Wet or green lumber used in a building, and so enclosed as to partially cut off air circulation, is very likely to be affected by dry rot. The fungi are sometimes found in dry wood because they can draw their needed moisture long distances. Wood in the advanced stages of dry rot is shrunken, discolored, brittle and powdery.

A *split* is a lengthwise separation of the wood due to the tearing apart of wood cells.

Stain is a discoloration, occurring on or in lumber, of any color other than the natural color of the piece on which it appears.

Wane is bark, or the lack of wood or bark, from any cause, on the edge or corner of a piece of lumber.

Care Of Lumber

The care that lumber is given after being delivered to the building site often determines, to a large extent, the rate of depreciation of the building. Green or partially dry lumber, when not properly piled, will twist and warp in drying and will retain this twisted and warped shape. Lumber should *not* be stored in tight piles without some type of protection. If lumber is not to be used for several days or a week, it should be unloaded on skids with a 6-inch clearance above the soil. The pile should then be covered with waterproof paper, canvas, or polyethylene so that it sheds water. However, the cover should allow air to circulate and should not enclose the pile to the groundline. In a tight enclosure, moisture from the ground may affect the moisture content of lumber. The use of a polyethylene cover over the ground before lumber is piled will reduce moisture rise. The same type of protection should be given to sheathing grade plywood.

Lumber that has been used for concrete forms, scaffolds and staging should be properly cleaned and inspected for defects caused by rough handling before it is used for permanent parts of the building.

Different Kinds Of Wood

You should be careful to select the proper kind of wood for the type of work you are doing. Different kinds of wood, such as Douglas fir, hemlock, several kinds of pine, western red cedar, cypress, and redwood, have different characteristics. One wood has great strength or the ability to support loads. Another wood may have these qualities in a lesser degree but is far superior in its ability to resist the weather or, in other words, is more satisfactory for outside use. Another wood will take finish better and have a finer appearance when used for interior purposes. Some of the more important varieties of timber will now be mentioned with a brief description of each to convey an idea of its characteristics and uses.

Douglas fir: Western Oregon and Washington is called the Douglas fir region of the United States. This wood is adapted to more different uses than any other kind. It is practically all heartwood, which is more durable than sapwood, is free from blue stain, and has little tendency to warp or twist. Its large heartwood content makes it second only in decay resistance to the most durable cedars, cypresses and redwoods, giving it equal rank with white oak and long leaf pine. In color the wood varies with conditions of growth, from yellow to a reddish brown.

Douglas fir is made into all forms of dimension sizes for house construction. It is used not only for framing and sheathing but for floors, doors, finish, siding, studding, joists and lath. It meets every item in government requirements for framing lumber: high stiffness, good bending strength, good nail-holding power, hardness and durability.

West Coast hemlock grows only in the Pacific Northwest from Oregon to Alaska. It has an attractive, straight, even grain, formed of long tough fibers. It is free from pitch and gum, light in weight when dry, naturally light in color, highly paintable, a good nail-holder. It is an excellent wood for general construction because of its stiffness and strength-to-weight ratio. It is capable of either a high polish or a satiny finish. Machined West Coast hemlock is beautifully light and bright in its natural state. It not only gets harder with age but is one of the few woods that does not darken with age. No other softwood is so nearly like the hardwoods in color, texture, aging, and finishing qualities.

Sitka spruce, largest of any spruces in the world, grows along the coast of the Pacific Northwest from northern California to Alaska. The wood is exceptionally light in weight, of moderate hardness and, because of its long fibers, possesses great toughness and strength in proportion to its weight. These characteristics make Sitka spruce a preferred wood for scaffolding and ladders. Its key quality is resiliency, the ability to take shock and to bend and recover after carrying a load. In house construction it makes a good finish, although it does not have as distinctive an appearance as fir, cedar or western hemlock. It is a good wood for siding and an easy wood to paint. It is used for drainboards and heavy doors for garages, residences, clubs, warehouses and freight sheds.

On account of its small, tight knots, its freedom from pitch and odor, and its ability to take nails without splitting, Sitka spruce is widely used for built-up fixtures and for the best kinds of crating boxes and food containers.

Western red cedar is the largest and finest of the cedars grown anywhere. It is often called "The Enduring Wood of the Ages" due to its ability to withstand decay. This wood grows principally in the moist regions of Washington and Oregon. No other wood is so light, so evenly grained, and so well able to resist decay. Its uniform structure and light cell walls provide a larger percentage of air spaces which give the wood exceptional insulating qualities.

Besides being the best wood for shingles, western red cedar is also an excellent wood for bevel siding. It has little tendency to warp, buckle, shrink, swell or check. It possesses an attractive grain and fixture for natural finishes and also takes paint and enamel very well. It is extensively used for both clear and knotty paneling interior as well as exterior finish, sash, doors and frames. It is particularly well adapted to use where mitered corners and tight joints are required. Its fiber permits it to be machined with remarkable precision. It is easy to work, takes a very smooth finish, cuts to sharp and true edges, fits accurately, and maintains its position and shape -- properties which make it particularly valuable for specialized uses such as patterns or moulds for metal castings. It is especially valuable for roof plank, for such structures as paper mills and textile plants where conditions of humidity are high, for boat and racing shell construction, and for trellises, garden furniture, greenhouses and equipment, posts and other items usually in contact with the soil or in damp locations.

Incense cedar lumber, one of the most durable and decay resistant of native American woods, is produced from a forest tree found in California, southern Oregon and western Nevada. The wood is non-resinous. The sapwood is white or cream-colored and the heartwood is light brown or light reddish-brown. Texture is fine and uniform with small, evenly arranged cells. The wood has a spicy odor characteristic of all cedars. Incense cedar has a low heat conducting property which makes it one of the finest wood insulators and leads to its use for sheathing, siding, floor and roof decking and other building uses where protection against heat and cold is important. The excellent decay resistance of the wood, coupled with its fine dimensional stability and high insulation quality, makes it one of the best siding, sheathing, and sub-flooring materials. Its light weight and easy workability lead to marked economics in handling on the job site, resulting in lower construction costs.

Its readiness to take paint and the smooth silken surfaces to which incense cedar is machined combine to form handsome exteriors. The wood may be painted or stained in any of the large variety of modern colors. The rich, brown-red color, small, sound knots and graceful grain combine to form in incense cedar an unusual and distinctive paneling and woodwork material. In its clear grades particularly, the wood has gained favor for interiors in the clean-lined modern style. A popular paneling article is "Pecky Cedar," manufactured from pieces containing small openings where the live tree had been attacked by parasites. The parasite dies and disappears when the tree is cut down, leaving no threat to the lumber or the home where it is used. The pleasant fragrance of incense cedar increases its value as closet lining.

Redwood grows extensively in California to heights of 275 to 300 feet, with diameters from 6 to 10 feet or more. Known chiefly for its durability, redwood lumber finds greatest use where this quality is important. A special grade is maintained for foundation and sub-structure use. Its exceptionally low shrinkage and high paint retention values have led to its widespread use for siding and exterior trim. It also has a wide industrial market, largely for tanks, vats, and mill roofs.

Red cypress is grown in the deep swamps of the coastal plains of the southeastern states, and along the Gulf of Mexico next to the tidewater lands. This wood is famous for its fine texture, beautiful figure and grain. Its natural color is neither too light nor too dark, therefore giving a very pleasing appearance for natural finishes.

The natural oil in red cypress prevents the growth of plant life which causes decay. Nature has done for this wood what man tries to do when he forces preservatives into wood to prevent decay. Red cypress is completely impregnated by nature with antiseptic or toxic oils. These oils not only prevent termites from destroying the wood but also preserve paint on cypress. Red cypress is highly resistant to a great many chemical solutions used in industrial plants. For this reason it is a preferred material for tanks, vats, pipes, troughs, and conveyors. It has long been used in corn products refineries, in breweries, in wineries and by others wishing to avoid contamination because it imparts neither color, taste nor odor to products coming in contact with it.

This wood is manufactured under the rules and specifications of the Southern Cypress Manufacturer's Association. Its grades and sizes conform with the American Lumber Standards.

Idaho white pine is grown in northern Idaho and the adjoining parts of Washington and Montana. The wood is light in color, distinctly straight-grained and soft. Though soft and light in weight, Idaho white pine lumber does not split easily and can safely be nailed right up to the ends of the pieces, a quality which makes the white pines a favorite with carpenters.

The clear wood is so much like that of northern white pine from New England and the Lake States that the two can hardly be distinguished. The knotted board grades are used for interior paneling, exterior trim, cornice lumber, shelving and siding, purposes for which it would be necessary to use select grades of most other softwoods.

In all of the soft pines, the board or common portion of the log is divided into five grades, instead of the three or four divisions made in the other kinds of softwoods. The names of these five grades in Idaho white pine, are "Colonial," "Sterling," "Standard," "Utility" and "Industrial." Colonial and Sterling grades both are characterized by small, tight, smooth knots and are used for flush sidings, paneling, and for other finish and similar purposes requiring a high quality board. Standard is the highest grade ever used for sheathing and subfloors. Utility grade gives very satisfactory service when used for these items.

New England homes of white pine, some of them now over three centuries old, give ample evidence to the long life of white pine construction. All of the white pines are, because of their soft, even texture, exceptionally resistant to weathering. Paint adheres to them firmly. The Colonial and Sterling grades, which have numerous smooth knots, make excellent siding, window frames, and exterior trim if the knots are sealed with aluminum house paint or shellac before the finish coats of paint are applied.

Ponderosa pine grows throughout the twelve western states, mostly at altitudes of 2,000 to 6,000 feet. Like the other western pines, ponderosa pine is always well seasoned before surfacing and shipment. Its close, even texture and remarkable toughness for such a lightweight wood has made ponderosa pine valuable for heavy duty flooring, grain chutes, and auto beds which are subject to heavy wear. The three select grades of ponderosa pine are known as "1 & 2 Clear" (B & Btr), "C Select" and "D Select." The five board grades are numbered from one to five. There are three grades of dimension which should never be confused with the board grades. Boards are graded chiefly on the basis of appearance, while for dimension lumber strength is more important. Clear cuttings from the "factory" grades are used in large amounts by manufacturers of millwork and other items.

For many years ponderosa pine has been the leading wood used for doors, sash, window frames, screens and most other stock millwork items. It is used for a majority of the softwood doors produced. Because of its light weight, its freedom from warping and twisting, and its excellent gluing qualities, it is used for the cores of most hardwood veneered doors. A carpenter can hang several more doors a day than he can doors of heavier woods.

Much of the yard lumber of this wood is used for rough carpentry items like sheathing, sub-flooring and roof boards. The knotted material which usually goes into the upper board grades makes fine shelving. Selected Number 2 or even Number 3 is much used for knotty pine paneling. The knotty boards leaving the high speed planing machines are surfaced so smoothly that they have almost a polished appearance and are easily machined accurately in a variety of patterns.

Sugar pine is the largest of all the pines, occasionally reaching a diameter of 12 feet and a height of 250 feet. It is not at all uncommon for lumber operators to cut trees six or seven feet thick and which are 75 feet or more from the ground to the first limb. It grows in the Sierra Nevada of California and parts of southern Oregon. There is ample virgin timber available to continue for over a hundred years at the present annual production. Long before that time many of the seedlings now growing will have reached commercial size.

Sugar pine, like Idaho white pine and northern white pine is classified as a true white pine. The wood cannot be distinguished with certainty from the other two genuine white pines. It is a beautiful creamy white, which darkens to a pale brown sometimes tinged with pink as it ages. It is light in weight and has a uniformly soft, even texture which makes the wood firm, yet exceptionally easy to cut smoothly in any direction. Its corky texture is a joy to pattern makers, wood carvers, and others who must do precise work with hand tools. The wood fibers seem to be unusually strong and tough for a wood which is so light in weight and which cuts so smoothly and easily.

Sugar pine has one of the very lowest and most uniform rates of shrinkage, and that is the main reason sugar pine piano keys, organ pipes, foundry patterns, sash and other fabricated products hold their size and shape so well. The thorough seasoning of sugar pine lumber before shipment also is a factor.

Since the trees are so large and the wood develops few defects in seasoning, sugar pine is unequaled for the production of wide and thick clear pine lumber. Selects up to 22 and 24 inches wide and 3, 4 or occasionally even 6 inches in thickness are regularly produced and thoroughly seasoned by the mills. Clear pieces up to 48 inches in width are obtainable for special purposes, though ordinarily it is preferable for users to glue up such widths with several pieces.

Among the important construction uses are doors, sash, window frames, millwork, drainboards, mouldings, exterior trim and siding. In most parts of the country, sugar pine, though not used much for stock millwork, is often specified by architects for special doors, sash, built-ups and enameled interior trim on high class jobs. In some localities, especially in southern California where high temperatures and low humidity put the wood to a severe test, sugar pine is generally used in most doors and sash. Sugar pine may be used for wall paneling in either knotty material or clear stock offering great decorative value.

Southern or yellow pine is principally produced in eleven Southern States. There are several varieties but only two are recognized of importance - the long leaf and the short leaf. Long leaf pine is superior in strength and other qualities to short leaf pine. Southern pine is unexcelled as a framing material. It is extensively used for framework, siding, flooring, sheathing, wall panels, sash, doors, rails, newel posts and every variety of dressed or turned exterior and interior finish in homes and commercial buildings. It is approved for all types of heavy construction such as warehouses, factories, office buildings, grandstands, speedways, oil derricks, beams, girders, posts, columns, joists, rafters, roof trusses, factory flooring, and more.

Chief characteristics of southern pine are its great strength, durability, hardness, toughness, stiffness, nail and screw-holding power and its varied and attractive texture, grain and natural figures. It takes stains, varnishes, hard oil and paint if no resin is present. Southern pine lumber is graded and classified with reference to its suitability for general use. Southern Pine Association grading rules conform to American Lumber Standards.

Arkansas soft pine is a trade name for a certain short leaf pine with qualities which distinguish it from short leaf pine in general.

These qualities are extraordinary soft texture, close fiber, light weight, bright color and freedom from heavy pitch -- all of which are especially noticeable in short leaf timber growing in the mountains and hilly regions of the south and west central parts of Arkansas. In no short leaf timber elsewhere are these characteristics so fully developed. Its soft texture is both tough and resilient and it reduces splitting by nails and screws. Tight, firm joints in the framework and mirror-smooth surfaces on paneling and woodwork are provided by these characteristics.

Arkansas soft pine products include all standard items in lineal trim, "Trim Pak" finish mouldings, ceilings, siding, jambs, flooring, common lumber, and small and medium timbers. They are graded and sold under grading rules published by the Arkansas Soft Pine Bureau which conform to American Lumber Standards.

Identification Of Wood

Familiarity with most kinds of wood makes it possible to identify them by their general appearance. In the technical identification of wood, specific differences must be pointed out. Some woods, as black walnut, can readily be identified by their color; others, such as Douglas fir, cypress and the cedars, can be distinguished by their odor. Many woods have a pronounced difference in color between sapwood and heartwood, whereas in others, there is no difference in color. Pitch is normally found only in the pines, Douglas fir, spruce and tamarack. Frequently, for accurate identification, it is necessary to refer to the finer details of the structure, such as the size of the rays, the size and arrangement of the pores, and the presence or absence of resin ducts.

Grain And Texture

The terms *grain* and *texture* are commonly used rather loosely in connection with wood. In fact, they do not have any definite meaning. Grain is often made to refer to the annual rings, as in fine grain and coarse grain. It is also employed to indicate the direction of the fibers, as in straight grain, spiral grain, and curly grain. Painters refer to woods as open-grained and close-grained, meaning the relative size of the pores which determines whether the piece needs a filler. Texture is often used synonymously with grain. Usually it refers to the finer structure of the wood rather than to the annual rings.

The Seasoning Of Lumber

The first operation after the lumber is cut and sawed at the mill is drying. Before drying, all timber is termed "green", meaning it contains a great quantity of sap and water. This moisture must be driven from the wood before it can be used. There are two methods by which this can be done - (1) air seasoning and (2) kiln-drying.

In air seasoning the boards and timbers are stacked in piles in the storage yard, each layer being separated by strips so that the air can circulate between them. In a few months the wood is fairly dried by evaporation and can be used for framing timbers but not for interior finish.

Where the lumber is used for interior finish, sash, doors, frames, porches, cornices and in fact everything except framing, it should be kiln-dried.

Kiln-drying is done by placing the lumber in a chamber called a kiln. The moisture is reduced in three to four days under scientifically controlled conditions. The lumber may be removed from the kiln at any time the moisture is reduced to the desired amount. By weight, fresh-cut wood is from one-third to two-thirds water. Most of this has to be removed and many modern mills now do it with dry kilns. This offers the following advantages to both manufacturer and contractor: (1) Time is shortened between manufacture and shipping; (2) Orders can be filled on short notice; (3) Degrading losses are lessened; (4) Insects and organisms are killed; (5) Less space in the lumber yard is needed; (6) The process is controllable, regardless of season, temperature and weather.

For many years there was a misapprehension that air-dried lumber was stronger or better than kiln-dried lumber. Exhaustive tests have conclusively shown that good kiln-drying and good air-drying have exactly the same results upon the strength of the wood. Wood increases in strength with the elimination of moisture content. This may account for the claim on the other side that kiln-dried lumber is stronger than air-dried lumber. This has little significance because, in use, wood will come to practically the same moisture content whether it has been kiln-dried or air-dried.

The Grading Of Lumber

No two trees grow exactly alike. Therefore no two pieces of lumber can be exactly alike with the same width of growth rings, sapwood and other characteristics. It is necessary to judge each individual piece that comes from each log. Specific grade and size standards have for years been established by the American Lumber Standards, in cooperation with the U. S.

Department of Commerce and representatives of lumber manufacturers, distributors, wholesalers, retailers, engineers, architects, and builders. In each accredited sawmill belonging to a member of any of the lumber manufacturing associations, trained lumber graders sort lumber as it comes through the mill. Their work, in turn, is checked at regular intervals by association lumber inspectors whose job it is to keep grading standards as specified by American Lumber Standards and grading rules adopted by the different lumber manufacturing associations.

Lumber was a more familiar item to our parents and grandparents when locally grown species were used and when grades and species selections were minimal. Softwood lumber of several species and from a number of regions is currently available in many communities, and dimension lumber grades have been developed on a more precise engineering basis. For these reasons, any complete listing of all the grades and species of softwood lumber available throughout the nation tends to appear somewhat confusing at first. Not all the listed grades and species of lumber are available in every locality, however. The builder and craftsman will quickly become familiar with the most commonly marketed species and grades in his community.

At the retail level, lumber is classified primarily by use and by size. It is also differentiated by the extent to which it has been manufactured, that is, rough sawn, dressed and worked (tongued and grooved, shiplapped or patterned.)

Types of lumber by use are *Yard, Structural,* and *Factory and Shop.*

Yard lumber consists of those grades, sizes and patterns used in ordinary construction. It is broken down into *Select* and *Common* grades.

Select grades of yard lumber have the best appearance and are used where a clear or high grade of finishing is desired. Grade names such as "B and Btr" (Better), "C" and "D" are used for most species.

Common grades of lumber are suitable for general construction. You will find even the lower common grades of boards yield many clear cuttings which can be used for furniture, cabinetry and other projects. Common grades of lumber are "Number 1," "Number 2," "Number 3," and "Number 4." Alternate names for common lumber, such as "Sel Merch" (Select Merchantable), "Merch", "Const." (Construction), "Std" (Standard), and "Util" (Utility), may also be encountered.

The common grades will frequently be found at the retail level as combination grades where the grade requirements for many uses make it practical to inventory and sell a "No. 2 and Btr" ("No. 1" and "No. 2" grades, combined), or "No. 3 and Btr" ("No. 3" and "No. 2" and possibly "No. 1", combined). You can check the suitability of such combined grades for your purposes by inspecting several pieces of the material which is inventoried this way at the lumber yard.

Structural lumber is 2 inches or more in nominal thickness (1½" actual dressed thickness). It is also called stress graded lumber because each grade is assigned working stress values to permit its use in engineered structures.

Factory and Shop lumber is produced primarily for industrial purposes such as the manufacture of windows and doors. You will not ordinarily find it at the retail yard.

Lumber is separated by size into *boards, dimension,* and *timbers.*

Lumber sizes are usually referenced for convenience and tally as nominal sizes, such as 1" x 2", 2" x 4", 4" x 10", etc. Actual surfaced sizes are smaller in thickness and width. Lengths are actual lengths as specified or slightly longer. Table 1-14 contains common nominal sizes and corresponding surfaced sizes.

Boards are less than 2 inches in nominal thickness and are 1 inch and larger in width. Boards less than 6 inches in nominal width may be called strips. Boards are used for fencing, sheathing, subflooring, roofing, concrete forms, box material and as a source of many smaller cuttings.

Dimension is from 2 inches to, but not including, 5 inches in nominal thickness and 2 inches or more in width. Such lumber, depending upon use, may be called framing, studs, joists, planks, rafters, and the like.

Timbers are 5 inches or more in their least dimension. According to use in construction, they are classified as beams and stringers, girders, purlins and posts.

Nominal sizes (2 x 4, etc.) are widely used in lumber tallying and, for simplicity, by the construction trades. Board and dimension lumber is generally sold by the *board foot*. This is a volume unit 1" in thickness, by one foot in length and one foot in width. To determine the number of board feet in a piece of lumber, multiply the thickness (in inches) by the width (in feet), by the length (in feet). For example, a 2" x 4" twelve feet long contains 2" x 4"/12" x 12' = 8 board feet.

The exception to the rule occurs where the nominal thickness is less than 1 inch. In this case it is only necessary to multiply the width in

ITEM	THICKNESSES			FACE WIDTHS		
	NOMINAL	MINIMUM DRESSED		NOMINAL	MINIMUM DRESSED	
		Inches			Inches	
Select or Finish (19 per cent moisture content)	3/8	3/8		2	1-1/2	
	1/2	7/16		3	2-1/2	
	5/8	9/16		4	3-1/2	
	3/4	5/8		5	4-1/2	
	1	3/4		6	5-1/2	
	1-1/4	1		7	6-1/2	
	1-1/2	1-1/4		8	7-1/4	
	1-3/4	1-3/8		9	8-1/4	
	2	1-1/2		10	9-1/4	
	2-1/2	2		11	10-1/4	
	3	2-1/2		12	11-1/4	
	3-1/2	3		14	13-1/4	
	4	3-1/2		16	15-1/4	
		Dry Inches	Green Inches		Dry Inches	Green Inches
Boards	1	3/4	25/32	2	1-1/2	1-9/16
				3	2-1/2	2-9/16
				4	3-1/2	3-9/16
				5	4-1/2	4-5/8
				6	5-1/2	5-5/8
				7	6-1/2	6-5/8
	1-1/4	1-1/4	1-1/32	8	7-1/4	7-1/2
				9	8-1/4	8-1/2
	1-1/2	1-1/4	1-9/32	10	9-1/4	9-1/2
				11	10-1/4	10-1/2
				12	11-1/4	11-1/2
				14	13-1/4	13-1/2
				16	15-1/4	15-1/2
Dimension				2	1-1/2	1-9/16
				3	2-1/2	2-9/16
				4	3-1/2	3-9/16
	2	1-1/2	1-9/16	5	4-1/2	4-5/8
	2-1/2	2	2-1/16	6	5-1/2	5-5/8
	3	2-1/2	2-9/16	8	7-1/4	7-1/2
	3-1/2	3	3-1/16	10	9-1/4	9-1/2
				12	11-1/4	11-1/2
				14	13-1/4	13-1/2
				16	15-1/4	15-1/2
Dimension				2	1-1/2	1-9/16
				3	2-1/2	2-9/16
				4	3-1/2	3-9/16
				5	4-1/2	4-5/8
	4	3-1/2	3-9/16	6	5-1/2	5-5/8
	4-1/2	4	4-1/16	8	7-1/4	7-1/2
				10	9-1/4	9-1/2
				12	11-1/4	11-1/2
				14		13-1/2
				16		15-1/2
Timbers	5 & Thicker		1/2 Off	5 & Wider		1/2 Off

Nominal and minimum-dressed sizes of finish boards, dimension, and timbers
(The thicknesses apply to all widths and all widths to all thicknesses)
Table 1-14

feet by the length in feet to obtain the board foot tally of the piece.

Select or finish lumber sizes apply a 19 or lower percent moisture content. Most finish lumber is manufactured at a maximum moisture content of 15 percent. Board and dimension lumber may be surfaced "dry" at 19 percent maximum moisture content (marked S-DRY), at 15 percent maximum moisture content (marked MC15 or KD), or at the green condition (S-GRN). Dressed sizes for such lumber at the dry and green conditions are given in Table 1-14. Timbers (over 5 inches in thickness) are produced in the green condition, permitting them to season in service.

Lumber may be manufactured *rough, dress-*

ed, and *worked* (matched, shiplapped or patterned). *Rough* lumber has been sawed, edged and trimmed but not surfaced. *Dressed* lumber has been surfaced on one or more sides to remove saw marks and surface blemishes. The most common dressed lumber is surfaced on all sides or S4S. *Worked* lumber has been tongued and grooved (T&G), shiplapped or patterned, in addition to being surfaced.

With the issuance of the American Softwood Lumber Standard, PS 20-70, in September 1970, development of a National Grading Rule for dimension lumber became a reality. Prior to development of the standard published by the U.S. Department of Commerce, each regional grade writing agency developed structural grades for dimension lumber based upon regional species or species groups. These grades took into consideration the characteristics of the species i.e., knot size, size of timber, non-dimension uses of the species, etc. As a result, a great many different engineering stress levels were developed and an even greater number of allowable spans for joists and rafters were tabulated.

To simplify the multitude of stuctural grades and working stresses available to the designer and the user, the Softwood Lumber Standard PS 20-70 provided that a National Grading Rule Committee be established, "to maintain and make fully and fairly available grade strength ratios, nomenclature and descriptions of grades for dimension lumber". These grades and grade requirements, as developed, are now used by all regional grade writing agencies. Since development of the National Grading Rule and the adherence of all regional and species rules to its requirements, a great deal more uniformity has resulted and use of dimension lumber by the architect and engineer has been materially simplified. All softwood grade writing agencies which publish grading rules, certified by the American Lumber Standards Committee, adhere to this rule for dimension lumber. The rule provides for uniform grade names for all species and grades of structural lumber.

The National Grading Rule separates dimension lumber into two *width* categories. Pieces up to 4 inches wide are graded as "Structural Light Framing," "Light Framing," and "Studs." Pieces 6 inches and wider are graded as "Structural Joists and Planks." (See Table 1-15). For special uses where a fine appearance and high bending strength are required, the national rule also provides a single "Appearance Framing" grade.

"Structural Light Framing" grades are available for those engineered uses where the

GRADE	
2"-4" Thick, 2"-4" Wide	
Structural light framing	Sel Str (Select Structual)
	No. 1
	No. 2
	No. 3
Studs	Stud
Light Framing	Const (Construction)
	Std (Standard)
	Util (Utility)
2"-4" Thick, 6" and Wider	
Structural Joists and Planks	Sel Str (Select Structual)
	No. 1
	No. 2
	No. 3
2"-4" Thick, 2" and Wider	
Appearance Framing	A (appearance)

Dimension lumber grades
(National grading rule)
Table 1-15

higher bending strengths are required. The four grades included in this category are "Sel Str" (Select Structural), "No. 1," "No. 2," and "No. 3."

"Light Framing" grades are available for those uses where good appearance at lower design level is satisfactory. Grades in this category are called "Const" (Construction), "Std" (Standard), and "Util" (Utility).

A single "Stud" grade is also provided under the National Grading Rule. It is intended specifically for use as a vertical bearing member in walls and partitions and is produced to a maximum length of 10 feet.

"Structural Joist and Plank" grades are available in widths 6 inches and wider for use as joists, rafters, headers, built-up beams, etc. Grades in this category are "Sel-Str" (Select Structural), "No. 1," "No. 2," and "No. 3."

Not all grades described under the National Grading Rule and listed in Table 1-15 will be available in all species or regions. The "Sel-Str" and "No. 1" grades are frequently used for truss construction and other engineered uses where high strength is required. For general construction, the grades normally encountered at the retail yard are "No. 2," or "No. 2 and Btr," or "Std and Btr." The "No. 3" and "Util" grades are also available and provide important economies where less demanding strength requirements are involved. In selecting dimension lumber for load bearing purposes, it is prudent to have engineering assistance and, where required, a building permit.

Where severe use conditions require lumber to be frequently wet or exposed to damp soil, naturally durable species and pressure-treated lumber are available. Lumber used for sleepers or sills resting on a concrete slab which is in direct contact with the earth, or joists closer than 18 inches to the ground, should be of naturally durable species or pressure-treated lumber. Where lumber is imbedded in the ground to support permanent structures, pressure-treated lumber should be used.

Naturally durable species most frequently encountered at the retail yard are California redwood, western red cedar and tidewater red cypress. The foundation grades of redwood and red cedar should be selected for ground contact. In cypress, a heart structural grade should be selected for similar exposure.

Assume you have decided upon the species and grade of lumber required for a project, as well as the appropriate number of board feet you will need. Make a simple sketch showing the layout of framing members, furring strips, paneling, etc., and include it with the list of materials you will take to the lumber yard. At the yard, request the assistance of a clerk who can check your sketch, quantity and grade selections.

If at all possible, follow your order through the yard. Look at the types of material in stock, the grade marks and other identification. Inspect other wood items, such as windows, doors, moulding and trim. This is a part of the education process which will be most helpful in planning the next project and in providing a mental picture of the appearance of the various lumber grades and species.

Using Lumber Grades

The use to which lumber is put determines the number, size, and position of the defects it may contain and still be satisfactory. In siding, for example, a reasonable number of knots on the edges are covered when the siding is in place. In flooring, some knots and other defects on the underside are allowable, since they will not show when the flooring is in use. Sheathing and subflooring may have a considerable number of defects since both kinds of lumber are entirely covered by finishing material. Covered lumber, such as sheathing, should be free from decay, even if it does not show, as the decay is quite likely to spread rapidly.

The condition of defects may also influence the grade of a piece of lumber. Tight knots in certain grades of siding or ceiling may be allowed. Loose knots likely to drop out would be objectionable.

The grading rules in general use at present, with very few exceptions, have to do with defects and do not take into account the quality of the wood itself. That is, if two boards of the same kind of wood are clear or if they have similar defects, both boards are placed in the same grade regardless of the quality of the wood itself. The wood in one board may be dense, heavy, and strong. The wood in the other light and weak. For some purposes, such as siding, ceiling, or finish, it may not matter whether dense or light wood is used. For other purposes, such as structural timber or flooring, where strength or hardness is a prime requisite, the wood must be dense to give satisfactory service.

Grades Recommended For Framing

Durability is always important in the construction of homes. Economy is also to be considered due to the increasing costs of building. It is important that the right grade of lumber be used in the right place in house framing to avoid unnecessary cost and at the same time guarantee sturdiness and durability. The most economical construction is that which uses the lowest grade of lumber which is suitable for the purpose. Not all housing lumber needs to be "select structural" any more than all parts of an automobile need to be tempered steel. The chief consideration is right use. Each piece of lumber has a job to perform as shown in Figure 1-16. It must add stiffness, insulation, strength or pleasing appearance to the house.

Modern lumber grading rules and markings used by lumber manufacturers under American Lumber Standards assure uniform characteristics in any one grade and size. Each grade -- Number 1, Number 2, and Number 3, for example -- exists to meet certain requirements. In any well constructed house there are places where the builder can find uses for all three grades.

Number 1 and Number 2 Douglas fir and West Coast hemlock are stress grades used extensively for horizontal, load-bearing members such as joists and rafters. These grades meet superior strength requirements. Number 2 Douglas fir joists and rafters can be used conservatively on spans not exceeding three-fourths the maximum allowable span for Number 1 joists and rafters of the same thickness and depth. For example, in a dwelling or apartment house with a live load of 40 pounds per square foot, 2-inch Number 1 joists, 16 inches on centers, can be used safely on a span

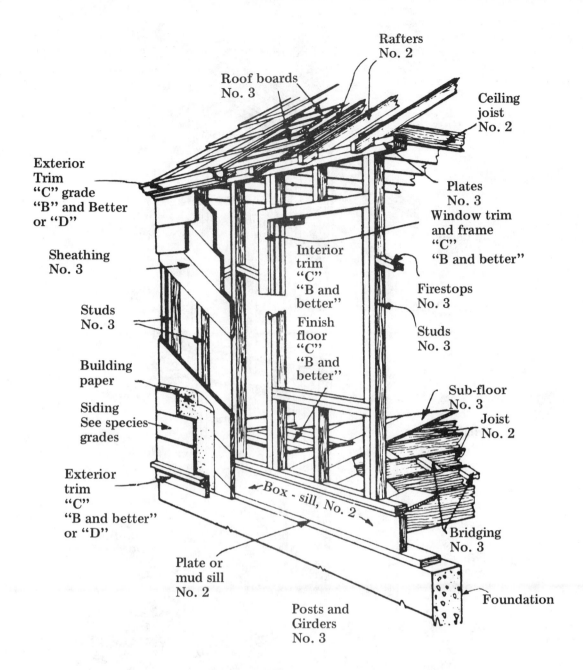

Grade use key for low cost one-story construction
Figure 1-16

in feet equal to 1½ times their nominal depth in inches; 2-inch Number 2 joists, 16 inches on centers can be used safely on a span in feet slightly more than their nominal depth in inches. For example, a 2 x 10 Number 1 can be used on a span of 15 feet; a 2 x 10 Number 2 on a span of eleven feet, three inches.

Number 3 studs may be used for walls and partitions in one-story structures and for non-bearing or minor partitions in other structures. In two-story construction Number 2 studs can be used for the first floor and Number 3 for the second floor. In each case experience has determined that the grade will give the

required strength with a dependable margin of safety to spare. Number 3 studs do not look as good as Number 1. They may contain large knots, small knot holes, short strips of bark along one edge, white specks or pitch pockets. They are enclosed in the walls where the owner is interested in having strength and not perfect appearance. Number 3 used where it belongs can give all the strength required with a margin to spare. See Figure 1-17.

Sheathing

Diagonal sheathing is much stiffer and stronger than horizontal sheathing with braces,

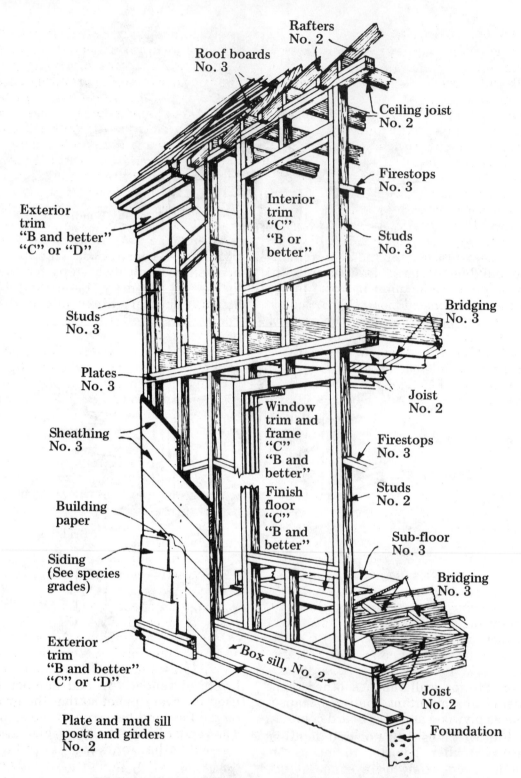

Rafters
No. 2

Roof boards
No. 3

Ceiling joist
No. 2

Firestops
No. 3

Exterior
trim
"B and better"
"C" or "D"

Interior
trim
"C"
"B or
better"

Studs
No. 3

Bridging
No. 3

Studs
No. 3

Plates
No. 3

Window
trim and
frame
"C"
"B and
better"

Joist
No. 2

Firestops
No. 3

Sheathing
No. 3

Studs
No. 2

Building
paper

Finish
floor
"C"
"B and
better"

Sub-floor
No. 3

Bridging
No. 3

Siding
(See species
grades)

Exterior
trim
"B and better"
"C" or "D"

Box sill, No. 2

Joist
No. 2

Plate and mud sill
posts and girders
No. 2

Foundation

Grade use key for low cost two-story construction
Figure 1-17

but it requires more material. Diagonal sheathing is most widely used in earthquake and tornado areas. Number 3 is economical and amply strong for use in both types. Where long lengths or unusual strength are required, such as for rafters and joists, use one grade higher. Lower grade boards, graded according to lumber association grading rules, are satisfactory for subflooring and sheathing. In buying and using lower grades for house framing, builders should make sure that the grades conform to those specified by the various regional lumber manufacturing associations under American Lumber Standards.

Modern Lumber Products

At one time there were only two kinds of lumber: clear and common. The buyer came to the sawmill and picked out what he wanted. Today, the sawmill operator sends his representative to the buyer and sells him the product of logs which have been sawn into as many as 2,000 items, each having its own grade, length, width, thickness and uses.

Many new products have been developed. Glued laminated timbers for rafters, beams, arches and trusses can now be obtained. This timber is made by gluing together, with a highly water resistant glue, a number of layers of wood and bending them into any shape required. It permits the building of large structures with long spans at great reductions in cost under modern methods of construction. Of particular interest to architects, engineers and contractors is the type of construction using small metal rings and plates. These modern connectors have simplified the joining of wood to wood and wood to steel. Less lumber, less hardware and a better joint are possible when modern timber connectors are used. Cut-to-size building parts, standardized sizes of millwork, and mass production of doors, windows and cabinet work were all developed to lower the cost of construction. Lumber today may be cut to exact length so it can be used with the least amount of cutting and fitting. It is trade-marked, packaged, bundled, tied and labeled like any other high class merchandise.

Use of the correct lumber products and new methods can be of great help in lowering construction costs. Exact length joists can be obtained, thus avoiding waste. Ceiling heights can be dimensioned to permit the use of an eight-foot exact length stud with squared ends. Each stud, as with the joists, may be used just as it comes from the dealer. With the increasing popularity of interior wall finishes other than plaster, this is an important matter because such finishes as plywood and fiberboard come in eight-foot lengths, and the eight-foot stud makes their use easier.

Another important item is the development of end-matched lumber which may be used for subfloors, wall sheathing, and roof sheathing. Its use offers advantages that contribute to sound, economical construction. Each piece is tongued and grooved on the ends as well as the edges. When applied, it meshes together to form strong, tight walls. The only cutting required is at the openings and corners. The piece that is cut off may be used to start the next run. Because the ends of the boards interlock in a firm tongue and groove joint, pieces do not have to be joined over framing members, as is required when plain square end lumber is used. The result is that this sheathing goes on faster, and is easier to apply, requiring no special skill on the part of the user -- and there is practically no waste. Nailing time is also considerably reduced because fewer nails are used.

Among many of the other improved lumber items that should be mentioned are packaged bevel siding and exterior finish, packaged mouldings, interior wall finishes, plywood, wood gutters, stepping, and shingles.

Just as the production of precut framing embraces the two main steps -- cutting and bundling -- construction with precut framing necessitates only two steps for the builder -- listing and assembly. Estimating is, therefore, an easy matter. Since one is dealing with bundled assemblies of framing members, it is only necessary for the contractor to tally the number of doors and windows of each size, count the studs, add in a bundle of diagonal bracing for each run, and list a fire block for each spaced stud to have a complete schedule of framing required between top plate and sole plate.

Selecting Plywood

For everything from subfloors to roof decks to siding and built-ins, plywood is an all-around building material. Plywood is a real wood. Although it is an engineered product, the natural wood is changed very little in the manufacturing process. When a log arrives at a plywood mill, it is peeled, placed in a giant lathe, and turned against a lathe knife. The veneer (thin sheet of wood) that flows from the lathe in a continuous ribbon is clipped into pieces of a convenient size for kiln drying and assembly into plywood panels. Every plywood panel is a built-up board made of kiln dried layers of veneer. An odd number of layers is used for every panel so that the grain direction on the face and back run in the same direction. The veneer layers are assembled at right angles to each other and are united under high pressure with an adhesive. The resulting glue-bonds become as strong or stronger than the wood itself. Since wood is stronger along the grain, this cross-lamination distributes the wood's strength in both directions. Peeling veneers from a log and reassembling them also provides a means for making panels much larger than those that could be produced by sawing. (Standard plywood panels are 4 x 8 feet, though wider and longer panels are also produced.)

One basic thing to remember in selecting plywood is to look for the grade-mark which identifies the product, subject to the quality inspection of an approved testing agency. The grade and type of a panel should also be considered in the selection of plywood. Type refers to the durability of the glueline or the degree of exposure the panel should be subjected to. Letter grades N, A, B, C, D, refer to the quality of the face and back veneers. N represents the highest veneer quality. A two letter combination, such as A-C, is used to indicate the quality of the panel face and back. ''A'' indicates the face quality and ''C'' describes the back.

Grades A-A, A-B, or A-D interior type sanded plywood are recommended for cabinet doors, furniture, built-ins, and other projects to be painted. B-D may also be painted. If you prefer a waxed, sealed, or varnished natural finish, select fine grain panels in A-A, A-B, or A-D grades. For exteror siding, interior paneling and ceilings, textured plywood panels are available in many different species, surfaces, and patterns. Textured plywood sidings are best finished with stains, although some species may be left to weather naturally. Good results for painted surfaces, interior or exterior, can also be achieved with Medium Density Overlaid (MDO) plywood. This grade has a smooth resin-treated fiber surface bonded to the panel face. It takes and holds paint well.

Plywood is available in both appearance and in engineered grades. Appearance grades are normally sanded. Engineered grades which are left unsanded are generally applied where a high degree of strength and rigidity are required. You'll notice that the approved grade-trademark on appearance grades includes a group number. That number stands for one of the more than 70 wood species from which plywood is manufactured. See Figure 1-18. Since species vary in strength and stiffness, they have been classified into five groups under PS 1 (Product Standard 1). The strongest woods are found in Group 1. Unsanded engineered grades of plywood bearing a grade-trademark carry an Identification Index which tells you the maximum support spacing to which the plywood can be applied in conventional construction. See Figure 1-19. The Identification Index appears as two number designations separated by a diagonal such as 24/0, 32/16, etc. The number to the left of the diagonal indicates the maximum spacing of supports in inches that should be used when the panel is applied as roof decking. The number designation to the right provides the same information for subflooring applications. Unsanded grades designated as *Structural I* and *Structural II* are recommended

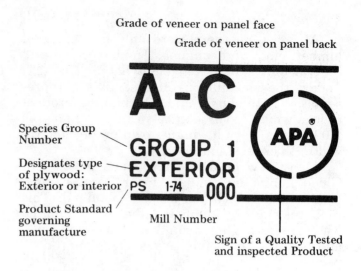

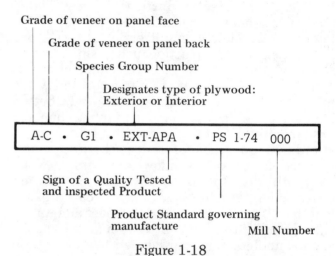

Figure 1-18

Engineered Grade-Trademark
Figure 1-19

for heavy load applications, where plywood's strength properties are of maximum importance.

Now that you've become a little better acquainted with plywood, we can move on to its role in residential construction. Starting with the basics, let's look at the part plywood plays in the area of flooring. In a double layer system the floor will be made up of plywood subflooring and a separate layer of underlayment. The underlayment plywood offers a high degree of dimensional stability that eliminates swelling and buckling. The result is a smooth, solid, stable base for any kind of finish flooring you desire. One good way to save money and time is to use one layer of plywood as a combination subflooring and underlayment material. The plywood serves both as a structural subfloor and as an excellent base for resilient floorings, carpeting, and for other nonstructural floorings. Plywood bearing the registered grade-trademark 2.4.1 can be used to provide both subflooring and an underlayment surface in a single 1-1/8 inch thick plywood panel. It's best to use 2 x joists spaced 32 inches on center or 4 x girders spaced 48 inches as the support system.

One excellent construction system you can use in building a floor is the APA Glued Floor System in which glue and nails are used to secure the structural underlayment to wood joists. This system was designed to produce floors that would be stronger and less apt to squeak. The glues needed for this system are elastomeric adhesives meeting performance specification AFC-01. These glues, which may be applied even in below freezing weather, are available in cartridges designed for conventional caulking guns. Specific recommendations on the APA Glued Floor System may be obtained from the American Plywood Association.

Plywood is also a key word in any discussion of wall construction. Performing structurally as wall sheathing, it covers large areas rapidly and supplies strength and rigidity. Neither let-in bracing nor building paper is required with it. Plywood wall sheathing may be installed either vertically or horizontally. Horizontal application of panels will give you greater stiffness under loads perpendicular to the surface. So, if you're going to be nailing siding such as shingles directly into your wall sheathing, it would be wise to apply the sheathing horizontally. The availability of many siding textures adds an aesthetic dimension to plywood's structural role as a wall material. Textured plywood can be used for interior paneling as well as exterior siding.

Although many builders still apply plywood siding in a two layer system, more and more are turning to the APA Single Wall System in which plywood siding is applied directly to studs. Accepted by the Federal Housing Administration, the Farmers Home Administration, and most local building codes, the single wall system is designed to offer tight wall construction. All horizontal and vertical joints are backed with lumber framing members. Nails around the edges secure panels to the framing and provide draft stops at all points. To insure weather tightness, single wall joints are shiplapped,

battened, or backed with building paper. When you're using the APA Single Wall System, you may apply plywood panel siding, or lap, or beveled siding. Specify 3/8, 1/2, or 5/8 inch, depending on your stud spacing. Be sure to seal plywood edges. If you're going to paint the surface, the prime coat can serve as your sealer. If the plywood is to be stained, seal it first with a water-repellent preservative that's compatible with the finish. Although plywood sidings are normally installed vertically, you may place panels horizontally with the face grain across supports.

Determining what is the allowable support spacing for the construction of single walls is a simple matter. Panels for single wall construction which are identified as 303 sidings have their maximum support spacing listed in their grade-trademark. A 303 siding, for example, bearing a "303-24 in. o.c." may be applied vertically to studs 16 or 24 inches on center, while panels marked "303-16 in. o.c." may be applied vertically over studs spaced no more than 16 inches apart. Texture 1-11, a 5/8 inch thick exterior siding panel with a 3/8 inch wide vertical grooving spaced 2, 4, 6, or 8 inches on center may be used vertically over studs spaced 16 inches on center.

All edges of panel siding should be backed with framing or blocking. And to keep from staining siding with nails, use hot dip galvanized, aluminum, or other nonstaining nails. No extra corner bracing is needed with plywood panel siding.

Moving along to roof construction, you'll find that plywood roof sheathing gives you the strength and rigidity you need while it makes a solid base for roofing material. Plywood sheathing bears an Identification Index which tells you the recommended rafter spacing for a specific plywood thickness. For example, for roof systems with a 24 inch span (distance between rafters), plywood with a marking of 24/0 will do the job. These Identification Index panels are available in thicknesses ranging from 5/16 through 7/8 inch.

Plywood roof sheathing has superior nail holding capabilities. Extensive laboratory and field tests have proven that even 5/16 inch plywood will hold shingle nails securely and permanently in place, in the face of hurricane force winds.

Your house plan will show either "open soffits" or "closed soffits". For a roof deck over closed soffits, you can use C-D Interior grade sheathing. To enclose the soffits, Medium Density Overlaid plywood is preferable because it has a superior painting surface. With open soffits, panels will be exposed at the overhang. Thus, you'll want to select an Exterior type plywood. In addition to being an Exterior type plywood, the plywood you choose for an open soffit application should be a high enough appearance grade to permit painting or staining. Textured plywood with the textured side down can be used for exposed soffits and ceiling applications. Staining is the only finishing required.

Wood Panel Products

Besides softwood plywood, you have several choices of panel products. Some provide needed strength and stiffness while others are primarily for finish, sound reduction, insulation, or other characteristics. Take time to look them over and select the materials that best fit your job.

Most of these products are partly or entirely of wood-based material -- hardwood plywood, insulation board, hardboard, laminated paperboard, particleboard, and gypsum board. Manufacturing and finishing methods vary greatly to provide materials with specific desirable properties. The materials don't look or feel alike and vary widely in properties, but all are manufactured in panel form. Thus they can cover large areas quickly and easily.

Most other panel products involve breaking the wood down into small portions and then reassembling the elements into boards. For such common materials as insulation board, hardboard, and laminated paperboard, the material is broken essentially into fibers, which are interfelted into panels. With particleboard, the wood is broken into particles that are bonded together with resin, heat, and pressure. Processes vary until sometimes the distinctions are blurred. The other main member of the panel group is gypsum board, which has a noncombustible gypsum core between faces of paper.

The insulation board-hardboard-paperboard group is customarily known as building fiberboard and may be called by such proprietary names as "Celotex," "Insulite," "Masonite," "Beaverboard," and "Homasote" without regard to the actual manufacturer.

Oldest of the boards is insulation board, made in two categories -- semi-rigid and rigid. Semi-rigid consists of the low-density products used as insulation and cushioning. The rigid type includes both the interior board used for walls and celings, as well as the exterior board used for wall sheathing.

Hardboard is a grainless, smooth, hard product. It is used for siding, underlayment and as prefinished wall panel.

Laminated paperboard serves as sheathing and other covering but is not used as much as the other building fiberboards.

Particleboards are often known by the kinds of particle used in their makeup, such as flakeboard, chipboard, chipcore, or shavings board. New products have greater strength, stiffness, and durability than those made a few years ago. Some higher quality products have recently been approved for subfloors and sheathing. Underlayment, however, is still the principal use of particleboard in houses even though more and more is going into shelving.

Gypsum board is used principally for interior covering. Builders like it because, unlike plaster, it is a dry-wall material. Edges along the length are usually tapered to allow for a filled and taped joint. It may be obtained with a foiled back which serves as a vapor barrier on the exterior walls and is also available with vinyl or other prefinished interior surface.

Each product will serve well if used as intended. Panels must be properly fastened to framing members and used in the right places under conditions for which they were designed. Manufacturers have found most instances of unsatisfactory performance directly related to improper application. Therefore, they usually include application and use instructions on each package or bundle of their material.

Insulation board has been and continues to be used extensively for wall sheathing, reportedly more than any other sheathing material for residential construction in the United States. It is inexpensive and it combines bracing strength, insulation, and a degree of noise control through walls.

Insulation board sheathing is available in three types -- nail-base, intermediate density, and regular density -- in order of decreasing density and strength. Thickness is ½ inch in the first two types and ½ or 25/32 inch in the last. Wood and asbestos shingles can be applied directly to nail-base with annular-grooved nails. Regular density, as the least dense, is the best insulator. The Federal Housing Administration recommends that all three types be applied vertically on the wall. The boards provide adequate racking resistance without corner bracing except that some specifying agencies require corner bracing with half-inch regular density board.

Insulation board and gypsum board sheathing are also available in 2 by 8 foot panels. They are applied horizontally and require corner bracing of the walls.

A special type of exterior particleboard sheathing panel, 4 feet wide, has been approved by some building codes. It must be not less than 3/8 inch thick and the studs must not be more than 16 inches on center. A 3/8-inch thick sheathing grade of laminated paperboard is also approved. Corner bracing is not required for either.

Roof sheathing provides the strength and stiffness needed for expected loads, racking resistance to keep components square, and a base for attaching roofing. Panels have several advantages over lumber, particularly in ease and speed of application and resistance to racking.

Softwood plywood is now used most extensively for roof sheathing, but other panel materials are satisfactory if they meet performance requirements. Special structural particleboards are under development for such purposes and the Federal Housing Administration has approved specific ones for some areas. Some insulation boards also meet the requirements.

Panel products are available in a wide variety of forms and finishes for interior coverings, and are far more common today than lath and plaster. The thin sheet materials, however, require that studs and ceiling joists have good alignment to provide a smooth appearance.

Finishes in bathrooms and kitchens have more rigid stain and moisture requirements than those for other living spaces. Be a bit more careful when selecting materials for these areas.

For ceilings, gypsum board is perhaps the most common sheet material, but large sheets of insulation board are also used. The panels are generally nailed or screwed in place and then joints and fasteners are suitably covered or trimmed. Insulation board in the form of acoustical and decorative ceiling tile is popular, especially for dens and recreation rooms.

Panels are typically 4 by 8 feet and applied vertically as single sheets so that the vertical joints butt at the stud. But there are many alternatives. For example, builders in some areas apply gypsum board horizontally in sheet sizes 4 feet wide and up to 16 feet long; or they may use more than one layer.

After ceiling and wall covering is complete, you need to prepare for the final wearing surface applied to a floor. Unless you apply wood strip or wood tile over the subfloor, you will usually need an underlayment. Underlayment is supplied in 4-foot wide panels and provides a smooth, uniformly thick base ideal for resilient flooring, carpeting, and other finish flooring. Panels are installed just before applying the finish flooring.

Hardboard and particleboard join plywood as the principal panel products used for floor underlayment. Hardboard comes in 3' by 4' or 4' by 4' panels a little less than ¼" thick. It is often used in remodeling because of the floor thicknesses involved. Particleboard comes in a 4' by 8' size in a variety of thicknesses from ¼" to ¾". Special insulation boards are also used as underlayment for carpeting as a resilient, noise-deadening material.

Before buying the underlayment, find out what is recommended for your specific covering and the techniques for installing it. You need to know about conditioning the underlayment, preparing the surface, panel arrangement, edge clearances to allow for shrinking and swelling, nailing, stapling or gluing, and possible filling and sanding. Do it right and avoid future problems.

Sound insulation, or minimizing unwanted noise, is accomplished to some extent in conventional construction. With increasing emphasis on the "quiet" home, it may be that you want to plan for better-than-normal sound barriers. Unwanted sounds, whether from a crying baby, a noisy party, or the flushing of a toilet can be reduced by special construction.

Airborne noises inside the house create sound waves that radiate outward until they strike a surface such as a wall, floor, or ceiling. Insulation board, especially with holes in it (acoustical tile), absorbs these waves and reduces room reverberation. Draperies and carpeting do too. The sound waves set the room surfaces in vibration and how much sound is transmitted to the next room depends on the construction.

If sound barriers are to be better than conventional walls and ceilings, they require extra thought and care in construction. They usually include "decouplers" to reduce transfer of sound waves. For a wall, the covering can be decoupled from the stud by *resilient channels*, ½" sound-deadening insulation board, or ¼" hardboard. Improved walls can be made with double-row-of-stud construction. The wall must, of course, also be suitably sealed against sound leaks through or around it.

Recommended Nailing Practices

Of primary consideration in the construction of a house is the method used to fasten the various wood members together. These connections are most commonly made with nails, but on occasions metal straps, lag screws, bolts, and adhesives may be used.

Joining	Nailing method	Nails		
		Number	Size	Placement
Header to joist	End-nail	3	16d	
Joist to sill or girder	Toenail	2	10d or	
		3	8d	
Header and stringer joist to sill	Toenail		10d	16 in. on center
Bridging to joist	Toenail each end	2	8d	
Ledger strip to beam, 2 in. thick		3	16d	At each joist
Subfloor, boards:				
1 by 6 in. and smaller		2	8d	To each joist
1 by 8 in.		3	8d	To each joist
Subfloor, plywood:				
At edges			8d	6 in. on center
At intermediate joists			8d	8 in. on center
Subfloor (2 by 6 in., T&G) to joist or girder	Blind-nail (casing) and face-nail	2	16d	
Soleplate to stud, horizontal assembly	End-nail	2	16d	At each stud
Top plate to stud	End-nail	2	16d	
Stud to soleplate	Toenail	4	8d	
Soleplate to joist or blocking	Face-nail		16d	16 in. on center
Doubled studs	Face-nail, stagger		10d	16 in. on center
End stud of intersecting wall to exterior wall stud	Face-nail		16d	16 in. on center
Upper top plate to lower top plate	Face-nail		16d	16 in. on center
Upper top plate, laps and intersections	Face-nail	2	16d	
Continuous header, two pieces, each edge			12d	12 in. on center
Ceiling joist to top wall plates	Toenail	3	8d	
Ceiling joist laps at partition	Face-nail	4	16d	
Rafter to top plate	Toenail	2	8d	
Rafter to ceiling joist	Face-nail	5	10d	
Rafter to valley or hip rafter	Toenail	3	10d	
Ridge board to rafter	End-nail	3	10d	
Rafter to rafter through ridge board	Toenail	4	8d	
	Edge-nail	1	10d	
Collar beam to rafter:				
2 in. member	Face-nail	2	12d	
1 in. member	Face-nail	3	8d	
1-in. diagonal let-in brace to each stud and plate (4 nails at top)		2	8d	
Built-up corner studs:				
Studs to blocking	Face-nail	2	10d	Each side
Intersecting stud to corner studs	Face-nail		16d	12 in. on center
Built-up girders and beams, three or more members	Face-nail		20d	32 in. on center, each side
Wall sheathing:				
1 by 8 in. or less, horizontal	Face-nail	2	8d	At each stud
1 by 6 in. or greater, diagonal	Face-nail	3	8d	At each stud
Wall sheathing, vertically applied plywood:				
3/8 in. and less thick	Face-nail		6d	6 in. edge
1/2 in. and over thick	Face-nail		8d	12 in. intermediate
Wall sheathing, vertically applied fiberboard:				
1/2 in. thick	Face-nail		1½ in. roofing nail	3 in. edge and
25/32 in. thick	Face-nail		1¾ in. roofing nail	6 in. intermediate
Roof sheathing, boards, 4-, 6-, 8-in. width	Face-nail	2	8d	At each rafter
Roof sheathing, plywood:				
3/8 in. and less thick	Face-nail		6d	6 in. edge and 12 in. intermediate
1/2 in. and over thick	Face-nail		8d	

Recommended schedule for nailing the framing and sheathing of a well-constructed wood-frame house
Table 1-20

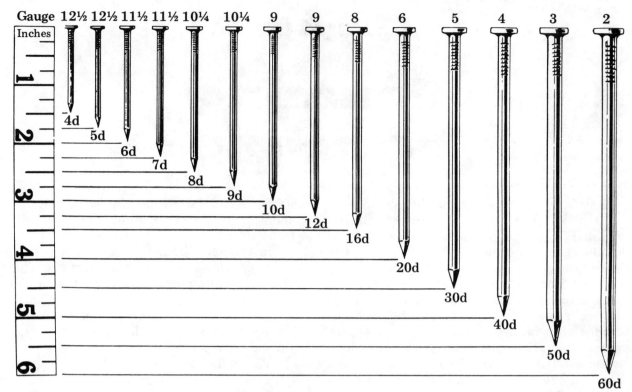

Sizes of common wire nails
Figure 1-21

Proper fastening of frame members and covering materials provides the rigidity and strength to resist severe windstorms and other hazards. Good nailing is also important from the standpoint of normal performance of wood parts. For example, proper fastening of intersecting walls usually reduces plaster cracking at the inside corners.

The schedule in Table 1-20 outlines good nailing practices for the framing and sheathing of a well constructed wood frame house. Sizes of common wire nails are shown in Figure 1-21.

Chapter 2

Framing Principles

All structural parts of a building which are hidden by sheathing or plaster are called *framing*. The frame provides the network to which the other materials are fastened as well as the strength and stiffness required to support loads. The frame includes the floor beams or joists, the roof supports or rafters, the uprights in the walls and partitions called the studs or studding, and the beams or girders which are used to support the ends of the joists and which are supported at their ends by posts or columns.

Each of the principal regions of the United States -- the Middle West, the Pacific Coast, the South, and the Atlantic Coast -- uses a different type of construction. However, the same basic principles of framing and bracing are used in all houses.

Buildings constructed entirely of wood above the foundations may be classified as follows: early braced frame, modern braced frame, western frame, balloon frame and plank-and-beam frame. The latter three framing methods have come about because of the rapid development of woodworking machinery, the improvements in the manufacture of building materials and the competition among contractors for low cost and high production.

Fundamentals Of Frame Construction

During the lifetime of a building, the lumber of which it is constructed undergoes many changes of temperature and humidity. Even well seasoned lumber will dry out further under the artificial heat of a house in winter. As the moisture content is reduced, the lumber will shrink somewhat. The amount of shrinkage depends on the moisture content of the structural member at the time it was set in place, the temperature and humidity of the building, and also the compression caused by the load on the structural member.

The shrinkage occurs primarily across the board, very little taking place lengthwise. Therefore, the more horizontal bearing members there are in a house frame, the more shrinkage there will be in the frame. This

fundamental fact is of utmost importance and should be used as a guide in the selection of framing methods.

In the early braced frame construction most of the bearing members were vertical. This type of construction presented very little shrinkage in the frame of the building. Some of the modern systems of framing have retained this principle as far as possible, while others have, to some extent, disregarded it.

The side and end walls of a frame are generally braced by inserting diagonal or knee braces at the corner posts of the building or by running the sheathing diagonally to the studs. The walls should be tied together by adequate joists, which should be braced to each other by bridging at intervals not exceeding 8 feet. Openings in wall partitions and floors should be provided with headers strong enough to carry the load of the studs or joists that have been cut to make the opening.

Early Braced Frame

In the early braced frame, each post and beam of the framework was mortised and tenoned together, and the angles formed by these posts or beams were held rigidly by diagonal braces. These in turn were held by wooden pins or dowels through the joints. This type of frame construction did not rely on the sheathing for rigidity; in fact, the sheathing or planking could run vertically up the sides of the structure. Barn construction was typical of this type of building. A more modern method of bracing the framework has replaced this type of construction, and it is used to a large extent where sheathing is not entirely relied upon for the rigid support of the building. This type is called the modern braced frame type of construction. The early braced frame will be given no further consideration.

Modern Braced Frame

Figure 2-1 shows the details of the modern braced frame. The studs of the side walls and center partitions are cut to exactly the same

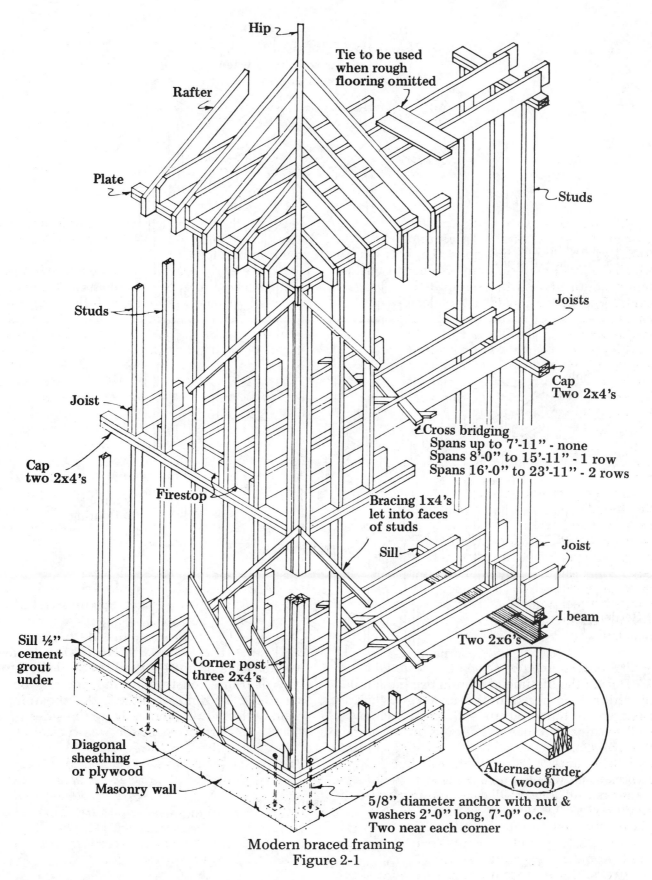

Hip

Rafter

Plate

Tie to be used
when rough
flooring omitted

Studs

Studs

Joists

Cap
Two 2x4's

Joist

Cross bridging
Spans up to 7'-11" - none
Spans 8'-0" to 15'-11" - 1 row
Spans 16'-0" to 23'-11" - 2 rows

Cap
two 2x4's

Firestop

Bracing 1x4's
let into faces
of studs

Sill

Joist

Sill ½"
cement
grout
under

I beam

Two 2x6's

Corner post
three 2x4's

Alternate girder
(wood)

Diagonal
sheathing
or plywood

Masonry wall

5/8" diameter anchor with nut &
washers 2'-0" long, 7'-0" o.c.
Two near each corner

Modern braced framing
Figure 2-1

length. The side wall studs are attached to the sill and plate, and the partition studs are attached to the girder and plate. This construction permits uniform shrinkage in the outside and inside walls if a steel girder is used as shown.

Diagonal braces are let into the studs at the corners of the building to provide rigidity. Any

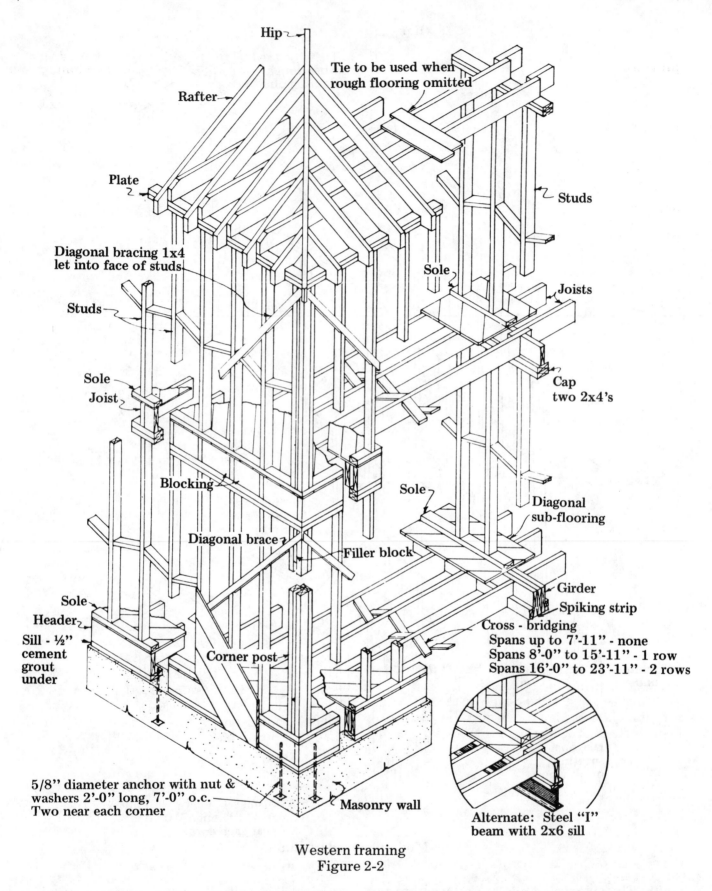

Hip

Tie to be used when
rough flooring omitted

Rafter

Plate

Studs

Diagonal bracing 1x4
let into face of studs

Sole

Joists

Studs

Sole
Joist

Cap
two 2x4's

Blocking

Sole

Diagonal
sub-flooring

Diagonal brace

Filler block

Girder

Spiking strip

Sole

Header

Cross - bridging
Spans up to 7'-11" - none
Spans 8'-0" to 15'-11" - 1 row
Spans 16'-0" to 23'-11" - 2 rows

Sill - ½"
cement
grout
under

Corner post

5/8" diameter anchor with nut &
washers 2'-0" long, 7'-0" o.c.
Two near each corner

Masonry wall

Alternate: Steel "I"
beam with 2x6 sill

Western framing
Figure 2-2

one of several systems of bracing may be used at
the corners, depending on the placement of
openings in the side or end walls.

The joists are lapped and spiked to the side

wall studs, and are either continuous between
the two opposite side walls, or are side lapped
and spiked at the center partition. In either case,
these joists form a tie for the two side walls. The

28

joists are bridged and are supported by their full width at the sill and girder. Fire stops are provided at the ends of all joists.

The flooring is laid diagonally to the floor joists in opposite directions on the upper and lower floors. This provides great rigidity to the building as well as to the floors.

The Platform Or Western Frame

Figure 2-2 shows the platform type of frame. Each story of the building is built as a separate unit, the floor being laid before the side walls are raised. This provides safer working conditions and could be used in the other types of framing. The other features of this type of construction are similar to those of the modern braced type, except that the ends of the joists form a bearing for the side and center wall studs. This disregards the principle of shrinkage, but permits equal shrinkage at the side and center walls.

With the modern use of kiln-dried lumber, this type of framing has become popular. It is probably the fastest and safest form of good construction and also allows greater use of short materials for studding. In this method the studding extends through the height of one story only and rests at the bottom on a "sole" plate nailed on top of the flooring. The joists of the second floor are carried on a two-piece girt or plate placed on top of the studding of the story below. As the rough floor extends to the outside edge of the building, no wooden fire stops are required.

In western framing a continuous header should be carried around the entire building. This header is the same size as the joists. At the first floor, the bottom edge of the header rests on the sill with its outside face flush with the outside edge of the sill and forms what is known as a *box-sill*. At the second floor, the bottom edge of the header rests on the girt which carries the joists, the outside face being flush with the outside edge of the girt. At the second floor, instead of using a continuous header, the joists may be extended flush to the outside of the plate and blocking placed between the joists.

Although western framing is not so well known in all localities, it is the most scientific method yet devised. Partitions are built and supported in the same manner in nearly all wood frame buildings and require girders, double joists, sole pieces and plates for support. All of these members are placed in such a manner that the shrinkage will amount to as much as 2 inches in some two-story buildings. When the studs are full length in the outside walls, this shrinkage is largely in the partitions, and, therefore, the

floors and ceilings all tend to slope toward the interior of the building. This shrinkage causes the doors to stick, the plaster to crack, and ruins the appearance of the interior.

In western frame construction, the outside walls are built with members that correspond with members used in the building of partitions. This does not eliminate shrinkage, but it makes the shrinkage uniform in both outside walls and partitions, thereby keeping the floors and ceilings level regardless of the shrinkage. Where steel beams are used instead of wood girders, wood of the same cross-section size as the sills on the foundation should be used to give an equal amount of shrinkage. Shorter lengths of lumber are also used, resulting in lower costs for lumber. The lengths of the pieces of framing can be readily determined from plans and cut by modern power saws previous to starting the actual building. Large units of framing, including trusses, headers, trimmers, etc., can be put together before the studs are erected. The units are raised in place in one single operation.

Balloon Frame

Figure 2-3 shows the balloon frame. The sill construction of Figure 2-1 might also be used in this type of frame. The second floor joists are carried by a ribbon let into the studs, thereby allowing the use of long side wall studs but necessitating the use of more inserted fire stops. The other features are similar to those used in the types previously described.

This system of framing was used extensively to replace the old style braced frame until the modern braced and platform types became popular.

The balloon type of framing has many features to recommend it. The outside studs are made to extend the full two stories from the sill to the roof plate. The load bearing partitions can also be made in this manner. At the second floor level, the joists rest on a 1'' x 6'' ribbon or ledger board and are nailed against the studs. The ribbon board is cut into the supporting studs. The attic or ceiling joists rest on the doubled top plate.

The flue effect of the openings between the studs is much stronger in the balloon frame than in the other types and fire-stopping is, therefore, much more important. Without some such check, flames from the basement could sweep to the top plate or into the attic. Fire-stopping must therefore be plentiful with blocking both at the sills and half story heights.

The balloon frame is not as rigid as the braced type of framing until after the outside sheathing has been applied. After this sheath-

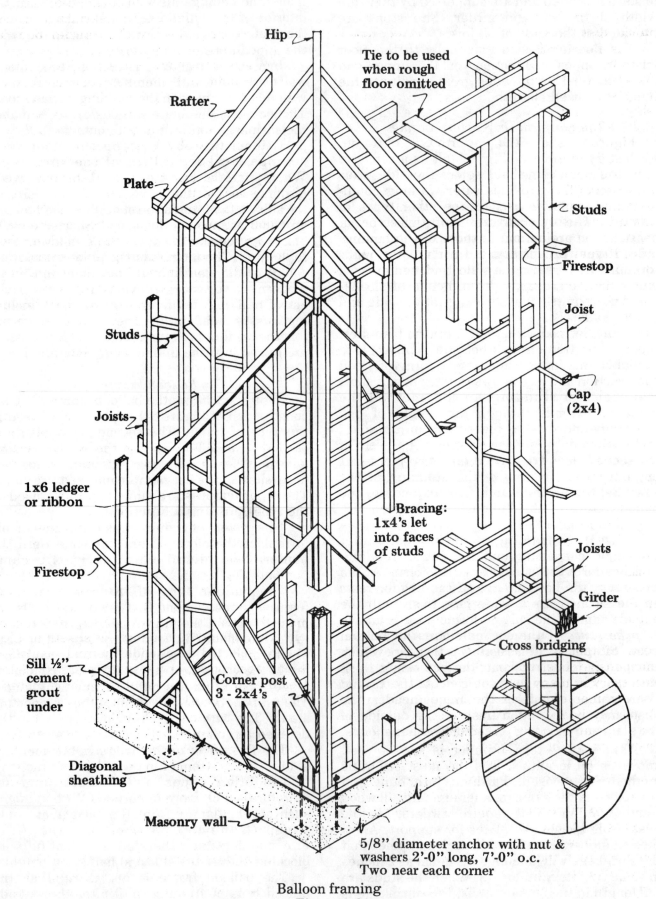

Hip

Tie to be used when rough floor omitted

Rafter

Plate

Studs

Firestop

Studs

Joist

Joists

Cap (2x4)

1x6 ledger or ribbon

Bracing: 1x4's let into faces of studs

Joists

Girder

Cross bridging

Sill ½" cement grout under

Corner post 3 - 2x4's

Diagonal sheathing

Masonry wall

5/8" diameter anchor with nut & washers 2'-0" long, 7'-0" o.c. Two near each corner

Balloon framing
Figure 2-3

ing is in place, a balloon frame building is as strong and stiff as the other types but the cost is much less because it can be erected faster. Where the sheathing is run diagonally from the corner post to the sill or if plywood or fibrous wall board is used, the 1 " x 4" diagonal brace may be omitted. The balloon frame also reduces shrinkage by reducing the amount of cross-section lumber to a minimum.

Plank And Beam Framing

The plank-and-beam system for the construction of floors and roofs in today's houses is uniquely adaptable to the modern design trends toward one-story structures, large glass areas, modular coordination, open-space planning, and, if desired, "natural finish" materials. Because this framing system is not familiar to many builders, it will be discussed in somewhat more detail here.

It is characterized by the concentration of structural loads on fewer and larger sized pieces than in conventional construction, resulting in rapid site assembly with fewer manhours. Dual function of materials and the use of planks continuous over two or more spans result in further economy.

Adapted from heavy timber construction, the plank and beam is used for floors and roofs in combination with ordinary wood stud or masonry walls, or curtain wall construction with skeleton frame. It is very popular for vacation-type homes.

Figure 2-4A illustrates the general principles and basic simplicity of the system. The similarity to structural steel framing practices is striking. Obviously, an engineered design is necessary for economy and safety. High quality materials and careful workmanship are inherent requirements. The resulting discipline produces a high quality product, sometimes at a lower cost. Simplicity of assembly extends even to inexpensive construction of the roof overhang so necessary in many parts of the country to keep the summer sun from the large glass areas. Assembly is easy because there are no rafter tails, no soffit, fascia, moldings, or eave vents. Insulation, roofing, and, in this case, gravel are stopped by a simple wood member, covered with metal to provide a drip.

Figure 2-4B shows the plank-and-beam roof applied to masonry walls. If you assume that Figures 2-4A and B represent a 24 foot wide house, you will note, with either a flat or a low-pitch roof, that 14 foot planks automatically provide an approximate 2 foot overhang. In any case, with a plank-and-beam roof the planned overhang is, for all practical purposes, complet-

ed when the carpenters have sheathed the roof.

Planning a house on a modular basis is a proven moneysaver for the builder because materials are used with a minimum amount of cutting and fitting. The plank-and-beam system fits naturally with this modular design. Thus, if you use the plank-and-beam system, you must decide what the beam spacing is to be. Your decision will naturally and logically be influenced by efficient use of collateral materials such as the planks and drywall sheets. For example, you might decide on an 8 foot spacing for beams and columns so that 16 foot planks will be continuous over two spans, and 4 by 8 sheets of drywall and plywood will work without cutting on the walls.

In addition to the economies resulting from adaptability to open planning, solar orientation, and modular coordination, the plank-and-beam system has a great potential for other savings in time, material, and money over conventional residential construction. These potential savings are due to the dual function of materials, fewer pieces to handle, and higher structural efficiency than in conventional construction.

The wide beam spacing and the 2 by 6, or 2 by 8 tongue-and-grooved dual-function planking provides the primary potential cost advantage—the elimination of the usual ceiling. Material and labor are saved in the ceiling joists, bridging, lath, and plaster or drywall. The planks and beams may be finished on the ground with "natural finishes" before erection. Interior scaffolding is unnecessary. The ceiling then would be completed when the planks are nailed in place. Too, there is the element of less time to build, which to the builder is money.

Applying a ceiling material to the underside of the planking between the beams, although not a very common practice, may be functionally, esthetically, or economically desirable. Such material, usually in the form of sheets or tiles, can be selected to provide thermal insulation, sound absorption, or a visually pleasant ceiling treatment. With this applied ceiling material you need not consider lumber appearance when you select the planks.

If the plank-and-beam system is used for a floor, it is possible to make the 2 inch planking serve both as the subfloor and finish flooring, although this practice may not be generally accepted. Further, the cost of bridging is eliminated and the 2 inch planking provides additional insulation for the floor.

Conventional Framing Vs. Plank-And-Beam

You can see the basic differences between conventional framing and the plank-and-beam system by carefully examining Figure 2-5A and

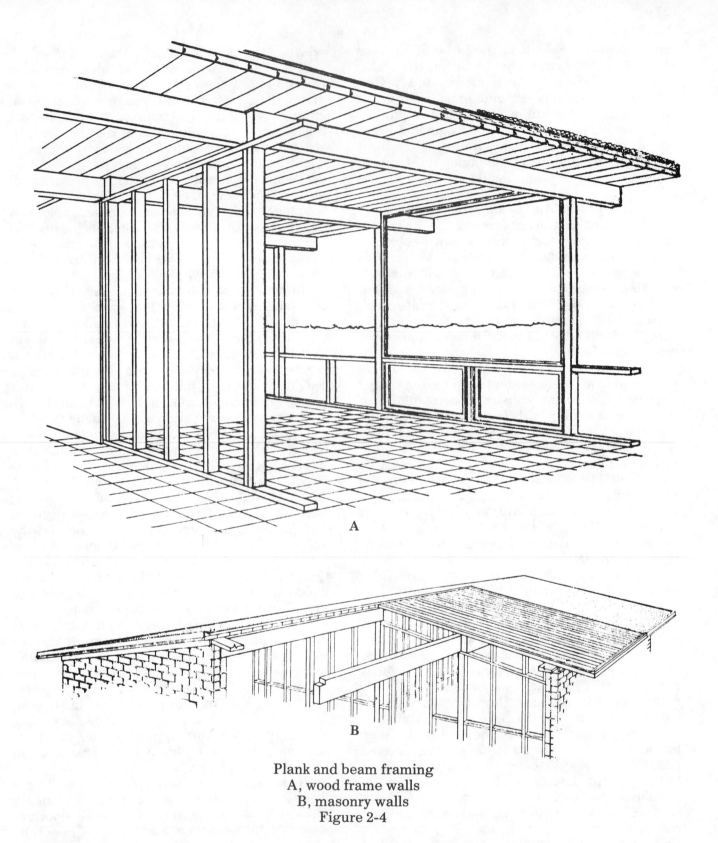

Plank and beam framing
A, wood frame walls
B, masonry walls
Figure 2-4

B. Each illustrates a typical 40 foot wall section with the same openings, the same floor to ceiling height, and the same height of the rough floor over the outside grade. The studs in Figure 2-5A are on 2 foot centers and those in Figure 2-5B on 16 inch centers. This was not done to bias the reader in favor of the plank-and-beam system but rather to compare the plank-and-beam system with the most commonly found conventional framing. The apparent relative complexity of the conventional framing would be somewhat less if the 2 foot module had been used.

At first glance, the most marked distinction

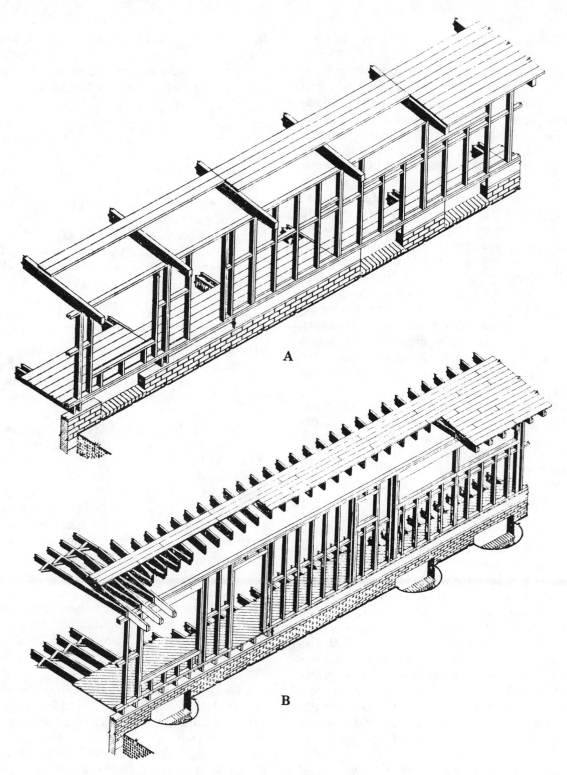

Plank and beam framing (A) and
conventional framing (B)
Figure 2-5

between Figures 2-5A and B is the large difference in the number of pieces of lumber. It is obvious from these illustrations that with the plank-and-beam system, the builder has an opportunity to save. The carpenters have fewer pieces to handle, saw, align, and nail. In addition, there are fewer points at which members must be nailed. What is not obvious, however, is that the plank-and-beam system demands very good workmanship. Because of the high load concentration, members must be cut true and square; and with so few points of contact between members, adequate nailing becomes very important.

Because ceiling heights are measured from the floor to the underside of the planking, you can save money in the construction of basements and walls with a plank-and-beam floor. Assume Figures 2-5A and B represent houses 24 feet wide by 40 feet long, with basement ceiling heights of 8 feet. Under this assumption, there is a saving on the basement construction in Figure 2-5A over that of Figure 2-5B of at least 85 square feet of 8 inch thick masonry wall, 30 cubic yards of excavation, and 9½ square yards of parging and waterproofing. Or to restate it, for the same amount of excavation and masonry you will have a greater volume in the plank-and-beam system basement than in the conventional basement. For the same reason, a saving can be realized on the wall studs in Figure 2-5A. With an 8 foot ceiling height, studs need be only 7 feet long in the plank-and-beam system. This means you can make full use of stock 14 foot 2 by 4's.

With the roof loads concentrated on columns spaced as much as 10 feet apart, the architect and builder are free to use large windows with no heavy lintels. Notice in Figure 2-5A how the window and door heads are brought directly to the top plate. There are no lintel beams or crippled studs as in Figure 2-5B. Although only a single top plate is indicated in Figure 2-5A, actually, a nailing strip approximately equal to a second plate must be provided on the underside of the planking in order to receive the interior finish and the exterior sheathing and siding.

Although the basement window openings in Figure 2-5A and B are the same size, and the height of the floor above grade is the same, window wells and lintels are necessary with the conventional construction. These same windows in the plank-and-beam system are above grade and without lintels. If you decide to use window wells, the planks-and-beam floor can be brought closer to the grade. Architecturally, this can be quite pleasing besides lowering the costs of entrance steps and rails.

As shown in Figure 2-5A, generally you need not provide double studs around openings because there is no roof load on the walls between columns. However, you must be careful about arbitrarily eliminating these double studs. When a wall has many large openings and low racking-strength sheathing is used, excessive deflection due to racking loads may cause windows and doors to bind.

If you build houses with pier foundations, you will find that the plank-and-beam construction lends itself well to that type of foundation.

In contrast to conventional construction where floor joists are supported by girders set in piers, the floor beams in the plank-and-beam system are directly supported by the piers.

Considering all the previously discussed advantages, you might wonder why all houses in this country are not of the plank-and-beam construction. Actually, technical problems concerning structural and architectural design limitations, materials (especially in certain parts of the country), and a lack of a thorough knowledge have limited the use of the system. None of these limitations restricts the designer excessively, and most of the problems involved can be readily solved, once they are understood.

The exposed plank ceiling with no attic space creates a problem in areas where the climate dictates the need for thick roof insulation. The insulation must be placed either between the planking and the roofing material, or be fastened to the underside of the planking. If the latter, then the eye appeal of the insulation as a ceiling becomes a major consideration. If it is placed over the planking, your choice of insulation is limited to a degree because it must have certain properties. It must be of the rigid type, able to bear the weight of the men applying the roofing, the roofing itself, and the snow load, without crushing or breaking. It must substantially maintain its insulating properties even when slightly wet. It must not deteriorate in time because of rot or other decomposition.

There are several insulating materials on the market which will satisfy these demands. However, while some can, many cannot be nailed to the planking. These are generally laid in mastics on roof decks limited to a maximum pitch of 3 in 12. Because of this limitation, you are restricted to a built-up roof with this insulation.

The initial installed unit cost of the rigid type of insulation for a plank-and-beam roof may often be higher than insulation for a conventional frame roof. This is not in itself a real cost disadvantage because the other economies of the system generally compensate for this higher insulation cost. However, it is possible, in very cold climates, where a great deal of insulation is required and where an adverse market for labor and materials exists, to find that the beam roof is uneconomical.

Reducing Framing Costs

The key to cost-saving in rough carpentry is in preplanning. The wise builder plans his job before layout begins and, in fact, before the framing is ordered.

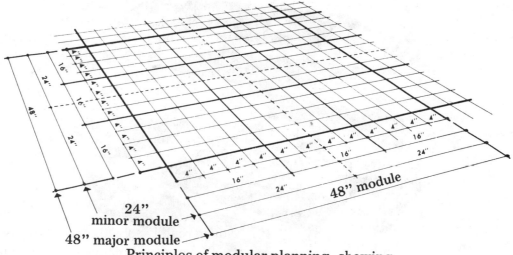

Principles of modular planning, showing
the planned grid
Figure 2-6

Home building in the United States is probably the most advanced of any nation in the world. Builder experience with various sizes and spacings of joists, studs and rafters together with standard 4' x 8' panel dimensions has resulted in a rapid and economical method of house framing. This method has contributed to the highest percentage of home ownership.

With the conventional method however, additional economies are possible. In 1974 a Washington, D.C. builder was able to demonstrate a labor and material savings of more than $43,000 for one 250-house tract as a result of preplanning. A Baltimore builder of lower cost homes was able to show a $16,500 savings for a 250-house project.

Such savings can be obtained without reducing the strength and rigidity of wood construction. They can stretch the nation's timber supply and help reduce building costs.

Preplanning

Expressions like modular coordination or modular dimensioning may sound overly involved. But they are today's terms for the preplanning which has historically been practiced by the most successful and cost-conscious contractors.

Using lumber framing and panel products on a preplanned (modular) basis in floors and roofs eliminates unnecessary waste of materials. Walls may also be preplanned to require a minimum of lumber in framing openings and corners and to accomodate insulation, plumbing, heating and drywall installation.

The Modular Plan

A modular plan involves three dimensions --

length, width and height. The modular length and width dimensions form the base and, in combination, are the two most important dimensions. The planning grid, Figure 2-6, is a horizontal plane divided into units of 4, 16, 24 and 48 inches. Overall dimensions are all multiples of 4 inches.

The 16-inch unit provides flexibility in spacing of windows and doors. Increments of 24 and 48 inches are used for the exterior dimensions of the house. Floor, ceiling and roof constructions are easily mated with these dimensions.

Modular plans for houses to be built of components of conventional framing are designed to exact size on the grid. The grid layout provides a rapid means of planning with accuracy.

Thickness and Tolerance Variables

Basic to preplanning of the house is the modular grid. The complete house is then divided into horizontal and vertical elements at regular intervals, as in Figure 2-7. Although the elements are shown as planes without thickness, allowance is made for wall thickness and tolerances based on fixed module lines at the outside faces of the studs.

Exterior walls and partitions may have many thickness variables. Floor and roof construction also vary in thickness, depending upon their structural requirements and types of framing and finishing.

Exterior Walls, Doors And Windows

An important feature of preplanning is the separation of exterior wall elements at natural division points. In Figure 2-8A, overall house

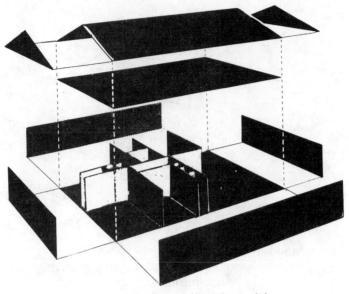

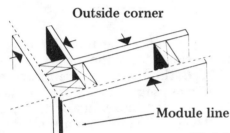

Outside corner   Exterior wall and partition   Inside corner

Module line

Division of houses into plane sections with
thickness and tolerance variables
Figure 2-7

dimensions are based on the 48-inch and 24-inch modules. However, maximum flexibility in placing door and window openings is achieved with the 16-inch module. Also, the precise location of wall openings on the 16-inch module eliminates extra wall framing frequently required in non-modular house construction.

Modular Roof Dimensioning

Increments for house depths are in 24-inch multiples. Six 48-inch module depths and five 24-inch module widths fulfill most roof span requirements. (See Figure 2-8B). Standard roof slopes combined with modular house depths provide all dimensions required for design of rafter, truss and panel roofs.

The pivotal point shown in Figure 2-9 is the fixed point of reference in the module line of the exterior wall. Modular roof design and construction dimensions are determined from this point.

Span tables for joists and rafters in this chapter permit selection of the most economical grades and sizes of locally available lumber. Floor joists are spaced 12, 16 or 24 inches on centers, depending upon the design floor loads and may be placed "lapped" or "inline." Joist spacings of 13.7 inches and 19.2 inches are alternatives which also divide the 8-foot length

of plywood subfloor panels into seven and five equal spaces, respectively.

Use of the 48-inch module on house depths permits greater use of full 4 x 8 foot plywood subflooring and minimizes cutting and waste.

Floor Framing

In many new homes, 6 to 17 percent of the cost of floor framing is wasted. If the front to back dimension of the floor plan measured between the outside surfaces of the exterior wall studs is not evenly divisible by 4, usable lengths of floor joists are being lost. See Figure 2-10. Check and determine the efficiency of your floor framing system. You may save money.

Lumber joists are produced in length increments of 2 feet with a tolerance of minus 0 inch and plus 3 inches. Joists 12 feet and longer typically have a plus tolerance of ½ inch or more to allow for cutting into shorter standard lengths.

In normal platform construction with lapped joists bearing on top of a center girder, the length of joist required is one-half the house depth minus the thickness of the band joist (1½ inch) plus one-half the required overlap at the support (1½ inch *under* Federal Housing Authority 1974 Minimum Property Standards).

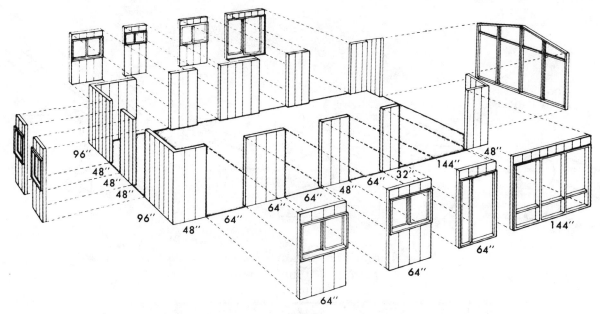

Exterior wall, door and window sections show
the need for modular planning of these
components. A 1/8-inch tolerance is
provided at each separation
Figure 2-8A

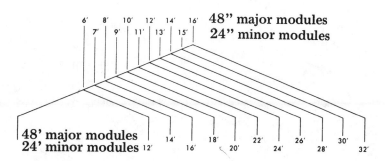

6' 8' 10' 12' 14' 16'   48" major modules
  7' 9' 11' 13' 15'   24" minor modules

48' major modules
24' minor modules 12'   14'   16'   18'   20'   22'   24'   26'   28'   30'   32'

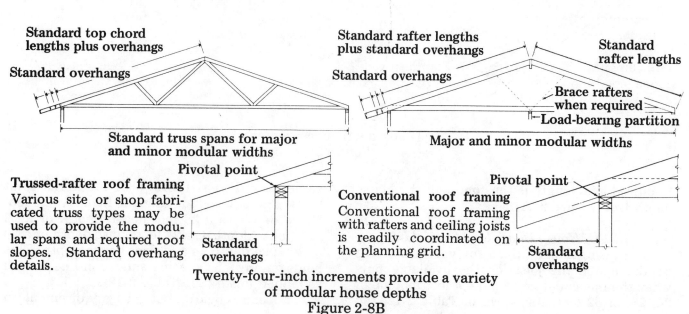

Standard top chord
lengths plus overhangs

Standard overhangs

Standard truss spans for major
and minor modular widths

Trussed-rafter roof framing
Various site or shop fabricated truss types may be
used to provide the modu-
lar spans and required roof
slopes. Standard overhang
details.

Standard rafter lengths
plus standard overhangs

Standard overhangs

Standard
rafter lengths

Brace rafters
when required
Load-bearing partition

Major and minor modular widths

Pivotal point

Conventional roof framing
Conventional roof framing
with rafters and ceiling joists
is readily coordinated on
the planning grid.

Pivotal point

Standard
overhangs

Standard
overhangs

Twenty-four-inch increments provide a variety
of modular house depths
Figure 2-8B

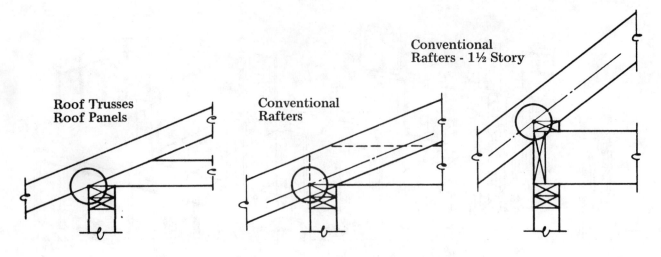

**Roof Trusses
Roof Panels**

**Conventional
Rafters**

**Conventional
Rafters - 1½ Story**

"Pivotal point" of reference for the modular
planning of house roofs
Figure 2-9

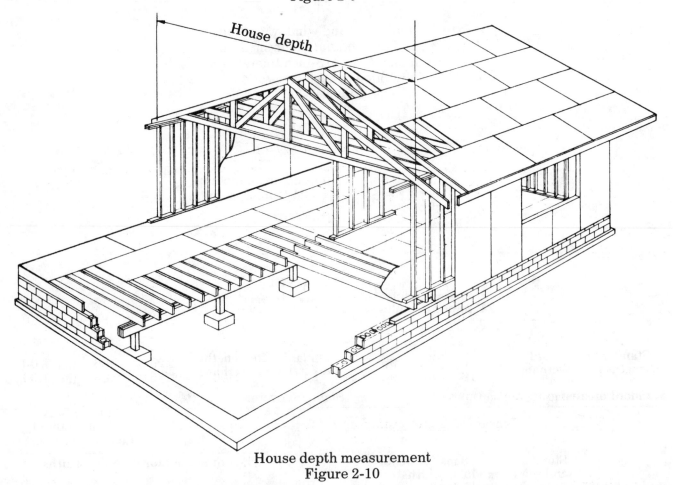

House depth

House depth measurement
Figure 2-10

Required joist length is thus one-half the house depth (see Figure 2-11).

Effect Of House Depth On Material Use

Joist lengths required for various house depths match the joist lengths available only when the house depth is on the 4 foot module of 24, 28, or 32 feet. As shown in Table 2-12, the same total lineal footage of standard joist lengths is required to frame the floor of a 25 foot house depth as a 28 foot house depth, or a 22 foot house depth as a 24 foot house depth or a 29 foot house depth as a 32 foot house depth. This holds for any joist spacing and for in-line as well as lapped joists (See Table 2-12).

The extra board foot joist requirement in

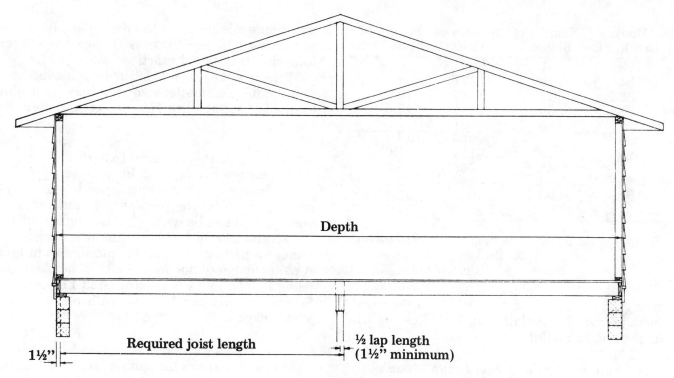

Depth

Required joist length | ½ lap length (1½" minimum)

1½"

Required joist length
Figure 2-11

Required footage of joists per 4' of house length based on standard joist length

House Depth Feet	Joist Required Feet	Length Standard Feet	Linear feet 16" joist Spacing	24" joist Spacing	Bd. ft. per sq. ft. of floor area* 16" spacing 2 x 8	2 x 10	24" spacing 2 x 10	2 x 12
23	10½	12	72	48	1.14	1.43	.95	1.14
22	11	12	72	48	1.09	1.36	.91	1.09
23	11½	12	72	48	1.04	1.30	.87	1.04
24	12	12	72	48	1.00	1.25	.83	1.00
25	12½	14	84	56	1.12	1.40	.93	1.12
26	13	14	84	56	1.08	1.35	.89	1.08
27	13½	14	84	56	1.04	1.30	.86	1.04
28	14	14	84	56	1.00	1.25	.83	1.00
29	14½	16	96	64	1.10	1.38	.92	1.10
30	15	16	96	64	1.07	1.33	.89	1.07
31	15½	16	96	64	1.03	1.29	.86	1.03
32	16	16	96	64	1.00	1.25	.84	1.00

*Floor area = House depth times 4'

Linear footage of joists required for various house depths
Table 2-12

non-modular house depths is taken up in lap over the center support. Extra overlap does not normally improve floor performance. Where in-line joists are used, the extra joist length is cut-off waste or must be recut to serve as blocking.

Savings

A 1970 survey of 272 builders by the National Association of Home Builders Research Founda-tion showed that more than 40 percent of the 14,248 housing units involved had front to back dimensions which were not on the 4 foot depth module of 24, 28 or 32 feet (see Table 2-13).

If the house plan involves a 22, 25, 26, 27, 29, 30 or 31 foot house depth measured out to out on the exterior studs, joist costs can be reduced up to 12 percent without changing joist size (labor and material costs for plywood subfloor can be reduced at the same time) by

House Depth, Ft.	Number of Builders	Number of Units	Percent of Total Units
22	1	22	0.15
24	18	1371	9.6
25	18	1445	10.1
26	50	2678	18.8
27	6	251	1.8
28	39	2059	14.4
29	10	134	1.0
30	32	1068	7.5
31	4	290	2.0
32	33	1947	13.7
33+	61	3083	21.6

House depths reported by builders in 1970 survey
Table 2-13

changing to the 4 foot module. Total floor area footage can be maintained by adjusting the house length module.

Changing To The 4 Foot Depth Module

If joist spans are increased in changing to the 4 foot module, the allowable span for the species and grade being used must be rechecked to determine adequacy on the new span. It will be satisfactory in many cases if you have been following traditional practice.

Where spans are reduced to achieve the 4 foot module, a smaller size joist may be usable and additional significant savings can be obtained.

Subfloor Panel Layout

Maximum savings can be achieved with the 4 foot house depth module if the layout of the subfloor panels are preplanned. Full 48 inch wide panels can be used without ripping for the 24, 28 and 32 foot house depths when joists are either lapped or trimmed for placement in line. A panel layout option for the 28 foot house depth using lapped joists is illustrated in Figure 2-14. Similar layouts can be used with other lapped joist designs.

Allowable Joist Spans

Allowable spans for lumber joists refer to the clear span between supports (see Figure 2-15). Such spans are determined by the working stresses for the joist grade and species, the joist

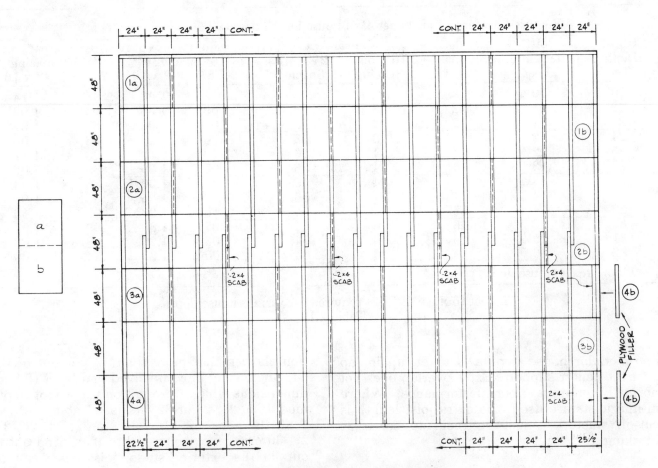

Subfloor panel layout for a 28' house depth
when joists are lapped over the center support
Figure 2-14

40

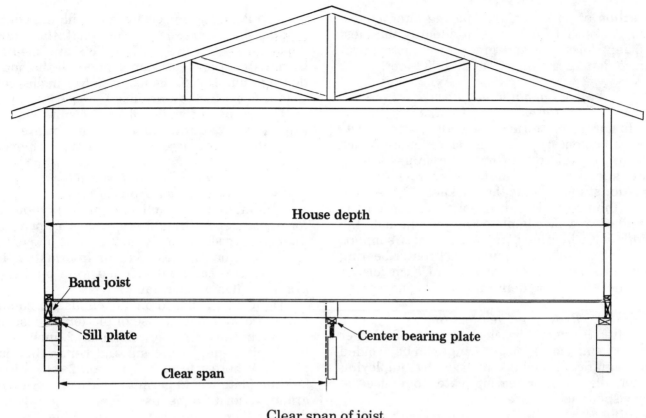

Clear span of joist
Figure 2-15

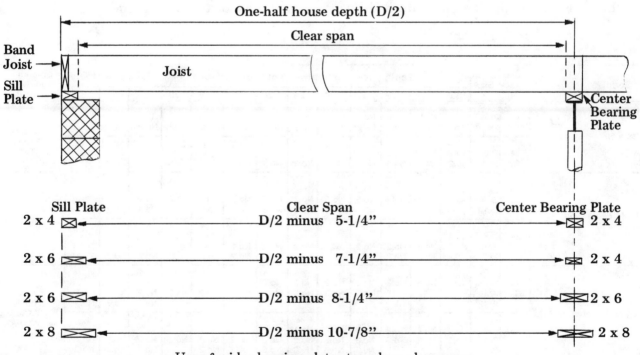

Sill Plate	Clear Span	Center Bearing Plate
2 x 4	D/2 minus 5-1/4"	2 x 4
2 x 6	D/2 minus 7-1/4"	2 x 4
2 x 6	D/2 minus 8-1/4"	2 x 6
2 x 8	D/2 minus 10-7/8"	2 x 8

Use of wider bearing plates to reduce clear spans
Figure 2-16

size and spacing and the design loads specified in the building code or Federal Housing Administration standard. Most requirements limit joist deflection to 1/360 of the span under a 40 pound per square foot (or smaller) uniform live load. Joist strength must be sufficient to carry the live load plus a dead load of 10 pounds per square foot.

Available lumber species and grades provide a broad range of span capabilities and a selection of joist size and spacing alternatives. The tables that follow help to identify the most efficient joist size and grade combinations for house depths ranging from 20 to 32 feet.

Span Capability Required For Different House Depths

In normal platform construction, where joists bear on a center girder, the clear span of each joist is equal to one-half the house depth (measured between outside surfaces of the exterior studs), minus the thickness of the band joist, minus the length of joist bearing on the foundation wall sill plate, minus one-half the width of the bearing plate on the center support. Where 2 x 4 sill plates and a 2 x 4 center bearing plate are used, the clear span of each joist is one-half the house depth minus 5¼ inches.

Reducing The Clear Span

Often the need for selecting joists of a larger size or using closer joist spacings can be avoided and significant savings realized by employing wider sills or center bearing plates to reduce the clear span (see Figure 2-16).

When 2 x 6 sill plates are used with a 2 x 4 center bearing plate, the clear span of each joist is one-half the house depth minus 7¼ inches, or 2 inches less than if a 2 x 4 sill plate had been

used. If 2 x 6's are used for both sill and center plates, the clear span is one-half the house depth minus 8¼ inches. If 2 x 8's are used for both plates, the clear span is one-half the house depth minus 10-7/8 inches, or 5-5/8 inches less than if 2 x 4 plates were used.

Where use of wider plates permits smaller joists, the added cost of the plates is insignificant compared to the saving achieved.

Minimum Joist Grades And Sizes For Different House Depths

Minimum grades and sizes of joists required for house depths from 20 to 32 feet at one foot intervals are shown in Table 2-17 for some of the most common species. Similar information for other species can be developed from published grading rules or span tables.

Table 2-17 is based on 40 pounds per square foot uniform live load, use of platform construction with center girder, nominal 2 inch thick band joists, and 2 x 4 sill and center bearing plates. Where use of a 2 x 6 or 2 x 8 sill or bearing plate would permit a particular size and grade of joist to be used for a larger house depth, this option is noted in the table.

Maximum efficiency is achieved by using the house depth which most fully utilizes the span capability of the joist grade and size selected.

Species	Joist Spacing	Joist Size	20	21	22	23	24	25	26	27	28	29	30	31	32	Grading Rule Agency*
Balsam Fir	16"	2x8	No.2	No.2	No.2	No.2	No.2[c]	No.1[c]								NeLMA NH&PMA
		2x10	No.3	No.3	No.3	No.3[a]	No.2	No.2	No.2	No.2	No.2	No.2	No.2[c]	No.1[a]		
	24"	2x8	No.2[c]	No.1[a]												
		2x10	No.2	No.2	No.2	No.2	No.2	No.2[c]	No.1	No.1[c]						
California Redwood (open grain)	16"	2x8	No.3[c]	No.2	No.2	No.2[a]	No.1[b]									RIS
		2x10	No.3	No.3	No.3	No.3	No.3	No.3[c]	No.2	No.2	No.2	No.2[a]	No.1[a]			
		2x12							No.3	No.3	No.3	No.3	No.3[b]	No.2	No.2	
	24"	2x8	No.2	No.1[a]												
		2x10	No.3	No.3[c]	No.2	No.2	No.2	No.2	No.2[c]	No.1[c]						
		2x12		No.3	No.3	No.3	No.3	No.3[c]	No.2	No.2	No.2	No.2	No.2	No.2[a]	No.1[b]	
Douglas Fir-Larch	16"	2x8	No.3	No.3	No.3	No.2	No.2	No.2	No.2	No.2	No.1[b]	No.2	No.2	No.2	No.2	NLGA WCLIB WWPA
		2x10				No.3	No.3	No.3	No.3	No.3	No.3[a]	No.3	No.3	No.3	No.3	
		2x12									No.3	No.3	No.3	No.3	No.3	
	24"	2x8	No.2	No.2	No.2	No.2	No.2[c]	No.1[c]								
		2x10	No.3	No.3	No.3	No.3[a]	No.2	No.2	No.2	No.2	No.2	No.2	No.2[b]	No.1[a]	No.1[c] DENSE	
		2x12				No.3	No.3	No.3	No.3	No.3	No.3[b]		No.2	No.2	No.2	

(Based on platform construction, nominal 2-inch thick band joists and, except where footnoted, nominal 2x4 sill and 2x4 center bearing plates)

Minumum joist grades for different house depths
40 psf live load
Table 2-17

Table 2-17 — Allowable joist grades by species, joist spacing/size, and house depth (House Depth measured between outside surfaces of exterior studs, feet)

Species	Joist Spacing	Joist Size	20	21	22	23	24	25	26	27	28	29	30	31	32	Grading Rule Agency*
Douglar Fir-South	16″	2x8	No.3	No.3	No.3[c]	No.2	No.2	No.2[a]	No.1[c]							WWPA
		2x10			No.3	No.3	No.3	No.3	No.3	No.3	No.2	No.2	No.2	No.2	No.2[c]	
		2x12									No.3	No.3	No.3	No.3	No.3	
	24″	2x8	No.2	No.2	No.2[a]	No.1[c]										
		2x10	No.3	No.3	No.3	No.3[c]	No.2	No.2	No.2	No.2	No.2[b]	No.1[c]				
		2x12					No.3	No.3	No.3	No.3[a]	No.2	No.2	No.2	No.2	No.2	
Eastern Hemlock-Tamarack	16″	2x8	No.3	No.3[c]	No.2	No.2	No.2[b]	No.1[a]								NeLMA, NLGA
		2x10		No.3	No.3	No.3	No.3	No.3	No.3[a]	No.2	No.2	No.2	No.2[a]	No.1	No.1[c]	
	24″	2x8	No.2	No.2[a]	No.1[a]											
		2x10	No.3	No.3	No.3[c]	No.2	No.2	No.2	No.2	No.2[c]	No.1[b]					
Eastern Spruce	16″	2x8	No.2	No.2	No.2	No.2	No.2[a]	No.1	No.1[c]							NeLMA, NH&PMA
		2x10	No.3	No.3	No.3	No.3	No.3[b]	No.2	No.2	No.2	No.2	No.2	No.2	No.2[c]	No.1	
	24″	2x8	No.2[b]	No.1	No.1[a]											
		2x10	No.3[c]	No.2	No.2	No.2	No.2	No.2[a]	No.1	No.1	No.1[b]					
Engelmann Spruce-Alpine Fir or (Engelmann Spruce-Lodgepole Pine)	16″	2x8	No.2	No.2	No.2	No.2	No.2[c]	No.1[a]								WWPA
		2x10	No.3	No.3	No.3	No.3[a]	No.2	No.2	No.2	No.2	No.2	No.2	No.2[c]	No.1	No.1[c]	
		2x12			No.3	No.3	No.3	No.3	No.3	No.3			No.2	No.2	No.2	
	24″	2x8	No.2[c]	No.1[a]												
		2x10	No.2	No.2	No.2	No.2	No.2	No.2[c]	No.1	No.1[c]						
		2x12	No.3	No.3	No.3	No.3[c]		No.2	No.2	No.2[c]	No.2	No.2	No.2[c]	No.1	No.1	
Hem-Fir	16″	2x8	No.3[a]	No.2	No.2	No.2	No.2	No.2	No.2[c]							NLGA, WCLIB, WWPA
		2x10		No.3	No.3	No.3	No.3	No.3	No.3[a]	No.2	No.2	No.2	No.2	No.2	No.2	
		2x12							No.3	No.3	No.3	No.3	No.3	No.3[c]	No.2	
	24″	2x8	No.2	No.2[a]	No.1	No.1										
		2x10	No.3	No.3[c]	No.2	No.2	No.2	No.2	No.2	No.2[c]	No.1	No.1[a]				
		2x12	No.3	No.3	No.3	No.3	No.3	No.3[a]		No.2	No.2	No.2	No.2	No.2	No.2	
Idaho White Pine or Western White Pine	16″	2x8	No.2	No.2	No.2	No.2[a]	No.1	No.1[a]								WWPA
		2x10	No.3	No.3	No.3	No.3[a]	No.2	No.2	No.2	No.2	No.2	No.2	No.1	No.1	No.1[b]	
		2x12			No.3	No.3	No.3	No.3	No.3	No.3	No.3[c]		No.2	No.2	No.2	
	24	2x8	No.1	No.1[b]												NLGA
		2x10	No.2	No.2	No.2	No.2	No.2[a]	No.1	No.1[b]							
		2x12	No.3	No.3	No.3	No.3	No.3[c]	No.2	No.2	No.2	No.2	No.2	No.2[a]	No.1	No.1[c]	
Lodgepole pine	16″	2x8	No.2	No.2	No.2	No.2	No.2	No.2	No.1[a]							WWPA
		2x10	No.3	No.3	No.3	No.3	No.3	No.2	No.2[c]	No.2	No.2	No.2	No.2	No.2[a]	No.1[c]	
		2x12							No.3	No.3	No.3	No.3	No.3[a]	No.2	No.2	
	24″	2x8	No.2	No.1	No.1[a]											
		2x10	No.3[a]	No.2	No.2	No.2	No.2	No.2	No.2[c]	No.1	No.1[b]					
		2x12			No.3	No.3	No.3	No.3[a]	No.2	No.2	No.2	No.2	No.2	No.2[b]	No.1	
Northern Pine	16″	2x8	No.3[c]	No.2	No.2	No.2	No.2	No.2[a]	No.1[c]							NeLMA, NH&PMA
		2x10	No.3	No.3	No.3	No.3	No.3	No.3[c]	No.2	No.2	No.2	No.2	No.2	No.2	No.2[c]	
	24″	2x8	No.2	No.2[b]	No.1	No.1[c]										
		2x10	No.3	No.3[c]	No.2	No.2	No.2	No.2	No.2[a]	No.1	No.1	No.1[c]				
Ponderosa Pine-Sugar Pine	16″	2x8	No.2	No.2	No.2	No.2	No.2[b]	No.1[c]								WWPA
		2x10	No.3	No.3	No.3	No.3	No.2	No.2	No.2	No.2	No.2	No.2	No.2[a]	No.1[a]		
		2x12							No.3	No.3	No.3	No.3	No.3[a]	No.2	No.2	
	24″	2x8	No.2[c]	No.1	No.1[c]											
		2x10	No.2	No.2	No.2	No.2	No.2	No.2[c]	No.1	No.1[a]						
		2x12	No.3	No.3	No.3	No.3[a]		No.2	No.2	No.2	No.2	No.2	No.2[b]	No.1	No.1	
Southern Pine	16″	2x8	No.3	No.3	No.3[a]	No.2	No.2	No.2	No.2[c]	MG No.2[b]	No.1[b]					SPIB
		2x10			No.3	No.3	No.3	No.3	No.3	No.3	No.3[c]	No.2	No.2	No.2	No.2	
		2x12									No.3	No.3	No.3	No.3	No.3	
	24″	2x8	No.2	No.2	No.2[c]	No.2	MG No.2[b]	No.1[c]								
		2x10	No.3	No.3	No.3	No.3[b]	No.2	No.2	No.2	No.2	MG No.2	MG No.2	MG No.2[b]	No.1[a]	No.1[c] DENSE	
		2x12				No.3	No.3	No.3	No.3	No.3	No.3[c]	No.2	No.2	No.2	No.2	

Table 2-17 (continued)

Species	Joist Spacing	Joist Size	House Depth (measured between outside surfaces of exterior studs), feet													Grading Rule Agency[*]
			20	21	22	23	24	25	26	27	28	29	30	31	32	
Southern Pine KD (15% mc)	16"	2x8	No.3	No.3	No.3	No.3[b]	No.2	No.2	No.2	MG No.2	No.1	No.1[b] DENSE				SPIB
		2x10				No.3	No.3	No.3	No.3	No.3	No.3	No.3[b]	No.2	No.2	No.2	
		2x12											No.3	No.3	No.3	
	24"	2x8	No.2	No.2	No.2	No.2[c]	MG No.2[a]	No.1[c]								
		2x10	No.3	No.3	No.3	No.3	No.3[b]	No.2	No.2	No.2	No.2	No.2[c]	MG No.2	No.1	No.1[c]	
		2x12					No.3	No.3	No.3	No.3	No.3	No.3	No.3[a]	No.2	No.2	
Spruce-Pine-Fir Coast Sitka Spruce or Sitka Spruce	16"	2x8	No.2	No.2	No.2	No.2	No.2[a]	No.1	No.1[a]							NLGA
		2x10	No.3	No.3	No.3	No.3	No.2	No.2	No.2	No.2	No.2	No.2	No.2	No.2[c]	No.1	
		2x12							No.3	No.3	No.3	No.3[a]		No.2	No.2	
	24"	2x8	No.2[b]	No.1	No.1[c]											WCLIB
		2x10	No.2	No.2	No.2	No.2	No.2	No.2[a]	No.1	No.1						
		2x12	No.3	No.3	No.3	No.3[a]		No.2	No.2	No.2	No.2	No.2	No.2[a]	No.1	No.1	
Western Hemlock	16"	2x8	No.3	No.3[a]	No.2	No.2	No.2	No.2	No.2	No.2[c]	No.1[b]					WWPA
		2x10		No.3	No.3	No.3	No.3	No.3	No.3	No.3	No.3[c]	No.2	No.2	No.2	No.2	
		2x12									No.3	No.3	No.3	No.3	No.3[a]	
	24"	2x8	No.2	No.2	No.2[a]	No.1	No.1[c]									
		2x10	No.3	No.3	No.3[b]	No.2	No.2	No.2	No.2	No.2	No.2	No.2[b]	No.1	No.1[b]		
		2x12										No.2	No.2	No.2	No.2	
White Woods (Western Woods)	16"	2x8	No.2	No.2	No.2	No.2[a]	No.1[b]									WWPA
		2x10	No.3	No.3	No.3	No.3[a]	No.2	No.2	No.2	No.2	No.2	No.2[a]	No.1[a]			
		2x12					No.3	No.3	No.3	No.3	No.3	No.3[c]	No.2	No.2	No.2	
	24"	2x8	No.1	No.1[b]												
		2x10	No.2	No.2	No.2	No.2	No.2[a]	No.1	No.1[a]							
		2x12	No.3	No.3	No.3	No.3[c]	No.2	No.2	No.2	No.2	No.2	No.2[a]	No.1	No.1	No.1[c]	

[a] Nominal 2x6 sill plate and 2x4 center bearing plate; or width of sill plate plus one-half width of center bearing equal to 7¼" or more.
[b] Nominal 2x6 sill plate and 2x6 center bearing plate; or width of sill plate plus one-half width of center bearing equal to 8¼" or more.
[c] Nominal 2x8 sill plate and 2x8 center bearing plate; or width of sill plate plus one-half width of center bearing equal to 10⅞" or more.
[*] NeLMA—Northeastern Lumber Manufacturers Association
NH&PMA—Northern Hardwood and Pine Manufacturers Association
RIS—Redwood Inspection Service
NLGA—National Lumber Grades Authority, a Canadian Agency
WCLIB—West Coast Lumber Inspection Bureau
WWPA—Western Wood Products Association
SPIB—Southern Pine Inspection Bureau

Minimum joist grades for different house
depths 40 psf live load
Table 2-17

For example, an existing floor design may call for Number 2 southern pine 2 x 10's, spaced 16 inches on center for a house depth of 30 feet. However, the same joist grade and size can be used for a 32 foot wide house. Changing the floor plan to a major 4 foot module of 32 feet eliminates wastage of available joist length while fully utilizing the span capability of the grade.

Alternatively, there are situations where a more economical grade or size may be used in an existing floor plan. A 28 foot house may be spanned at lower cost by using Number 3 Douglas fir-larch 2 x 10's, 16 inches on centers and 2 x 6 sill plates than with Number 2, 2 x 10's of the same species and spacing, but with 2 x 4 sill plates. Use of Number 1, 2 x 8's 16 inches on centers with 2 x 6 sill and center bearing plates may also be a more economical alternative.

In many cases, fuller utilization of span capability can be achieved with reduced costs by increasing joist spacing. For example, Number 2 Douglas fir-larch or southern pine 2 x 10's can span a 28 foot house when spaced 24 inches on centers as well as when spaced 16 inches on centers. In this instance the most economical joist spacing will depend upon a comparison of the added cost of the thicker plywood subfloor required for 24 inch spacing compared with savings in material and labor from reducing the number of joists required. Potential cost saving options such as these can be identified from Table 2-17.

Field-Glued Floor Systems

Maximum spans for many of the higher lumber grades are limited by stiffness. In such cases, additional spanning capability can be obtained by field-gluing plywood subfloor-underlayment to the joists. The glued-floor system also minimizes squeaks and nail pops, a major builder call-back problem.

Minimum field-glued floor systems required for various house depths are shown in Table 2-18. Only those joist grades and sizes which gain added span capability through field gluing are shown. Joist and plywood requirements in Table 2-18 are in agreement with maximum clear spans for glued plywood floor systems accepted by HUD/FHA and most building codes.

Use of field-gluing can reduce the joist size or grade required for particular floor designs. Such cost saving opportunities can be identified by comparing joist requirements in Tables 2-17 and 2-18 for the same house depth. For example, Number 1 Douglas fir-larch or southern pine 2 x 8's, 16 inches on centers, with 2 x 6 sill and center bearing plates are required for a 28 foot house. When field gluing is employed, Number 2 joists of the same size and species on the same spacing and with 2 x 4 bearing plates would be satisfactory.

Savings achieved by use of a lower, more readily available grade, or smaller size joist, often will be much larger than the additional cost of gluing (nails are spaced only 12 inches on centers in field gluing compared to 6 inches and 10 inches for conventional nailed construction).

It will pay to check out the field-gluing option.

Some species and sizes listed in Tables 2-17 and 2-18 are produced and used only in certain regions. Consult your local lumber supplier for information on available species and grades.

Exterior Wall Framing

Field observations indicate that builder costs for exterior wall framing can be as much as 25 percent higher than necessary. Some common ways wall framing costs can be reduced, while still building a high quality wall, are described in this manual. If you are not already taking full advantage of these techniques, significant additional earnings are possible on future construction.

Preplan For Savings

The key to economical wall framing is preplanning to eliminate unnecessary materials and labor. The wise builder plans his job before construction begins, and even before the lumber is ordered.

Builders frequently decide on a house design, estimate the required quantity of framing lumber and then turn the job over to the carpenter foreman or subcontractor to frame the house with little in the way of detailed instructions. As a carpenter you will do your best to follow a modular 16 inch or 24 inch spacing of studs, but you will usually find that

Species	Joist Spacing	Joist Size	Plywood Underlayment Thickness	House Depth (measured between outside surfaces of exterior studs), feet												
				20	21	22	23	24	25	26	27	28	29	30	31	32
Balsam Fir	16"	2x8	1/2	—	—	—	—	—	No.1	No.1[c]	—	—	—	—	—	—
		2x10	1/2	—	—	—	—	—	—	—	—	—	—	—	No.1	No.1
California Redwood (open grain)	16"	2x8	1/2	—	—	—	No.2	No.2	No.2[a]	No.1[b]	—	—	—	—	—	—
			19/32	—	—	—	—	—	—	—	No.1[c]	—	—	—	—	—
			23/32	—	—	—	—	—	—	—	No.1[a]	—	—	—	—	—
		2x10	1/2	—	—	—	—	—	—	—	—	—	—	No.2	No.2	No.1[a]
			19/32	—	—	—	—	—	—	—	—	—	—	—	—	No.2[b]
	24"	2x8	23/32	No.2	No.2	No.1	No.1[b]	—	—	—	—	—	—	—	—	—
		2x10	23/32	—	—	—	—	—	—	No.2[a]	No.1	No.1	No.1[b]	—	—	—
		2x12	23/32	—	—	—	—	—	—	—	—	—	—	—	No.2	No.2
Douglas Fir-Larch	16"	2x8	1/2	—	—	—	—	—	—	—	—	No.2	No.2[c]	No.1[c]	—	—
			19/32	—	—	—	—	—	—	—	—	—	—	No.1[a]	No.1[c]	—
			23/32	—	—	—	—	—	—	—	—	—	—	—	No.1[c]	—
	24"	2x8	23/32	—	—	—	—	—	No.1	No.1[c]	—	—	—	—	—	—
		2x10	23/32	—	—	—	—	—	—	—	—	—	—	—	No.1	No.1

Joist grades having additional house depth spanning capability through field-gluing of plywood subfloor-underlayment to joists
40 psf live load
Table 2-18

Species	Joist Spacing	Joist Size	Plywood Underlayment Thickness	20	21	22	23	24	25	26	27	28	29	30	31	32
Douglas Fir South	16"	2x8	1/2	—	—	—	—	—	No.2	No.2	No.2[a]	No.1[c]	—	—	—	—
			19/32	—	—	—	—	—	—	—	No.2[c]	—	—	—	—	—
			23/32	—	—	—	—	—	—	—	No.2[b]	No.1[c]	—	—	—	—
	24"	2x10	1/2	—	—	—	—	—	—	—	—	—	—	—	—	No.2
		2x8	23/32	—	—	No.2	No.2[b]	No.1	No.1[a]	—	—	—	—	—	—	—
		2x10	23/32	—	—	—	—	—	—	—	—	No.2	No.2[b]	No.1	No.1	No.1[b]
Eastern Hemlock-Tamarack	16"	2x8	1/2	—	—	—	—	—	No.2	No.2	No.2[b]	No.1[a]	—	—	—	—
			19/32	—	—	—	—	—	—	—	No.2[a]	No.1[c]	—	—	—	—
			23/32	—	—	—	—	—	—	—	—	No.1[a]	—	—	—	—
		2x10	1/2	—	—	—	—	—	—	—	—	—	—	No.2	No.2	No.2[a]
			19/32	—	—	—	—	—	—	—	—	—	—	—	—	No.2
	24"	2x8	23/32	—	No.2	No.2[c]	No.1	No.1[a]	—	—	—	—	—	—	—	—
		2x10	23/32	—	—	—	—	—	—	—	No.2	No.1	No.1	No.1	No.1[c]	—
Eastern Spruce	16"	2x8	1/2	—	—	—	—	—	—	No.1	No.1[b]	—	—	—	—	—
Engelmann Spruce-Alpine Fir	16"	2x8	1/2	—	—	—	—	—	No.1	No.1[c]	—	—	—	—	—	—
		2x10	1/2	—	—	—	—	—	—	—	—	—	—	—	—	No.1
Hem-Fir	16"	2x8	1/2	—	—	—	—	—	—	—	No.1	No.1[a]	—	—	—	—
			19/32	—	—	—	—	—	—	—	—	No.1	—	—	—	—
	24"	2x10	23/32	—	—	—	—	—	—	—	—	—	No.1	No.1[c]	—	—
Lodgepole Pine	16"	2x8	1/2	—	—	—	—	—	—	No.1	No.1[a]	—	—	—	—	—
		2x10	1/2	—	—	—	—	—	—	—	—	—	—	—	—	No.1
	24"	2x8	23/32	—	—	No.1	No.1[c]	—	—	—	—	—	—	—	—	—
		2x10	23/32	—	—	—	—	—	—	—	—	No.1	No.1[c]	—	—	—
Northern Pine	16"	2x8	1/2	—	—	—	—	—	—	No.1	No.1	No.1[c]	—	—	—	—
			19/32	—	—	—	—	—	—	—	—	No.1[a]	—	—	—	—
			23/32	—	—	—	—	—	—	—	—	No.1	—	—	—	—
		2x10	1/2	—	—	—	—	—	—	—	—	—	—	—	—	No.2[b]
	24"	2x8	23/32	—	—	—	No.1	—	—	—	—	—	—	—	—	—
		2x10	23/32	—	—	—	—	—	—	—	—	—	No.1	No.1[c]	—	—
Ponderosa Pine-Sugar Pine	16"	2x8	1/2	—	—	—	—	—	No.1	No.1[a]	—	—	—	—	—	—
		2x10	1/2	—	—	—	—	—	—	—	—	—	—	—	No.1	No.1
	24"	2x10	23/32	—	—	—	—	—	—	—	No.1	—	—	—	—	—
Southern Pine	16"	2x8	1/2	—	—	—	—	—	—	No.2[a]	MG No.2	MG No.2	MG No.2	No.1[c]	—	—
			19/32	—	—	—	—	—	—	—	—	—	—	No.1[a]	No.1[c]	—
	24"	2x8	23/32	—	—	—	—	—	No.1	No.1[b]	—	—	—	—	—	—
		2x10	23/32	—	—	—	—	—	—	—	—	—	—	—	No.1	No.1
Southern Pine KD (15% mc)	16"	2x8	1/2	—	—	—	—	—	—	—	No.2	MG No.2	MG No.2[a]	No.1[a]	—	—
			19/32	—	—	—	—	—	—	—	—	—	—	No.2[c]	No.1[c]	—
			23/32	—	—	—	—	—	—	—	—	—	—	—	No.1[b]	—
	24"	2x8	23/32	—	—	—	—	MG No.2	MG No.2[c]	No.1	No.1[a]	—	—	—	—	—
		2x10	23/32	—	—	—	—	—	—	—	—	—	—	—	MG No.2[c]	No.1
Western Hemlock	16"	2x8	1/2	—	—	—	—	—	—	No.2	No.2[b]	No.1	No.1[c]	—	—	—
			19/32	—	—	—	—	—	—	—	—	—	No.1[a]	—	—	—
			23/32	—	—	—	—	—	—	—	—	—	—	No.1[c]	—	—
	24"	2x8	23/32	—	—	—	—	No.1	No.1[c]	—	—	—	—	—	—	—
		2x10	23/32	—	—	—	—	—	—	—	—	—	—	No.1	No.1[b]	—
White Woods (Western Woods)	16"	2x8	1/2	—	—	—	—	No.1	No.1[a]	—	—	—	—	—	—	—
		2x10	1/2	—	—	—	—	—	—	—	—	—	No.2	No.1	No.1	No.1[c]

[a] Nominal 2x6 sill plate and 2x4 center bearing plate; or width of sill plate plus one-half width of center bearing equal to 7¼" or more.
[b] Nominal 2x6 sill plate and 2x6 center bearing plate; or width of sill plate plus one-half width of center bearing equal to 8¼" or more.
[c] Nominal 2x8 sill plate and 2x8 center bearing plate; or width of sill plate plus one-half width of center bearing equal to 10⅞" or more.

Joist grades having additional house depth spanning capability through field-gluing of plywood subfloor-underlayment to joists 40 psf live load
Table 2-18 (continued)

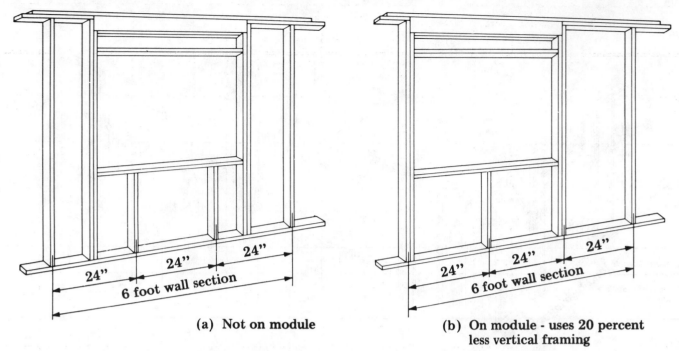

(a) Not on module
(b) On module - uses 20 percent less vertical framing

Windows located on module can save framing
Figure 2-19

extra pieces are required here and there to accommodate doors, windows and partitions. You may also follow traditional training by adding other things, such as extra studs where partitions meet outside walls, blocking at mid-height of walls, double studs and headers for opening in nonbearing walls and similar practices. Each of these is avoidable through proper attention in the planning stage. They add significantly to cost without corresponding benefit to the structure or the home buyer.

Modular Plans

Cost savings will be greatest when the overall size of the house plus the size and location of wall openings coincide with standard modular stud spacings. For the greatest possible savings of materials and labor, detailed plans showing size, grade, quality and location of all wall components should be prepared before construction begins. Preplanning the actual locations of studs will not only save framing lumber but will also permit better utilization of sheathing or siding. An important side benefit will be education of the work crew. When the crew has constructed a number of units in accordance with the cost saving techniques presented here, less detailed instructions and fewer modifications on the job will be required in the future. Note Figure 2-19.

Reducing Wall Framing Costs

Exterior wall framing which does not incorporate cost saving techniques is shown in Figure 2-20. In contrast, optimum wall framing is shown in Figure 2-21. A comparison of these two illustrations indicates the substantial savings that can be achieved through proper application of efficient framing principles. Estimated labor and material savings for a typical one-story house, using Figure 2-21 techniques, are given in Table 2-22. Savings were calculated by the National Association of Home Builders Research Foundation and are based on time and materials studies of actual house construction in 1974.

Details of individual methods for reducing exterior wall framing costs are described below.

Stud Spacing

For one-story houses, or the top floor of multi-story units, 24 inch stud spacing can be used for virtually all houses where exterior and interior facing materials can span 24 inches. Tests and structural analysis show that, for many installations, 24 inch spacing can also be used for walls supporting the upper floor and roof of two-story units.

Use of 24 inch stud spacing can result in significant savings. For a 1660 square foot one-story example house, with normal door and window areas, potential labor and material savings of about $36.00 are achievable when exterior wall framing is on 24 inch centers rather than 16 inch centers. In many cases, the same thickness of exterior sheathing and/or exterior

47

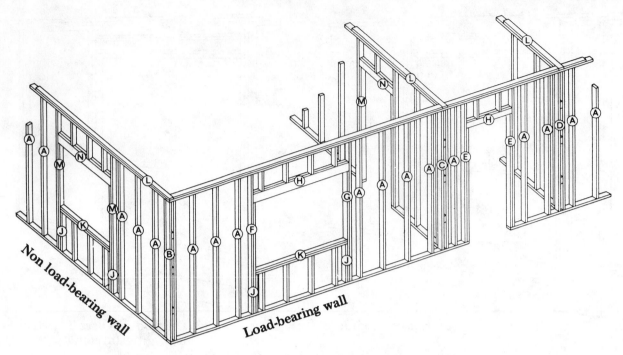

Wall framing: Cost saving principles not applied
Figure 2-20

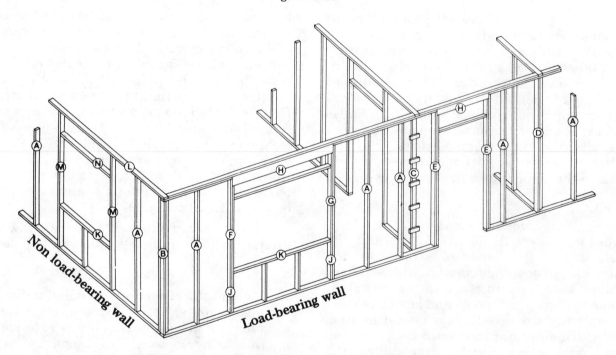

Wall framing: Incorporated cost saving principles
Figure 2-21

finish used for 16 inch stud spacing will be acceptable as well on 24 inch spacing. Assuming no change in materials, an additional $36.00 potential saving could be realized in reduced labor for drywall, sheathing, insulation and wiring application.

Studs should be located on a regular spacing, whether 16 inch or 24 inch, to minimize cutting and fitting of sheathing or cladding panels. Use of 16 inch or 24 inch modular stud spacing also minimizes labor costs for special cutting and fitting of insulation and vapor barriers. Where placement of the window or door opening is critical, use of a 16 inch module provides maximum flexibility.

Chapter 24 demonstrates the significant advantages of using 2 x 6 studs spaced at 24 inches to allow for insulation.

Cost Saving Principle	Illustration	Potential Cost Saving
1. 24" stud spacing versus 16" spacing; wall framing costs plus application of cladding, insulation, electrical, etc.	Figures 2-20 & 2-21	$36.05 (Framing) 36.30 (Other)
2. One side of door and window openings located at regular 16" or 24" stud position.	Figures 2-20 & 2-21, Details E,F,M	14.00
3. Modular window sizes used, with both side studs located at normal 16" or 24" stud position (Savings additional to No. 2, above).	Figure 2-19 Figures 2-20 & 2-21, Details G,M	9.75
4. Optimum three-stud arrangement at exterior corners.	Figures 2-20 & 2-21, Detail B Figure 2-24 (C)	4.70
5. Cleats instead of backup nailer studs where partitions intersect exterior walls.	Figures 2-20 & 2-21, Detail C Figure 2-23 (D)	13.15
6. Single sill member at bottom of window openings.	Figure 2-19 Figures 2-20 & 2-21, Detail K Figure 2-25	5.35
7. Support studs under window sill eliminated.	Figure 2-19 Figure 2-20 & 2-21, Detail J	7.55
8. Window and door headers located at top of wall; short in-fill studs (cripples) eliminated.	Figure 2-19 Figures 2-20 & 2-21, Detail H,N	12.55 (16" o.c.) 9.65 (24" o.c.)
9. Single top plate for non-bearing end walls.	Figures 2-20 & 2-21, Detail L	11.70
10. Single 2x4 header for openings in non-bearing walls.	Figures 2-20 & 2-21, Detail N	5.20
11. Header support studs eliminated for openings in non-bearing walls.	Figures 2-20 & 2-21	5.75
12. Mid-height wall blocking eliminated.		26.45 (16" o.c.) 25.35 (24" o.c.)
Potential cost saving if all cost saving principles are applicable in one house.		$188.45

Notes: 1. This example is based on a one-story house having 1660 square feet of floor area, 9 windows, 4 doors, 6 exterior corners, 8 partitions intersecting exterior walls, 2 non-bearing end walls and a total of 196 linear feet of exterior wall.
2. Estimated savings in labor and materials were calculated by the National Association of Home Builders Research Foundation based on time and materials studies of actual construction. Potential cost savings are based on labor at $6.50 per hour and lumber at $175 per thousand board feet.

Exterior wall framing: Potential cost saving for a one-story house
Table 2-22

Plan For Openings And Intersections

Window and door locations affect wall framing costs. When architectural considerations permit, one edge of window or door rough openings should be located at a regular stud position, thus using fewer studs. In a 1660 square foot one-story house with normal opening areas, location of one side of each door and window opening at a regular 16 inch or 24 inch stud position results in estimated savings of approximately $14.00. Details E, F, and M of Figure 2-21 illustrate this modular layout principle.

Wherever possible, window and door sizes should be selected so that the rough opening width is a multiple of the stud spacing. Thus both sides of the opening can be on the stud module. Not only does this reduce the number of studs required, but sheathing and siding can be better located to minimize wastage due to cut-outs for openings. Examination of Figures 2-19 and 2-21 will show how this can be accomplished. Locating both side studs of window openings at the normal 16 inch or 24 inch position results in an *additional* saving in the exterior wall framing cost of the example house of about $9.75. (See Table 2-22).

It is also advantageous to locate partitions at

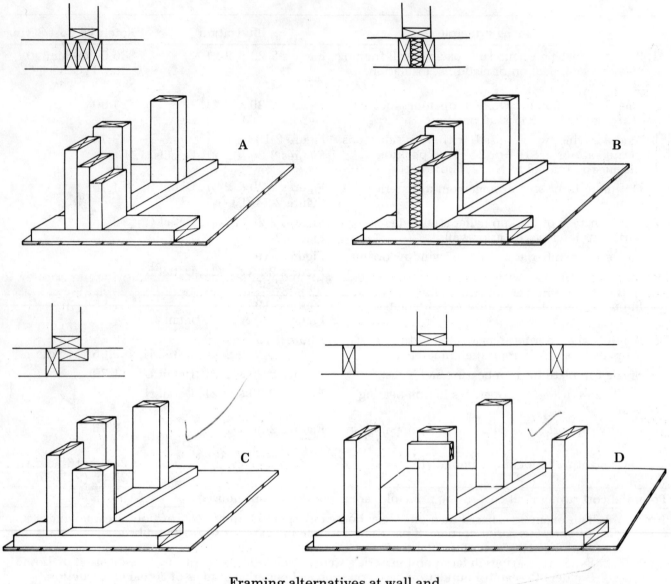

Framing alternatives at wall and
partition intersections
Figure 2-23

a normal wall stud position when backup cleats are not used at intersections of partitions and walls (See Figure 2-21, Detail D).

Studs At Wall Corners And Intersections

Traditional framing methods for exterior corners and intersections of partitions and walls use three studs. This is to provide support and a nail base for interior and exterior cladding. Such framing is shown in Details B, C and D of Figure 2-20 and in Figures 2-23(b) and 2-24(b). This arrangement requires that the space between studs be insulated prior to application of sheathing. One alternative, shown in Figure 2-23(a) and 2-24(a), uses an extra stud to provide the desired insulation. While this is thermally effective, it is an expensive use of wood.

A preferred three-stud arrangement is illustrated in Figure 2-21, Details B and D, and in Figures 2-23(c) and 2-24(c). This stud placement eliminates spacer blocks, provides the same backup support for wall linings as the more traditional arrangement, and insulation can be installed after the sheathing is applied.

Another alternative is shown in Figure 2-21, Detail C, and in Figure 2-23(d). Studs serving only as a backup for interior cladding are eliminated by attaching backup cleats to the partition stud. These cleats can be 3/8 inch plywood, 1 inch lumber or metal clips. A similar arrangement of cleats can be used to eliminate the back-up nailer stud at exterior corners. Figure 2-24(c).

Labor and material savings in the neighborhood of $18.00 are achievable in a 1660 square

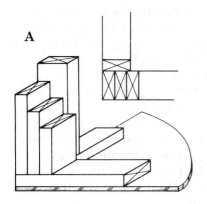
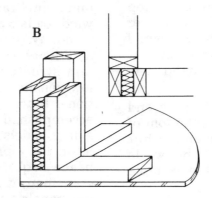
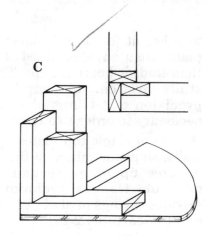

A   B   C

Stud arrangements at exterior corners
Figure 2-24

foot one-story house with common room sizes using the optimum three-stud exterior corner arrangement and backup cleats at all intersections of partitions and exterior walls (compare Details B and C, Figures 2-20 and 2-21).

Framing Around Windows And Doors

Perhaps the most commonly over-built wall framing is around window openings. Details J and K of Figure 2-20 show doubled sills under windows, plus short support studs at the ends of the sill. As can be seen from Figure 2-25, however, vertical loads at windows are transferred downward by the studs that support the header. Thus the sill and wall beneath it are nonload-bearing. Details J and K of Figure 2-21, and Figures 2-19 and 2-25, show elimination of the second sill and of the two sill support studs. Ends of the sill are supported by end nailing through the adjacent stud. Use of a single sill member at the bottom of all window openings and elimination of all window sill support studs in the 1660 square foot one-story example house result in savings of almost $13.00.

Further economics are usually possible by moving the header beam up to replace the lower top plate, as seen in Figures 2-20 and 2-21, Detail H, and Figures 2-19 and 2-25. While this requires longer support studs, it can eliminate most of the short in-fill studs between the top plate and the header since vertical loads no longer need be transferred to the top of the window opening. Short studs below the header would be needed only when required to support sheathing or cladding. If sheathing or cladding can span the distance between the header beam and the rough opening top plate, no short studs need be installed. Reduction in labor and materials for cutting and fitting of studs is further complimented by savings in cutting and

fitting insulation. The same principle applies to header beams across doorway openings. Potential savings of $9.50 to $12.50 in wall framing costs for the example house are achievable through elimination of short in-fill studs below all window and door headers in load bearing walls.

Framing Nonload-Bearing Walls

For most houses, not all exterior walls are load-bearing. Generally there are two or more walls parallel to joists and roof trusses that effectively support only their own weight. Such nonload-bearing exterior walls can be framed in the same way as interior partitions, as shown in Details K, L, M and N of Figure 2-21. Since there are no loads to transfer, doors and windows can be framed with single members. A single top plate is adequate provided metal straps or plates are used to tie top plates

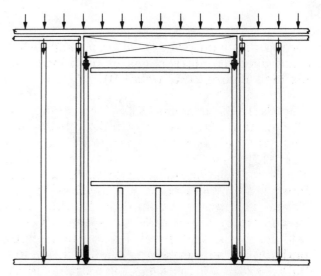

Load distribution through header and support studs at opening in load-bearing wall
Figure 2-25

together at joints. Of course, when single top plates are used, studs must be 1½ inches longer than studs in bearing walls. This is no problem if studs are cut to length on the job; if precision end-trimmed studs are used it is necessary to order two lengths.

Use of single top plates in non-bearing end walls, single members at the top and bottom of window openings in end walls and other optimum framing techniques in end walls produce savings of about $22.50 in the example house (See Table 2-22).

Mid-Height Wall Blocking

Mid-height blocking between wall studs is almost always unnecessary and should be eliminated. Top and bottom plates provide adequate fire stopping, and ½ inch gypsum board applied horizontally does not require blocking for support of taped joints when studs are 24 inches on centers or less. For the example house, mid-height wall blocking would cost approximately $26.00 more depending on stud spacing used (See Table 2-22).

Wall Bracing

Most codes require that exterior walls have a minimum "racking" strength for stability under wind loads. Sufficient racking strength is provided by panel type sheathing or siding; no additional let-in corner bracing is needed. If a non-structural sheathing is used, such as low-density fiberboard or gypsumboard, additional bracing may be necessary. This can be accomplished by using structural sheathing or siding panels at corners of the structure, or by let-in 1 x 4 bracing. Since installing the 1 x 4 brace requires additional time of a skilled craftsman, it generally is more cost-effective to use panels for bracing.

Lumber Grades

Economy is achieved by selecting the lowest grade of lumber that will accomplish the required task. Tests have proven that Utility grade 2 x 4's are more than adequate to support most wall loads. Utility grade studs at 24 inches on centers are accepted by the Federal Housing Administration (FHA/HUD) for load-bearing exterior walls supporting roof and ceiling loads. For other residential bearing walls, the next higher grade, i.e. Number 3, Standard or Stud grade, is acceptable. Calculations indicate these grades will support design loads for almost all conventional wood-frame residences.

Chapter 3

Sills

The sill plate is the bearing member placed directly on the foundation wall. The accuracy with which this member is framed is of great importance to the successful framing of the entire building. The dimensions given on floor plans and elevations should be closely studied so the sill may be properly placed on the foundation.

Wood such as southern yellow pine, Douglas fir, redwood, or cypress may be used for sill construction. Most sills are treated with a wood preservative and are sold by lumber yards as ''sill'' material.

In wood frame construction, the sill plate should be anchored to the foundation wall with ½-inch bolts hooked and spaced about 8 feet apart (Figure 3-1). In some areas, sill plates are fastened with masonry nails, but such nails do not have the uplift resistance of bolts. In high wind and storm areas, well anchored plates are very important. A sill sealer is often used under the sill plate on poured walls to take care of any irregularities which might have occured during curing of the concrete. Anchor bolts should be embedded 8 inches or more in poured concrete walls and 16 inches or more in block walls with the core filled with concrete. A large plate washer should be used at the head end of the bolt for the block wall. If termite shields are used, they should be installed under the plate and sill sealer.

The two general types of wood sill construction used over the foundation wall conform either to platform or balloon framing. The box is commonly used in platform construction. It consists of a plate anchored to the foundation wall over a sill sealer which provides support and fastening for the joints and header at the end of the joists (Figure 3-2). Some houses are constructed without benefit of an anchored sill plate although this is not entirely desirable. The floor framing should then be anchored with metal strapping installed during pouring operations (Figure 3-3).

Platform Construction

The box type sill provides the advantages of the single sill, namely the fire stop and the solid bearing upon which to nail the subfloor. The disadvantage is that the side wall studs rest on horizontal grain of considerable thickness. This thickness is equal to that of the shoe, subfloor, joist and sill. Notice that the subfloor runs to the outside header of the sill and must be laid before the side wall partition is raised. Another disadvantage is that there is no other means of tying the sill plate to the joists other than toenailing.

Figure 3-4 shows another type of box sill with a sill plate of the single type. The header rests directly on the foundation, thus affording good nailing into the edge of the sill plate. The plate upon which the side wall studs rest is let into the top of the joists, thereby providing a good nailing surface for tying the header and joist together.

This type of sill illustrates the practice of notching the bottom of the joists where they rest on the sill plate, thereby reducing the strength of the joist at this point. This is sometimes avoided by inserting a double sill plate as shown in Figure 3-5. This provides for full width

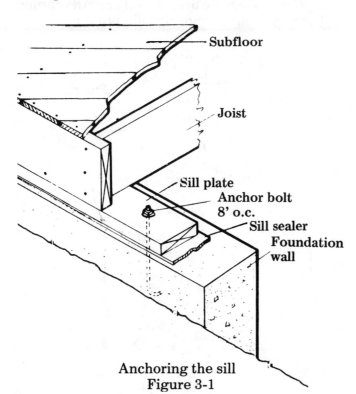

Anchoring the sill
Figure 3-1

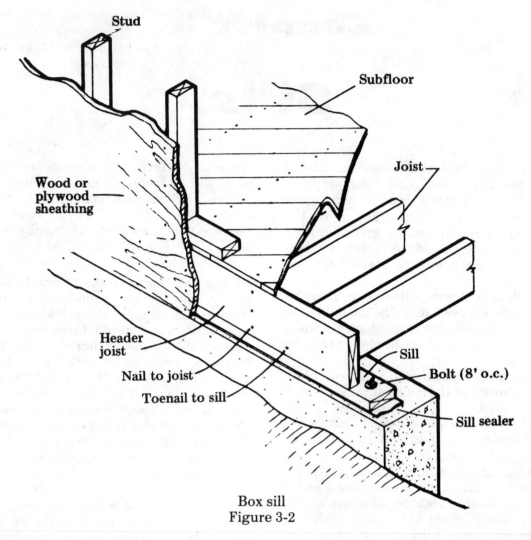

Stud

Subfloor

Joist

Wood or
plywood
sheathing

Header
joist

Nail to joist

Toenail to sill

Sill

Bolt (8' o.c.)

Sill sealer

Box sill
Figure 3-2

bearing of the joist on the sill. It also makes it possible to nail the header solidly to the double sill plate and permits the anchoring of a double sill plate to the masonry instead of the weaker method of anchoring only a single sill plate. The double sill plate may also be straightened more easily than a single plate and provides for the lapping of one plate over the other at the corners of the building as shown at Figure 3-6.

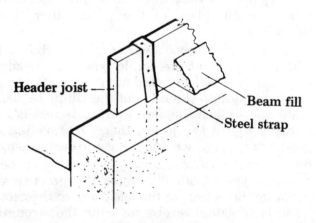

Header joist

Beam fill

Steel strap

Anchoring floor system to concrete or
masonry walls: Without sill plate
Figure 3-3

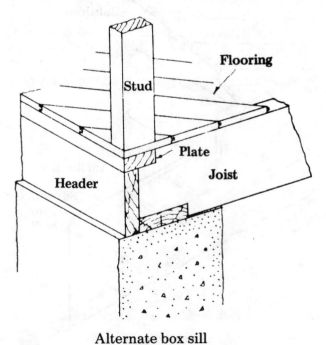

Flooring

Stud

Plate

Header

Joist

Alternate box sill
Figure 3-4

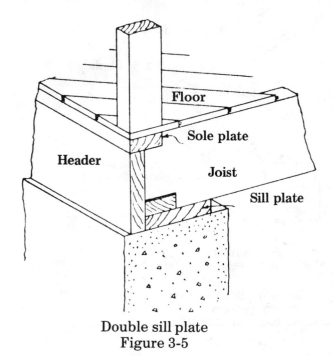

Double sill plate
Figure 3-5

Balloon Framing

Because there is less potential shrinkage in exterior walls with balloon framing than with platform framing, balloon framing is usually preferred over platform in full two-story brick or stone veneer houses.

Balloon frame construction uses a wood sill upon which the joists rest. The studs also bear on this member and are nailed both to the floor joists and the sill. The subfloor is laid diagonally or at right angles to the joists and a firestop is added between the studs at the floorline (Figure 3-7). When diagonal subfloor is used, a nailing member is normally required between joists and studs at the wall lines.

The sill plate should be so constructed that it will form a straight and true surface on which the joists and side wall studs may rest. The corner joints should be lapped, or strapped if

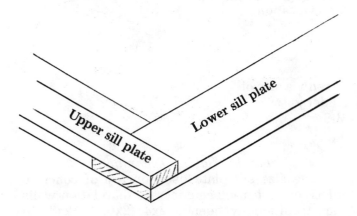

Lap of double sill plate
Figure 3-6

butted together. The side wall studs for balloon framing rest directly on the sill plate, thus reducing the bearing members of horizontal grain to a minimum.

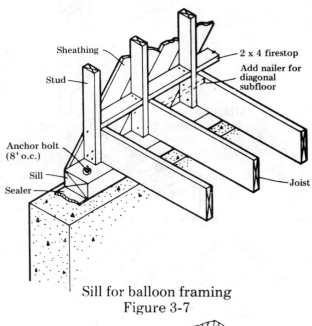

Sill for balloon framing
Figure 3-7

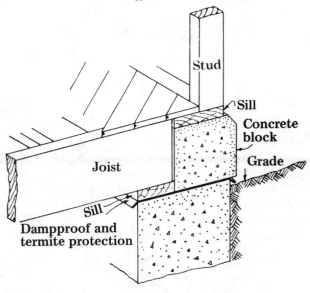

Solid sill at grade line
Figure 3-8

The header at the end of the joists prevents drafts between the studs and joists. It also provides solid sill construction and a firm base upon which to nail the subfloor. This header may be cut in between the joists or may be spiked to the ends. The method used would depend on the width of the sill plate. The joists should have a bearing seat on the sill plate of at least 4 inches.

The type of solid sill shown in Figure 3-8 may be used where the outside grade line is nearly as

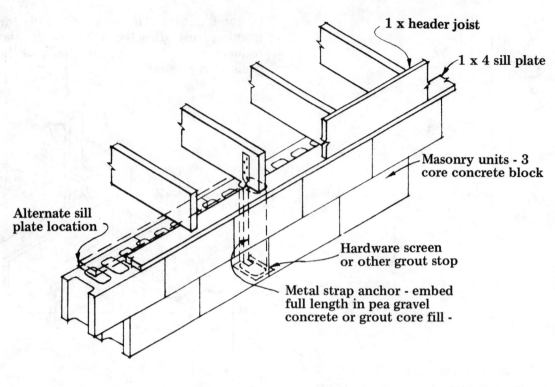

1 x header joist

1 x 4 sill plate

Masonry units - 3
core concrete block

Alternate sill
plate location

Hardware screen
or other grout stop

Metal strap anchor - embed
full length in pea gravel
concrete or grout core fill -

A

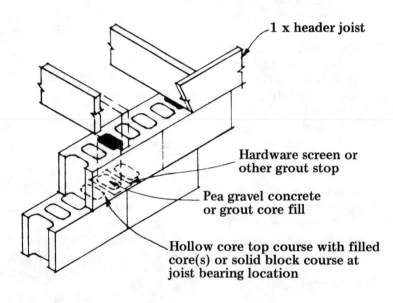

1 x header joist

Hardware screen or
other grout stop

Pea gravel concrete
or grout core fill

Hollow core top course with filled
core(s) or solid block course at
joist bearing location

B

Eliminate sill plate or reduce size of
sill plate and header joist
Figure 3-9

high as the inside floor line. This type of sill is generally associated with the balloon and modern braced frame. In this type of foundation, where the sill is within one foot of the ground, there should be some means of dampproofing in the masonry walls.

Lumber Saving Techniques

The size sill plate used on top of concrete block or cast-in-place concrete foundation walls has traditionally been a 2x4, 2x6, or 2x8. In many cases, the sill plates either could be eliminated or be a 1x4 and still provide adequate

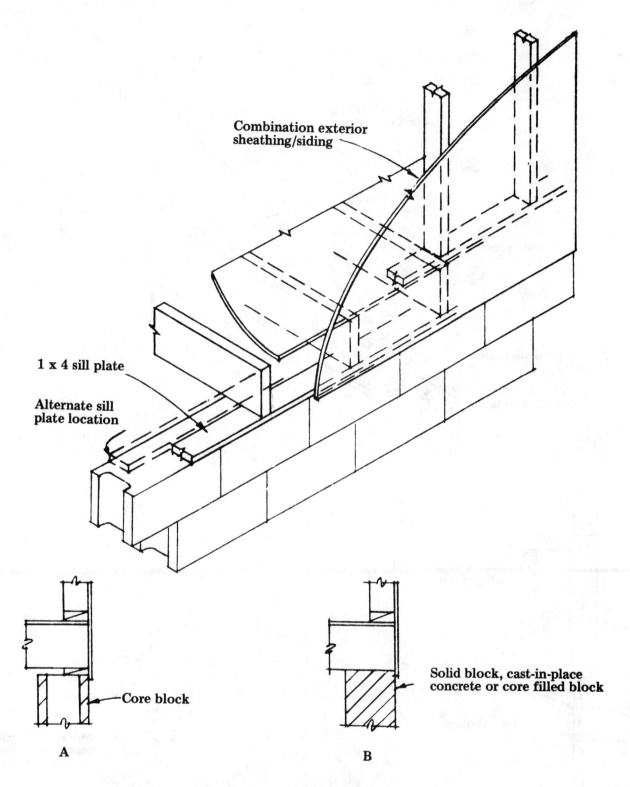

Combination exterior sheathing/siding

1 x 4 sill plate

Alternate sill plate location

Core block

Solid block, cast-in-place concrete or core filled block

A

B

Eliminate sill plate and header joist or reduce size of sill plate when roof live loads are 20 PSF or less and combination sheathing/siding is used

Figure 3-10

bearing. Also, it is wise to position the sill plate with the inside edge flush with the interior face of the foundation wall. This decreases the design span of the joist and in some cases will allow using a smaller size joist. Use appropriate sheathing or siding details, sill sealer, insulation or caulking to prevent entry of wind, rain, dust, or snow when the sill plate is recessed back from

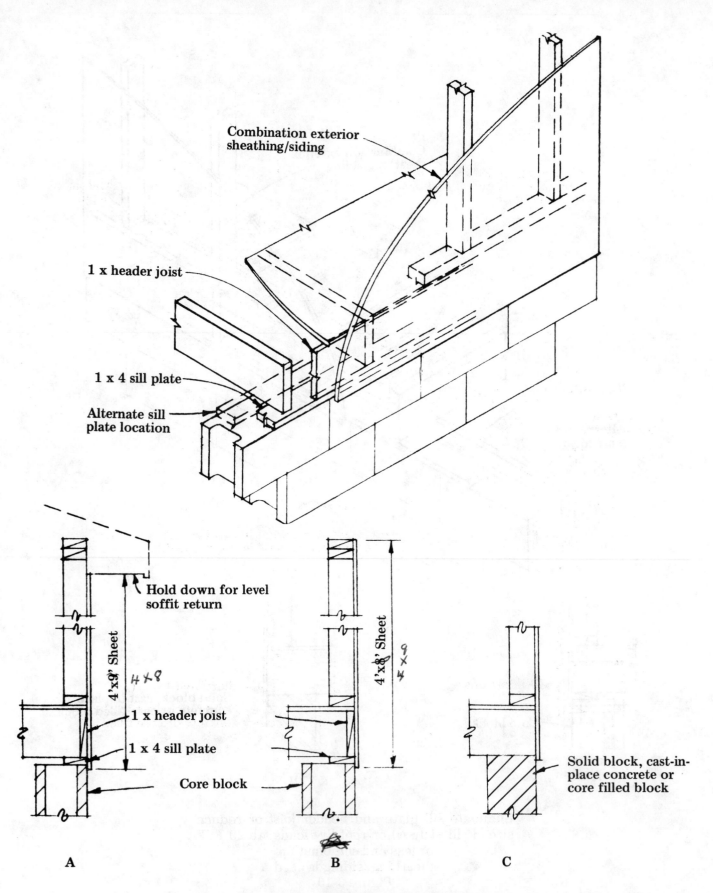

Combination exterior
sheathing/siding

1 x header joist

1 x 4 sill plate

Alternate sill
plate location

Hold down for level
soffit return

4'x9' Sheet

4x8

1 x header joist

1 x 4 sill plate

Core block

4'x8' Sheet

4 x 9

Solid block, cast-in-
place concrete or
core filled block

A   B   C

Eliminate sill plate or reduce size of sill plate
and header joist when combination
sheathing/siding is used
Figure 3-11

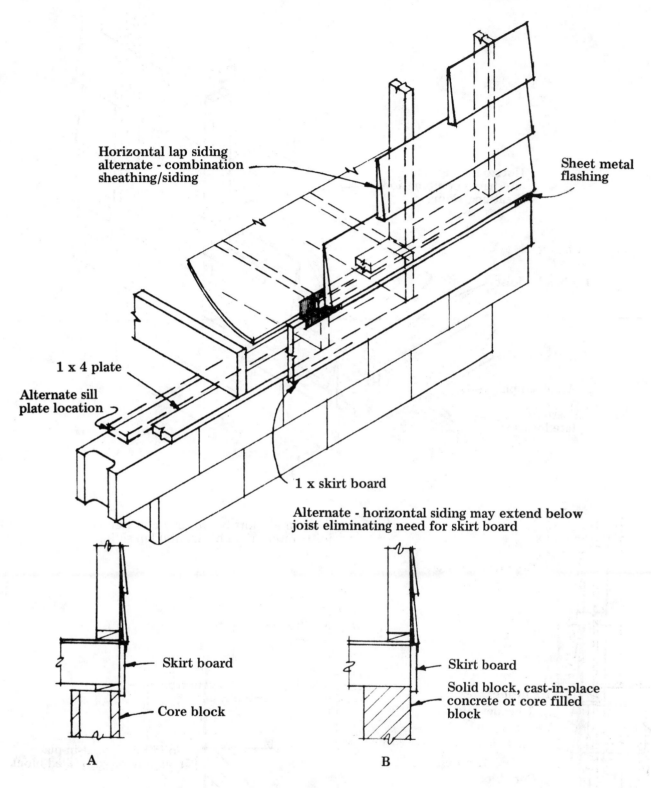

Horizontal lap siding
alternate - combination
sheathing/siding

Sheet metal
flashing

1 x 4 plate

Alternate sill
plate location

1 x skirt board

Alternate - horizontal siding may extend below
joist eliminating need for skirt board

Skirt board

Core block

A

Skirt board

Solid block, cast-in-place
concrete or core filled
block

B

Eliminate sill plate and header joist or reduce
size of sill plate when horizontal
lap siding is used
Figure 3-12

the outside edge of the foundation wall.

When a joist rests on top of a hollowcore masonry block wall, as shown in Figure 3-9A, a 1x4 sill plate is required to provide adequate bearing on the cross webs and face shells of the block.

When a joist bears directly on the cross web or on cores that have been filled with grout,

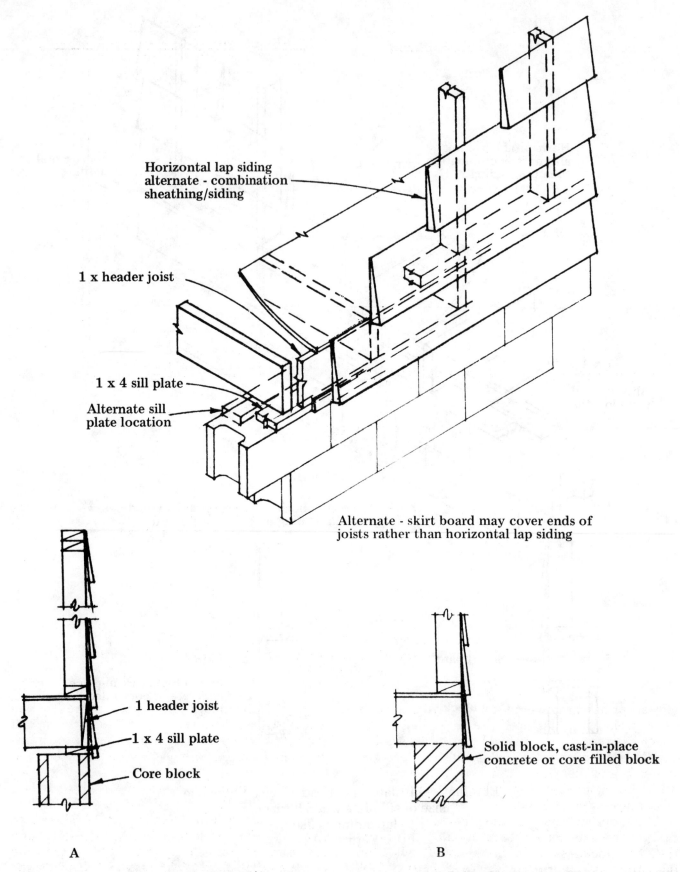

Horizontal lap siding
alternate - combination
sheathing/siding

1 x header joist

1 x 4 sill plate

Alternate sill
plate location

Alternate - skirt board may cover ends of
joists rather than horizontal lap siding

1 header joist

1 x 4 sill plate

Core block

Solid block, cast-in-place
concrete or core filled block

A

B

Eliminate sill plate or reduce size of sill plate
and header joist when horizontal
lap siding is used
Figure 3-13

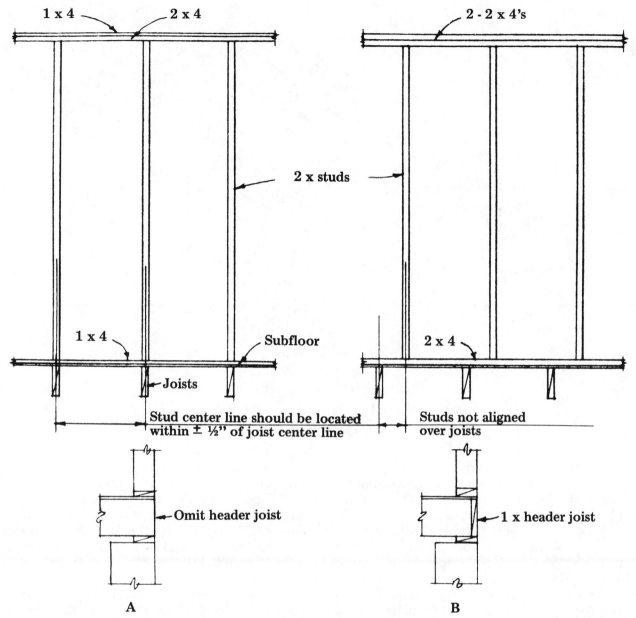

Use 1 x 4 bottom plate when exterior wall
studs are directly over floor joists
Figure 3-14

mortar, or pea gravel concrete, the sill plate can be completely eliminated (see Figure 3-9B). When the top course is solid block or the wall is cast-in-place concrete, no sill plate is required. When a joist bears directly on the top of the wall, the clear span is reduced. Reducing the span may reduce the joist size required.

A header joist used at the ends of floor joists that rest on exterior walls traditionally has been the same thickness as the floor joist. The header joist can be nominal 1-inch (see Figure 3-9) rather than 2-inch thick lumber and still perform its function properly, except where the header joist serves as a lintel over openings.

The header joist helps to transmit stud loads on the bottom exterior wall plate to the foundation. When the roof loads are not abnormally great and combination siding/-sheathing is used, the header joist can be eliminated completely when not required for edge support of subfloor such as diagonal boards (see Figure 3-10). Where roof loads are light, the header joist can be eliminated as shown in Figure 3-10A for hollowcore foundations, and the header joist plus the 1x4 sill can be eliminated as shown in Figure 3-10B when the joist has solid bearing.

Several combinations of framing that take advantage of the savings mentioned are possible. When single-layer combination

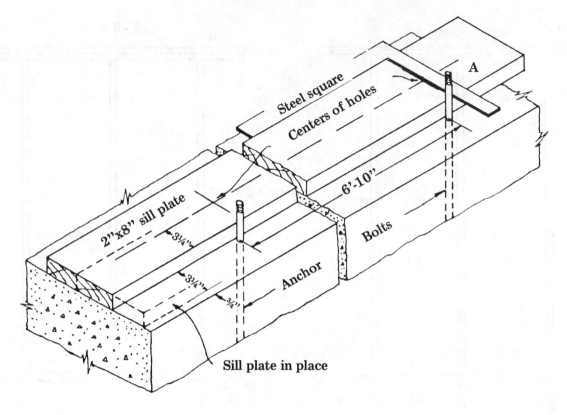

Method of marking sill for anchor bolt holes
Figure 3-15

sheathing/siding as shown in Figure 3-11 is used, method A or B can be used depending on the length of plywood sheet (4x8 or 4x9). Figure 3-11A shows how a 4x8-foot sheet can be held below a roof overhang with a level soffit return. Figure 3-11B shows a 4x9-foot sheet running to the wall top. When using the method in Figure 3-11A, provide an anchorage, as necessary, of the top plate to the top of the wall studs. In both Figures 3-11A and B, a 2x4 sill plate and a one-inch thick header joist would be acceptable when the foundation is core block. If the foundation wall is solid block, cast-in-place concrete, or has hollowcore masonry block with cores filled with concrete under the joist, the 1x4 sill plate and header joist can be eliminated as illustrated in Figure 3-11C.

In the case of horizontal lap siding, the header joist can be eliminated on hollowcore block as shown in Figure 3-12A. Both the header joist and sill plate can be eliminated (see Figure 3-12B) when the joist rests on solid block, cast-in-place concrete, or core-filled block.

In another case, the horizontal siding as shown in Figure 3-13 extends over the header joist. Figure 3-13A shows use of a one-inch thick header joist and a 1x4 sill plate on hollowcore block. Figure 3-13B is the same except that the joist rests on solid block, cast-in-place concrete,

or core-filled block which allows elimination of the sill plate and header joist.

If the exterior load-bearing wall stud spacing is the same as the floor joist spacing and the studs fall directly over the joist, it is possible to reduce the bottom plate size from a 2x4 to a 1x4 (see Figure 3-14).

To Install The Sill Plate On The Wall

1. Place the sill plates along the top of the four walls of the foundation in approximately their permanent positions. Use 16 or 18 foot lengths of stock if possible. Place the sill so that no joints occur over an opening in the foundation wall, and square the butt joints.

2. Lay the sill plate against the inside of the anchor bolts (Figure 3-15). The ends of the sills may be left projecting over the wall until after the sill plate has been fitted over the anchor bolts.

3. Square lines from each anchor bolt across the sill plate (Figure 3-15A).

4. Measure the correct distance along each line from the outside edge of the sill to locate the holes for the anchor bolts.

Note: The outer edge of the sill should be at least ¾ inch in from the outside edge of the foundation wall so the sheathing will come flush with the foundation wall. If a

62

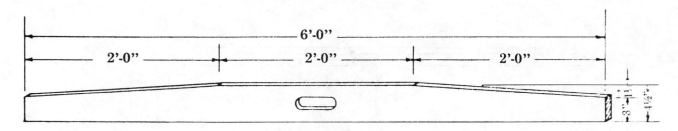

A straight edge
Figure 3-16

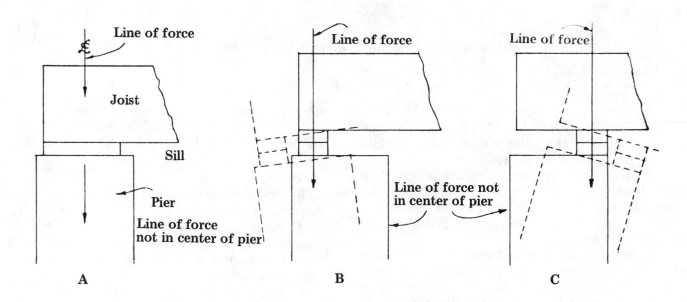

Placement of sills on piers
Figure 3-17

box sill is to be used, allowance for the header must also be made.

5. Mark these points and drill the holes about 3/8 inch larger than the diameter of the bolts. This size hole will allow for straightening the sill and will make it easy to place the sill plate over the bolts.

6. Place the sills over the anchor bolts.

Note: If a double sill is to be used, fit it in the same manner as the single sill plate, but make allowance for the lap joints at the corners of the building.

7. Cut the sill plates at the corners of the building so they butt together closely.

8. Place washers and nuts on the anchor bolts and draw the nuts to bring the sill plate down temporarily to the masonry wall.

Note: If termite shields are to be used, it is best to insert them before the sill plate is placed over the bolts.

9. Square the plate at the corners, using the 6-8-10 rule, and brace the sill in this position.

10. Level and straighten the sill over the bearing walls by using a straight edge and level. If shims are needed, place them every 4 feet between the bottom of the sill plate and the top of the foundation.

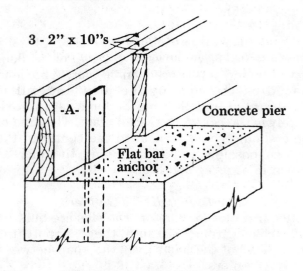

Anchoring a beam to a pier
Figure 3-18

63

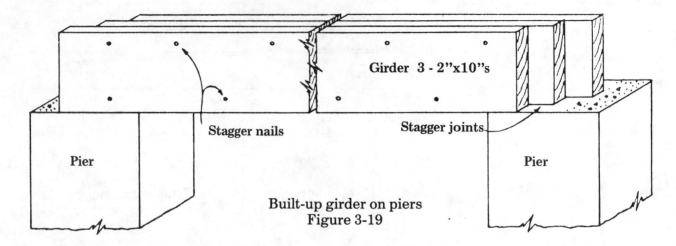

Girder 3 - 2"x10"s

Stagger nails

Stagger joints

Pier

Pier

Built-up girder on piers
Figure 3-19

Note: Some carpenters prefer to do this after the floor joists have been placed so the weight of the joists will hold the sill plate down against the wall.

11. After the top of the sill plate is level and straight and the outside edge is straight with the face of the wall, tighten the nuts on the anchor bolts, but not so much that they will pull the sill out of alignment.

12. Place the header and spike it to the sill so it stands squarely with the face of the wall.

Note: The remaining processes of sill assembly will be taken up under floor joists in Chapter 6.

How To Make A Straight Edge

1. Select a straight grained piece of white pine about 1¼ inch x 5 inches x 6 to 10 feet.

2. Plane the two edges until they are straight and parallel.

3. Taper the board and cut a handle in it (Figure 3-16).

Placement Of Sills On Piers

Figure 3-17A shows a section of a floor platform and sill placed on a pier. The load placed on the pier by the joist and sill is concentrated and the line of force is approximately in the center of the sill plate. This line of force should come as near to the center of the pier as possible to avoid tipping the pier as shown in Figures 3-17B and 3-17C.

How To Install Sills On Piers

Note: In a pier foundation, the sills are built up like girders because they support the load of the joists over the span between the piers.

1. Align and nail together the two outside members of the sill. Be sure the tops of the members are even and the crowned side is up.

2. Place the sill on the piers and against the iron strap anchors in the piers. Do not put any joints of the built-up sill over the span between the piers.

3. Spike the anchors to the side of the sill as at A, Figure 3-18.

4. Place the remaining member of the sill against the opposite side of the anchor and mark the location and size of the anchor on this remaining member.

5. Notch out this piece so that it will fit up tightly against the other two members.

6. Spike the third piece to the other pieces as shown in Figure 3-19.

7. Place the sill on the piers being sure the bearing seat of the top of the pier is level.

8. Level the sill from pier to pier.

Note: If shingles are used to shim up the sill, be sure they are at least 3 inches wide. They should be inserted from opposite sides of the sill so that the thin edges overlap each other, thus forming a shim of uniform thickness. See Figure 3-20.

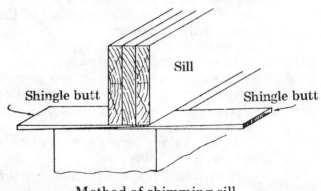

Sill

Shingle butt

Shingle butt

Method of shimming sill
Figure 3-20

Chapter 4

Girders

A girder, as used in house construction, is a large beam used at the first floor line to support the inner ends of the floor joists. Where the width of the building is greater than 14 feet, and it is not desirable to use an interior crossbearing wall, a girder should be used for furnishing added support near the center. The use of rather large joists is then unnecessary.

Both wood girders and steel beams are used in present day house construction. The standard I-beam and wide flange beam are the most commonly used steel beam shapes. Wood girders are of two types--solid and built up. The built-up beam is preferred because it can be made up from drier dimension material and is more stable. Commercially available glue-laminated beams may be desirable where exposed in finished basement rooms.

The built-up girder (Figure 4-1) is usually made up of two or more pieces of 2-inch dimension lumber spiked together, the ends of the pieces joining over a supporting post. A two-piece girder may be nailed from one side with tenpenny nails, two at the end of each piece and others driven stagger fashion 16 inches apart. A three-piece girder is nailed from each side with twentypenny nails, two near each end of each piece and others driven stagger fashion 32 inches apart.

Ends of wood girders should bear at least 4 inches on the masonry walls or pilasters. When wood is untreated, a ½-inch air space should be provided at each end and at each side of wood girders framing into masonry (Figure 4-1). In termite-infested areas, these pockets should be lined with metal. The top of the girder should be

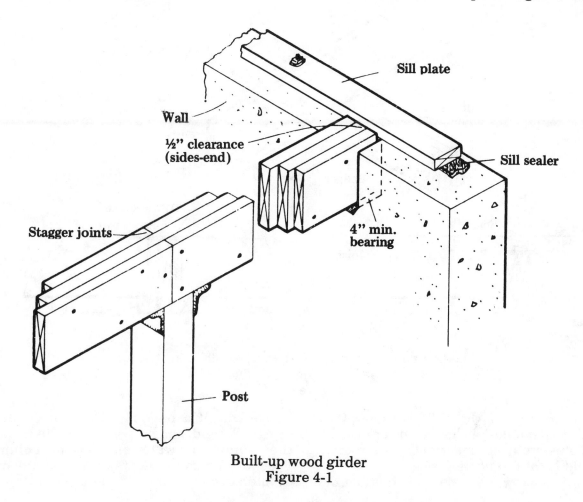

Sill plate

Wall

½" clearance (sides-end)

Sill sealer

Stagger joints

4" min. bearing

Post

Built-up wood girder
Figure 4-1

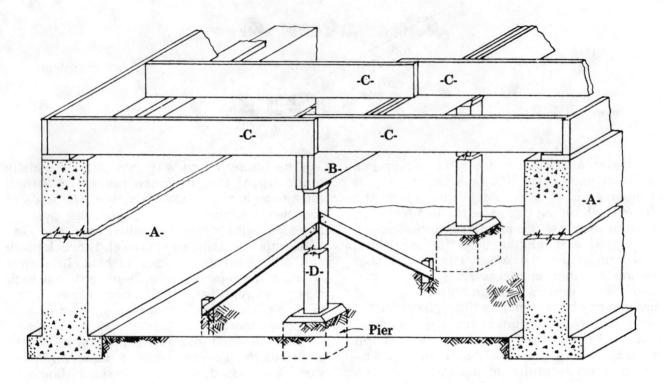

Built-up girder in place
Figure 4-2

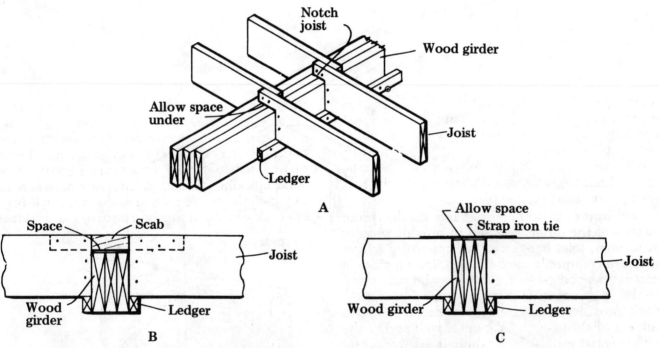

Ledger on center wood girder: A, Notched joist;
B, scab tie btween joist; C, flush joist
Figure 4-3

level with the top of the sill plates on the foundation walls, unless ledger strips are used. If steel plates are used under ends of girders, they should be of full bearing size.

The girders carry a very large porportion of the weight of a building. They must be well designed, rigid, and properly supported, both at the foundation walls and on the columns. Precautions must be taken to avoid or to counteract any future settling or shrinking that

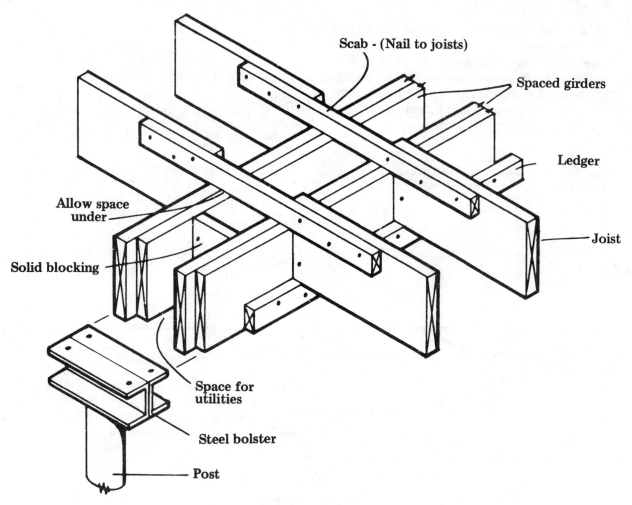

Scab - (Nail to joists)

Spaced girders

Ledger

Allow space under

Joist

Solid blocking

Space for utilities

Steel bolster

Post

Spaced wood girder
Figure 4-4

might cause distortion of the building. The girders must also be so installed that they will properly support the joists. Figure 4-2 shows two outside masonry walls (A). The built-up girder (B) supports the joists (C). The pier foundations support the columns (D) which in turn support the girder (B).

For more uniform shrinkage at the inner beam and the outer wall and to provide greater headroom, joist hangers or a supporting ledger strip are commonly used. Depending on sizes of joists and wood girders, joists may be supported on the ledger strip in several ways (Figure 4-3). Each provides about the same depth of wood subject to shrinkage at the outer wall and at the center wood girder. A continuous horizontal tie between exterior walls is obtained by nailing notched joists together (Figure 4-3,A). Joists must always bear on the ledgers. In Figure 4-3, B, the connecting scab at each pair of joists provides this tie and also a nailing area for the subfloor. A steel strap is used to tie the joists together when the tops of the beam and the joists are level (Figure 4-3, C). It is important

that a small space be allowed above the beam to provide for shrinkage of the joists.

When a space is required for heat ducts in a partition supported on the girder, a spaced wood girder is sometimes necessary (Figure 4-4). Solid blocking is used at intervals between the two members. A single post support for a spaced girder usually requires a bolster,

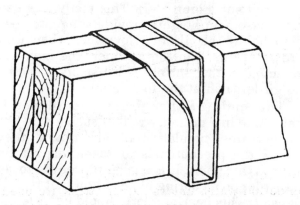

Girder with joist hanger
Figure 4-5

preferably metal, with sufficient span to support the two members.

Figure 4-5 shows a girder over which joist hangers have been placed to carry the joists. This type is used where there is little headroom, or where the joists carry an extremely heavy load, and spiking cannot be relied upon.

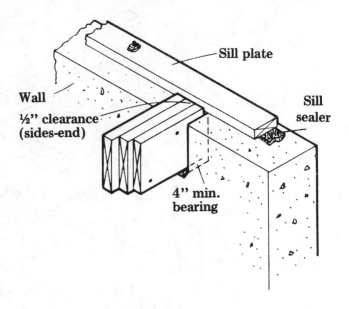

Built-up wood girder
Figure 4-7

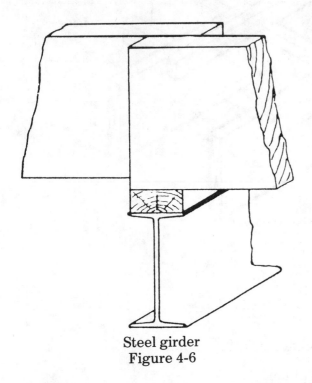

Steel girder
Figure 4-6

Figure 4-6 shows the steel I-beam used instead of the wooden girder. A 2 x 4 is sometimes used on top of the girder for the joists to rest on, to provide a surface to which to nail the joists, and to provide the same amount of horizontal grain wood to support the load as there is in the outside wall.

Perhaps the simplest method of floor-joist framing is one where the joists bear directly on the wood girder or steel beam, in which case the top of the beam coincides with the top of the anchored sill (Figure 4-7). This method is used when basement heights provide adequate headroom below the girder. However, when wood girders are used in this manner, the main disadvantage is that shrinkage is usually greater at the girder than at the foundation.

The ends of the girder should not be embedded in a masonry wall in such a way as to prevent free circulation of air around these ends. If this precaution is not taken, dry rot may occur at this point. It is a good procedure to rest the girder on a bearing plate of iron that is set in the wall. This bearing plate should be at least 2 inches wider than the girder on each side. Bearing plates range from ½ inch to 1 inch in

thickness and help distribute the load of the girder over a larger wall area. Plates of this type are particularly important where built-up beams are supported by masonry walls.

Spacing Of Girders

The length and depth of the joists and the location of bearing partitions of the floor above must be considered in determining the spacing of girders. It is good procedure to keep the distance between girders 14 feet or less. Thus, in a span of 25 feet, one girder is sufficient if it is placed halfway between each of the other supports. If the distance is 35 feet, two girders equally spaced are necessary.

Girders should be placed below bearing partitions. This restriction sometimes changes the spacing of girders between side wall supports.

A girder under joists having no bearing partitions above is shown in Figure 4-8, A. Figure 4-8, B shows a bearing partition placed midway between two side walls. The girder is directly beneath this partition. The bearing partition in Figure 4-8, C is 8 feet from one side wall and 16 feet from the other. Again, the girder is located directly underneath the partition. If there are two bearing partitions as shown in Figure 4-8, D, there should be two girders regardless of the span.

Effect Of Dimensions On Strength Of Girders

The girder must be large enough to support whatever load may be imposed on it, but not so large that it is wasteful. Before determining the

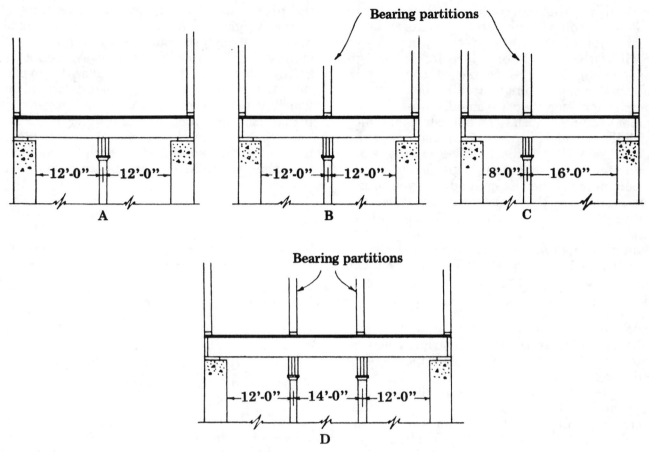

Placement of girders under partitions
Figure 4-8

size of a girder or beam the carpenter should understand three important relationships.

1. The effect of length of a girder on its strength.

2. The effect of width of a girder on its strength.

3. The effect of depth of a girder on its strength.

A. Length

If a plank supported at the ends carries an evenly distributed load throughout its entire length, it will bend to some extent. A plank twice as long, with the same load per foot of length, will bend much more and may break. If the length is doubled, the safe load will be reduced not to one-half, as might be expected, but to one-quarter. However, for a single concentrated load at the center, doubling the length decreases the safe load by only one-half. The greater the unsupported length of the girder, the stronger the girder must be. The strength may be increased by using a stronger material or by using a larger beam. The beam may be enlarged by increasing the width or depth or both.

B. Width

Doubling the width of a girder doubles its strength. This double width girder will have a load carrying capacity equal to two single width girders placed side by side.

C. Depth

Doubling the depth of a girder increases its carrying capacity four times. A beam 3 inches wide and 12 inches deep will carry four times as much as one 3 inches wide and 6 inches deep. Therefore, to secure additional strength, it is more economical to increase the depth of a beam than the width. However, it is well to avoid increasing the girder depth to much more than 10 inches, since a deeper girder will cut down the headroom in the basement. Instead, the girder could be made wider, additional supports could be put in to reduce the span, or a stronger material could be used for the girder.

Load Calculations

On many jobs a carpenter will be called on to select the size or placement of load bearing members. Normally, licensed structural engineers are retained to develop a structural plan

69

for more complex structures. However, in residential construction little structural engineering is done. Even if the plans were developed with full consideration to design loads, changes in the course of construction might make the loading very different than the designer or engineer assumed. Most of the lumber grading associations publish span tables which list permissible spans and loading for the type of trees they inspect. However, there are seven major lumber grading agencies and over 25 commercial species graded. If you don't happen to have the right span tables for the lumber you are using you may be left without guidance on how large and how many girders, columns, joists and rafters you need. Building codes are of little help. Most codes only make reference to how much load bearing capacity you must provide. They don't tell you whether you should use two 2 x 8's or two 2 x 12's for the girder you are installing. As a result, you probably install far larger and heavier structural members than are necessary, thus wasting lumber and labor.

The tables in this book will allow you to design the correct girder, column, joist, header and rafter combination for your job and with the lumber you have available. This manual won't make you a structural engineer, but it will explain a few simple principles so you can use the tables to arrive at the right timber size. The tables in this book are based on recommendations of the National Forest Products Association and are accepted by almost all modern building codes and the Federal minimum property standards.

Loading Principles

The load area of a building is carried by both foundation walls and the girder. Because the ends of each joist rest on the girder, there is more weight on the girder than there is on either of the walls. Before considering the load on the girder, it may be well to consider a single joist. Suppose that a 10-foot plank weighing 5 pounds per foot is lifted by two men. If the men were at opposite ends of the plank, they would each be supporting 25 pounds.

Now assume that one of these men lifts the end of another 10-foot plank with the same weight as the first one, and a third man lifts the opposite end. The two men on the outside are each supporting one-half of the weight of one plank, or 25 pounds apiece, but the man in the center is supporting one-half of each of the two planks or a total of 50 pounds.

The two men on the outside represent the foundation walls, and the center man represents

the girder; therefore, the girder carries one-half of the weight, while the other half is equally divided between the outside walls (Figure 4-9). However, the girder may not always be located halfway between the outer walls. To explain this, the same three men will lift two planks which weigh 5 pounds per foot. One of the planks is 8 feet long and the other is 12 feet long. Since the total length of these two planks is the same as before and the weight per foot is the same, the total weight in both cases is 100 pounds.

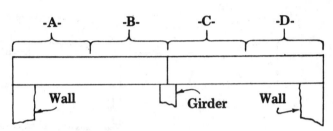

Distribution of weight
Girder in center of building
Figure 4-9

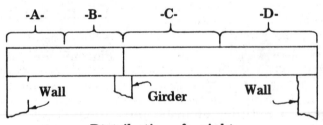

Distribution of weight
Girder off center of building
Figure 4-10

One of the outside men is supporting one-half of the 8-foot plank, or 20 pounds. The man on the opposite outside end is supporting one-half of the 12-foot plank, or 30 pounds. The man in the center is supporting one-half of each plank, or a total of 50 pounds. This is the same total weight he was lifting before. A general rule that can be applied when determining the girder load area is that a girder will carry the weight of the floor on each side to the midpoint of joists which rest upon it (Figure 4-10).

It has been assumed that the joists are butted or lapped over the girder. Loaded joists have a tendency to sag between supports (Figure 4-11) and when they are butted or lapped, there is no resistance to this bending over the girder. An exaggerated amount of this sag is shown by D in Figure 4-11.

If the joists are continuous, however, under load they will tend to assume the shape shown in

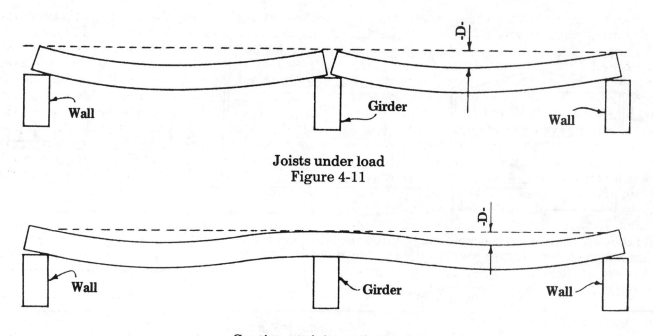

Joists under load
Figure 4-11

Continuous joist under load
Figure 4-12

Figure 4-12. Being in one piece, they resist bending over the center support, thereby forcing the girder to carry a larger proportion of the load than if the joists were cut. A girder at the mid-point of continuous joists will take five-eights instead of one-half the floor load.

After the girder load is known, the total floor load per square foot must be determined in order to select a safe girder size. Both dead and live loads must be considered in finding the total floor load.

The first type of load consists of all weight of the building structure. This is called the dead load. The dead load per square foot of floor area, which is carried to the girder either directly or indirectly by way of bearing partitions, will vary according to the method of construction and building height. The structural parts included in the dead load are:

Floor joists for all floor levels
Flooring materials, including attic if it is floored
Bearing partitions
Attic partitions
Attic joists for top floor
Ceiling lath and plaster, including basement ceiling if it is plastered

For a building of light-frame construction similar to an ordinary frame house, the dead load allowance per square foot of all the structural parts must be added together to determine the total dead load. The allowance for average subfloor, finish floor, and joists without basement plaster should be 10 pounds per square foot. If the basement ceiling is plastered, an additional 10 pounds should be allowed. When girders (or bearing partitions) support the first floor partition, a load allowance of 20 pounds must be allowed for ceiling plaster and joists when the attic is unfloored. If the attic is floored and used for storage, an additional 10 pounds (per square foot) should be allowed.

The second type of load to be considered is the weight of furniture, persons, and other movable loads which are not actually a part of the building but are still carried by the girder. This is called the live load. The live load per square foot will vary according to the use of the building and local weather conditions. The allowance for the live load on floors used for living purposes is usually 40 pounds per square foot. If the attic is floored and used for light

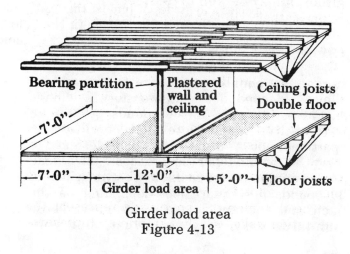

Girder load area
Figure 4-13

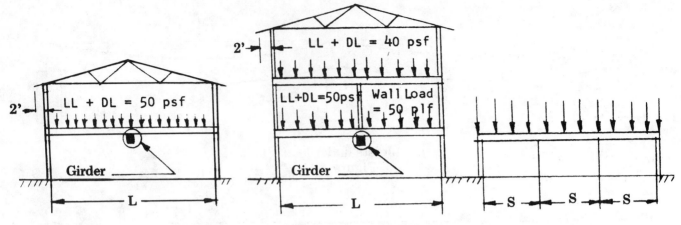

Note: Girder may be offset from centerline of house width up to one foot.

Note: End splits may not exceed one times girder depths.

Maximum span designs for girders supporting one-story floor loads

Girder Spans = S in Feet*

Nominal Lumber Sizes	House Widths = L in Feet								
	20	22	24	26	28	30	32	34	36
2 - 2x6	4'-5"	4'-0"	—	—	—	—	—	—	—
3 - 2x6	5'-6"	5'-3"	5'-0"	4'-10"	4'-8"	4'-5"	4'-2"	—	—
2 - 2x8	5'-10"	5'-3"	4'-10"	4'-5"	4'-2"	—	—	—	—
3 - 2x8	7'-3"	6'-11"	6'-7"	6'-4"	6'-2"	5'-10"	5'-5"	5'-1"	4'-10"
2 - 2x10	7'-5"	6'-9"	6'-2"	5'-8"	5'-3"	4'-11"	4'-8"	4'-4"	4'-1"
3 - 2x10	9'-3"	8'-10"	8'-5"	8'-1"	7'-10"	7'-5"	6'-11"	6'-6"	6'-2"
2 - 2x12	9'-0"	8'-2"	7'-6"	6'-11"	6'-5"	6'-0"	5'-8"	5'-4"	5'-0"
3 - 2x12	11'-3"	10'-9"	10'-3"	9'-10"	9'-6"	9'-0"	8'-5"	7'-11"	7'-6"

Maximum span designs for girders supporting two-story floor loads

Girder Spans = S in Feet*

Nominal Lumber Sizes	House Widths = L in Feet								
	20	22	24	26	28	30	32	34	36
3 - 2x8	4'-7"	4'-2"	—	—	—	—	—	—	—
2 - 2x12	4'-9"	4'-4"	4'-0"	—	—	—	—	—	—
3 - 2x10	5'-10"	5'-4"	4'-11"	4'-7"	4'-3"	4'-0"	—	—	—
3 - 2x12	7'-1"	6'-6"	6'-0"	5'-6"	5'-2"	4'-10"	4'-6"	4'-3"	4'-0"

*Linear interpolation within the above table for house widths not given is permitted.

Maximum spans using lumber having an allowable bending stress not less than 1000 psi. Some species and grades of lumber that can be used with this table: Douglas fir-larch No. 2, Douglas fir south No. 2, hem-fir No. 2, lodgepole pine No. 1, and southern pine No. 2

Table 4-14

storage, 20 pounds per square foot should be allowed. Second floors in two story dwellings should be designed for 30 pounds per square foot of live load. Note that corridors in apartment houses usually have to be designed to carry 80 pounds per square foot. Live load design minimums for roofs vary between 12 and 20 pounds per square foot depending on the pitch and the anticipated snow load. These standards may vary somewhat, but your building code probably closely follows the figures above.

When the total live and dead load per square foot of floor area is known, the load per linear foot on the girder is easily figured. Assume that the girder load area of the building shown in Figure 4-13 is sliced into 1-foot lengths across the girder. Each slice represents the weight supported by 1 foot of the girder. If the slice is divided into 1-foot units, each unit will represent

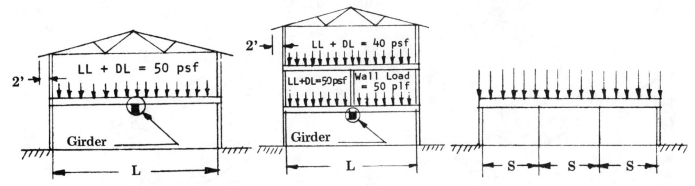

Note: Girder may be offset from centerline of house width up to one foot.

Note: End splits may not exceed one times girder depths.

Maximum span designs for girders supporting one-story floor loads

Girder Spans = S in Feet*

Nominal Lumber Sizes	House Widths = L in Feet								
	20	22	24	26	28	30	32	34	36
2 - 2x6	5'-3"	4'-10"	4'-5"	4'-1"	—	—	—	—	—
3 - 2x6	6'-9"	6'-5"	6'-2"	5'-11"	5'-8"	5'-3"	4'-11"	4'-8"	4'-5"
2 - 2x8	7'-0"	6'-4"	5'-10"	5'-4"	5'-0"	4'-8"	4'-4"	4'-1"	—
3 - 2x8	8'-11"	8'-6"	8'-1"	7'-9"	7'-5"	7'-0"	6'-6"	6'-2"	5'-10"
2 - 2x10	8'-11"	8'-1"	7'-5"	6'-10"	6'-4"	5'-11"	5'-7"	5'-3"	4'-11"
3 - 2x10	11'-4"	10'-10"	10'-4"	9'-11"	9'-6"	8'-11"	8'-4"	7'-10"	7'-5"
2 - 2x12	10'-10"	9'-10"	9'-0"	8'-4"	7'-9"	7'-2"	6'-9"	6'-4"	6'-0"
3 - 2x12	13'-9"	13'-2"	12'-7"	12'-1"	11'-7"	10'-10"	10'-2"	9'-6"	9'-0"

Maximum span designs for girders supporting two-story floor loads

Girder Spans = S in Feet*

Nominal Lumber Sizes	House Widths = L in Feet								
	20	22	24	26	28	30	32	34	36
3 - 2x6	4'-2"	—	—	—	—	—	—	—	—
2 - 2x10	4'-8"	4'-3"	—	—	—	—	—	—	—
3 - 2x8	5'-6"	5'-0"	4'-7"	4'-3"	—	—	—	—	—
2 - 2x12	5'-8"	5'-2"	4'-9"	4'-5"	4'-1"	—	—	—	—
3 - 2x10	7'-0"	6'-5"	7'-6"	5'-6"	5'-1"	4'-9"	4'-6"	4'-3"	4'-0"
3 - 2x12	8'-6"	7'-9"	7'-2"	6'-8"	6'-2"	5'-9"	5'-5"	5'-1"	4'-10"

*Linear interpolation within the above table for house widths not given is permitted.

Maximum spans using lumber having an allowable bending stress not less than 1500 psi. Some species and grades of lumber that can be used with this table: Douglas fir-larch No. 1, Douglas fir south select structural, mountain hemlock select structural, southern pine No. 1

Table 4-15

1 square foot of the total floor area. The load per linear foot of girder is determined by multiplying the number of units by the total load per square foot. Note in Figure 4-13 that the girder is off center. Therefore, the joist length on one side of the girder is 7 feet (one-half of 14 feet) and the other side is 5 feet (one-half of 10 feet), for a total distance of 12 feet across the load area. Since each slice is 1 foot wide, it has a total floor area of 12 square feet. Now, if we assume

that the total floor load for each square foot is 70 pounds, multiply the length times the width (7' x 12') to get the total square feet supported by the girder (7' x 12' = 84 square feet).

84 sq. ft.
x 70 lb. per sq. ft. (live and dead load)
5,880 lb. total load on girder

Wood Girder Designs
Selection of wood girders that support floor

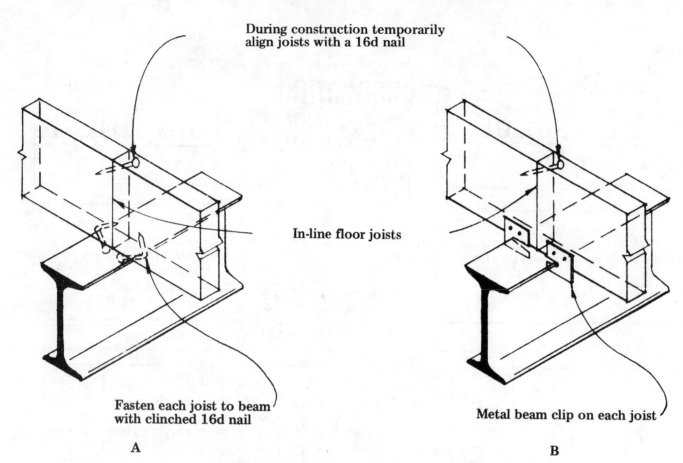

During construction temporarily
align joists with a 16d nail

In-line floor joists

Fasten each joist to beam
with clinched 16d nail

A

Metal beam clip on each joist

B

Methods of fastening in-line joists to beam
Figure 4-16

joists should be based on local building code requirements. Tables 4-14 and 4-15 have been prepared for one-story and two-story houses of varying depths where trusses transfer the roof loads to the outer walls. The girder, therefore, carries only floor loads. Some codes specify larger girders than are necessary for center floor joist supports because it would require extensive tables and calculations to provide information necessary to cover the many cases involved. Tables 4-14 and 4-15 can be used in many code areas and are based on F.H.A. recommendations. The allowable girder span "S" shown may also be used for designs where the girder is supported by two or more columns. Use the design data in Appendix A where roof trusses are not used.

Wood Sill Plates On Top Of Steel Girders

Many building codes require that a 2 x 4 or 2 x 6 continuous wood nailer be placed on top of steel beams used to support floor joists near the center of the house. This continuous wood nailer can be eliminated where the floor joist can be held down by nails driven directly into the bottom of the floor joist and clinched to the steel beam flange as shown in Figure 4-16,A.

In-line joists may be temporarily aligned over the center support by toenailing butt ends with a 16d nail.

How To Make A Built-up Girder

1. Select straight lumber free from knots or major defects. Use long lengths of stock so no more than one joint will occur over the span between the footings.

2. If 2 inch by 10 inch by 16 foot planks are used, cut one plank 8 feet long. Mark the crowned edge of the plank, and place this edge at the top of the girder.

3. Spike one 16 foot and one 8 foot plank together as shown in Figure 4-17. Place two 20d spikes about 6 inches from the end of the plank and about 2 inches from the top and bottom. Drive them at an angle of about 10 degrees into the two planks. Do not drive them home so the ends protrude as this makes it difficult to place the third plank on the girder.

4. Stagger the spikes over the entire length of the girder, working from the end and keeping the top edges of the planks flush with one another as they are spiked.

5. Stagger the joints in the beam as shown in Figure 4-17. Be sure the planks are squared at

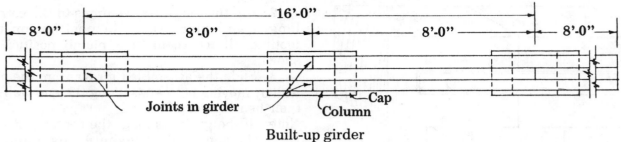

8'-0"   8'-0"   16'-0"   8'-0"   8'-0"

Joints in girder   Cap   Column

Built-up girder
Figure 4-17

each joint and butted tightly together.

6. After the first two members of the girder have been spiked, place the third member against the opposite side of the girder from which the spikes were driven.

7. Place the spikes in the same manner and proceed as when spiking the first two members together. Do not place the spikes directly opposite the ones on the other side of the girder. Drive the spikes in solidly on both sides of the girder.

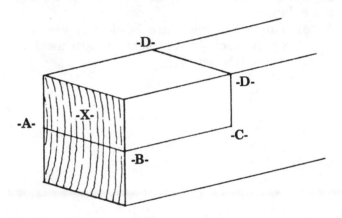

Layout of half-lap joint
Figure 4-18

How To Make Solid Girder Joints

Note: Figure 4-18 shows the half-lap crowned joint sometimes used to join solid beams.

1. Place the beam on one edge, crowned side up so the annual rings run from the top to the bottom of the beam (Figure 4-18).

2. Lay out a center line across the end of the beam as at AB, and on both sides of the beam as at BC. Make line BC about the same length as the depth of the beam, but not less than 6 inches. Square lines through points C and D on both sides of the beam and connect them across the top (DD). Mark an X on one of the sections as shown. This shows the portion of the beam to be cut out.

3. Use a rip saw to cut along lines ABC and a cross cut saw to cut along DDC.

4. Test the cut surfaces with a steel square.

If they are not square, pare the surfaces with a firmer chisel until they are square and true.

5. To make the matching joint on the other beam, proceed in exactly the same manner, but be sure the crowned side of the beam is down. When the joint is finished, turn the crowned side up and match the joints. Be sure all surfaces of the joint are square and tight.

6. Tack a temporary strap across the joint to hold it tightly together. Drill a hole through the joint with a bit about 1/16 inch larger than the bolt that is to be used. See Figure 4-19.

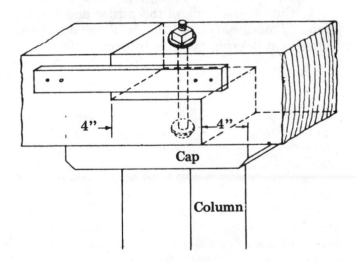

4"   4"   Cap   Column

Bearing cap
Figure 4-19

7. Place a bolt and washers as shown and tighten them.

Note: Provide a bearing cap at least 8 inches longer than the length of the joint (Figure 4-19). If metal is used, it should be ¾ inch thick. If wood is used, it should be not less than 2 inches thick. Another type of joint is shown in Figure 4-20. The straps are generally about 18 inches long and are bolted to each side of the beams.

How To Secure Ledger Boards On Girders

Nail the ledger board to the girder, staggering the nails as described in nailing the

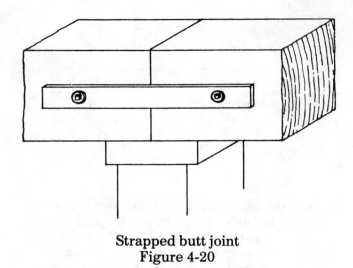

Strapped butt joint
Figure 4-20

beam. If steel beams are used, the boards are secured to the beam by bolts.

How To Support And Erect Girders

Note: Figure 4-2 shows a girder in place. Build the girder on top of the side wall, using the wall as a platform upon which to rest the girder while it is being assembled.

1. Cut off the ends of the girder and frame them, if necessary, so they will fit on their permanent bearing seats in the end walls.

2. Cut the temporary columns to the correct length and place them in position on the footings. Brace them in a plumb position as shown.

3. Slide the assembled girder from the side wall by moving both ends along the end walls to their bearing seats.

Note: If the girder is long, the center will sag. This may be prevented by building staging from the side wall, on which the girder was built, to the columns. The staging should be of such a height and width that it will allow men to support the girder while it is being slid across the end walls and placed on top of the columns.

Caution: Be sure to build the staging strong enough to support not only the weight of the men but also the weight of the girder.

4. Nail the girder in place on the columns and securely brace the columns. If a slight crown is desired in the center of the length of the girder, shingles may be used at the tops of the columns.

Note: The same procedure is used in erecting small steel girders such as are used in frame houses.

Chapter 5

Columns

If you do not have architectural or engineering guidance in selecting sizes and placement of columns, it is important that you know something of timber design principles. However, it is not necessary that the carpenter understand the mathematical basis of the design of these members. In describing supporting columns, this chapter will merely consider some of the elementary principles involved, and will acquaint you with methods of using simple charts and data.

Girders are supported by wooden posts, brick or concrete block piers, hollow steel pipe, or lally columns spaced 8 to 10 feet apart. No matter what the column may be, it must have a footing large enough to carry the load and a cap of some kind upon which the girder rests. The footing is usually concrete and about 2 feet square by 1 foot thick. For large masonry piers these dimensions are increased.

Brick piers must be at least 12 inches square unless the loads are very light and the piers are short. For heavy loads and high piers, the size should be 16 inches square. Concrete block piers are seldom less than 16 inches in each direction due to the size of the blocks available. Wood columns, if used, should not be less than 3 inches square and their lower ends must be raised off the floor about 3 inches so that any water standing on the floor will not cause decay to set in.

Steel pipe or lally columns have the advantage of being fireproof, shrink-proof, and less subject to deterioration. They are fitted with cap and base plates to obtain the proper bearing. Caps for columns of all kinds are usually flat steel plates from ¼ to ½ inch thick. Masonry piers are occasionally capped with 4 inch cut stone caps. Where wooden posts are as wide as the girder above, the cap plate is often omitted.

Figure 5-1 shows a solid wood column with a metal bearing cap, drilled to provide a means of fastening it to the column and to the girder. The bottom of this type of column may be fastened to the masonry footing by a metal dowel inserted in a hole drilled in the bottom of the post and in the masonry footing. The base at this point is sometimes coated with asphalt or pitch to prevent rust and rot.

It is best to avoid spans of more than 10 feet between columns that are to support girders. The farther apart the columns are spaced, the heavier the girder must be to carry the joists over the span between the columns.

Distribution Of Loads On Columns

The distribution of loads on columns or posts is shown in Figure 5-2. In this case, the girder is joined over the post B and rests on the outer posts A and C. The post B takes one-half the total girder load on each side so the load on the post is one-half the girder weight itself plus

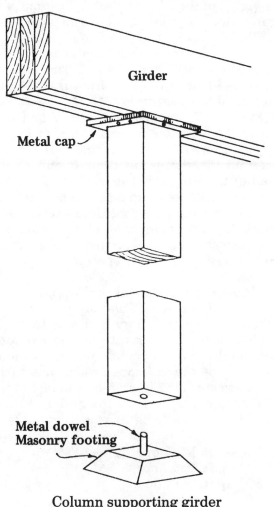

Girder

Metal cap

Metal dowel
Masonry footing

Column supporting girder
Figure 5-1

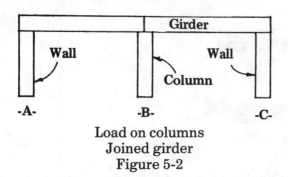

Load on columns
Joined girder
Figure 5-2

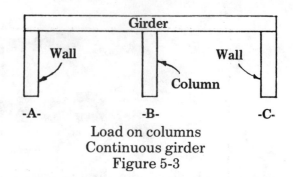

Load on columns
Continuous girder
Figure 5-3

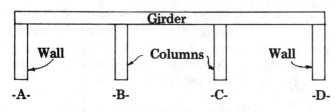

Load on columns
Continuous girder
Figure 5-4

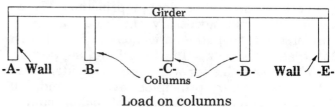

Load on columns
Continuous girder
Figure 5-5

one-half the load carried by each length of girder.

In Figure 5-3 the post *B* carries approximately five-eights of the load on the girder from *A* to *C* because the girder is continuous above post *B*.

In Figure 5-4, the load on post *B* is five-eights of the total load from *A* to *B*, plus one-half of the load from B to C. The load on post *C* is one-half of the load from *B* to *C*, plus five-eights of the load from *C* to *D*.

In Figure 5-5, the load on posts *B* and *D* is similar to the loads on posts *B* and *C* of Figure 5-4. Post *C* will bear one-half the weight imposed on the girder between *B* and *C* and one-half of the weight between *C* and *D*.

In general, where more than one post is to be used, each post must be considered separately. Although the above mentioned rules are not exactly correct in all cases, they are practical where the posts are evenly spaced.

An Example Of A Method Of Determining The Size Of Columns

Assume that a two story house is to be 24 feet wide and 40 feet long with an average total load on the floors of 40 pounds per square foot. The spacing of the columns, which are to be arranged as shown in Figure 5-5, is to be 10 feet.

1. Find the total load on the girder. 12 feet (half width) x 40 feet (length) = 480 square feet.

```
  480 sq. ft. 1st floor
  480 sq. ft. 2nd floor
  960 sq. ft. both floors
 x 40 lbs. per sq. ft.
38,400 lbs. total load on girder
```

2. Find the load per linear foot of girder. 38,400 pounds ÷ 40 feet = 960 pounds per linear foot.

3. Find the load on column *B*, Figures 5-5 and 5-6

This equals
{
5/8 of the total load from A to B
10 ft. x 960 lbs. per linear foot x 5/8 = 6,000 lbs.

Plus

1/2 of the total load from B to C
10 ft. x 960 lbs. per linear foot x 1/2 = 4,800 lbs.
}

Load on column B   10,800 lbs.

4. Find the load on column D. Same as column B.

5. Find the load on column C.

This equals
{
1/2 of the total load from B to C
10 ft. x 960 lbs. per linear foot x 1/2 = 4,800 lbs.

Plus

1/2 of the total load from C to D
10 ft. x 960 lbs. per linear foot x 1/2 = 4,800 lbs.
}

Load on column C   9,600 lbs.

Column B must support 10,800 lbs.
Column D must support 10,800 lbs.
Column C must support 9,600 lbs.

The loads imposed on columns or posts are called compressive and tend to bend the column unless the cross section is a certain proportion of

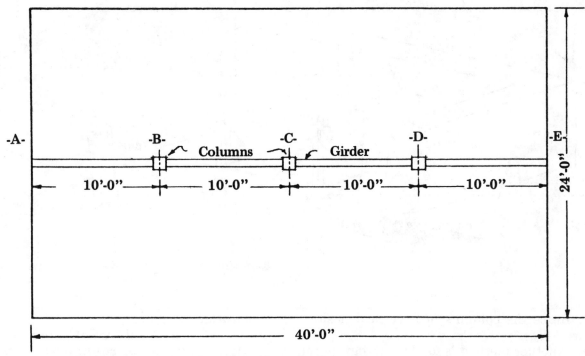

Location of columns
Figure 5-6

Height of column inches	Imposed Load In Pounds					
	3" x 4"	4" x 4"	4" x 6"	6" x 6"	6" x 8"	8" x 8"
4-0	8,720	12,920	19,850	30,250	41,250	56,250
5-0	7,430	12,400	19,200	30,050	41,000	56,250
6-0	5,630	11,600	17,950	29,500	40,260	56,250
6-6	4,750	10,880	16,850	29,300	39,950	56,000
7-0	4,130	10,040	15,550	29,000	39,600	55,650
7-6	---	9,300	14,400	28,800	39,000	55,300

For Douglas fir or southern pine No. 1 grade
Table 5-7

the length. Table 5-7 gives the maximum loads and lengths of wood columns. In the previous example, column B, which would have to support 10,800 pounds, would need to be a 4 x 4 according to the table, if it is to be 6 feet 6 inches high. The compressive capacities of steel beams may be found in Appendix A.

To Install Permanent Wooden Columns

1. Snap a chalked line across the footings to locate the center line of the columns on the footings.

2. Square the bottom ends of the wood columns or fit them to the top surfaces of the footings so the columns will stand in a vertical position. It may be necessary to hold the posts in position with stays.

3. Sight a line from the bearing seat on the masonry walls across the edges of the tempora-

rily placed columns (Figure 5-8) and mark them at these points.

Note: This may be done in several ways. Sometimes the builder's transit is used to sight a level line from some convenient point (A, Figure 5-8) above the bearing seat of the girder. The distance of A above the bearing seat is then deducted from the level points that have been marked on each column. This will determine the finished length of the columns.

4. Bore a hole in the bottom of each column and drill a hole in the top of each footing with a star drill to allow the metal dowel to be inserted (Figure 5-1).

5. Place the dowel in the footing and place the column on the footing so the dowel enters the hole.

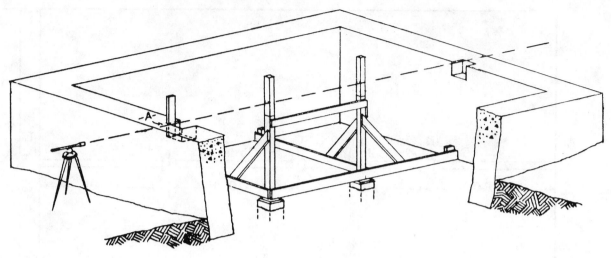

Placing columns
Figure 5-8

6. Brace the columns in a vertical position (Figure 5-8).

Note: Another method is to place the girder on temporary posts, line up the top of the girder to form a level or slightly crowned line between the bearing walls, and then cut the permanent columns to a length determined by the distance between the bottom of the girder and the top of the footing. This method may be used if temporary columns are replaced by permanent columns after the building has been framed and sheathed.

To Install Metal Columns

If circular steel columns are used, the carpenter generally places them. The same general procedure is followed but temporary wood columns are often placed on the footings to support the girder. After the framing of the building has been completed, the girder is checked for straightness and the steel columns put in place. Columns made of steel I-beams or channels generally come cut to the proper length.

Chapter 6
Floor Joists And Bridging

Floor joists are perhaps the most common area of failure in frame construction. The floor joists must be stiff to avoid squeaking floors, sagging partitions, sticking doors and cracked wallboard. The bearing surfaces of the joists should lie squarely on the sill and girder and they should be uniformly framed at these points so the tops of the joists form a straight and level floor line.

Floor joists support the loads of the rooms they span. For example, the load on the joists under the kitchen of a house would be the weight of the joists, subfloor, finish floor, lath, plaster, sink, cupboards, and all other immovable objects carried by the joists. In addition, it would include the stove, refrigerator and other movable objects in addition to any persons in the room. The joists also act as a tie to bind together and to stiffen the frame of the building.

Floor joists are selected primarily to meet strength and stiffness requirements. Strength requirements depend upon the loads to be carried. Stiffness requirements place an arbitrary control on deflection under load. Stiffness is also important in limiting vibrations from moving loads -- often a cause of annoyance to occupants. Other desirable qualities for floor joists are good nail holding ability and freedom from warp.

Wood floor joists are generally of 2 inch (nominal) thickness and of 8, 10, or 12 inch (nominal) depth. The size depends upon the loading length of span, spacing between joists, and the species and grade of lumber used. As previously mentioned, grades in species vary a great deal. For example, the grades generally used for joists are "Standard" for Douglas fir, "Number 2 or Number 2KD" for southern pine,

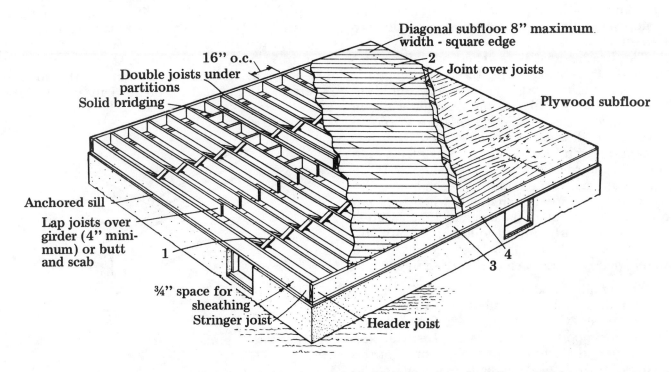

Floor framing: (1) Nailing bridging to joists; (2) nailing board subfloor to joists; (3) nailing header to joists; (4) toenailing header to sill
Figure 6-1

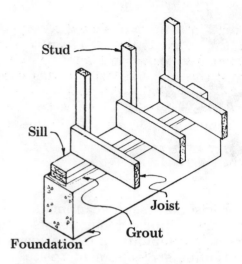

Sill construction without a header
Figure 6-2

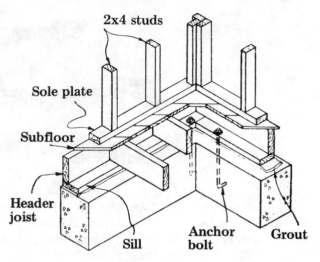

Box-sill construction
Figure 6-3

and comparable grades for other species.

Span tables for floor joists, which are published by the Federal Housing Administration or local building codes, can be used as guidelines. These sizes are of course often minimum, and it is sometimes the practice in medium and higher priced houses to use the next larger size than those listed in the tables.

Any joists having a slight bow edgewise should be so placed that the crown is on top. A crowned joist will tend to straighten out when subfloor and normal floor loads are applied. The largest edge knots should be placed on top, since knots on the upper side of a joist are on the compression side of the member and will have less effect on strength.

The header joist is fastened by nailing into the end of each joist with three sixteenpenny nails. In addition, the header joist and the stringer joists parallel to the exterior walls in platform construction (Figure 6-1) are toenailed to the sill with tenpenny nails spaced 16 inches on center. Each joist should be toenailed to the sill and center beam with two tenpenny or three eightpenny nails; then nailed to each other with three or four sixteenpenny nails when they lap over the center beam. If a nominal 2 inch scab is used across butt-ended joists, it should be nailed to each joist with at least three sixteenpenny nails at each side of the joint.

The simplest and in many ways the best way to frame joists into their supports is to set them on top as shown in Figures 6-2, 6-3 and 6-4. When so supported, the joists are easily spiked to the sill and stud and are easily lapped at the girders, spiked together and toenailed to the girder. However, joists resting on girders rest on wood in a direction perpendicular to the grain

and settle if the wood is not well seasoned. This is especially serious at the girders. Dropped girders interfere with headroom in the basement, require that heating pipes be carried rather low, and may therefore increase the depth in the basement or require the heater to be placed in a pit to provide clearance for any risers. For these reasons girders should be kept as high as possible. This is best accomplished by the ledger strip as illustrated in Figures 6-5 and 6-6. The joists notched over the girder in Figure 6-6 should be carefully noted. Bearing must not occur at the notch; it must all occur at the bottom of the joist where it rests on the ledger. Therefore, the notch must be made high enough to clear the girder by at least ½ inch so that any future shrinkage in the joist will not cause bearing at the notch. Otherwise the joist is certain to split at that point. This is a general

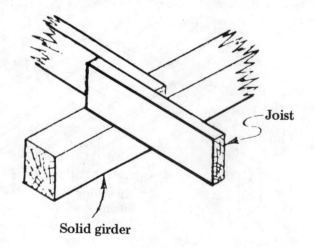

Framing joist on solid girder
Figure 6-4

82

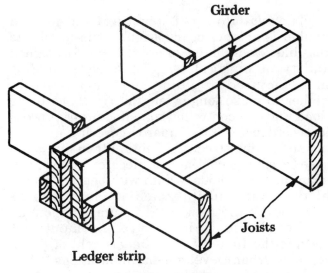

A built-up girder
Figure 6-5

practice covering all deep notches. When it is necessary to have girders and joists flush at the bottom, joists may be notched as shown in Figure 6-5. Notches at this depth are not objectionable on deep joints of long span which carry fairly light loads.

To avoid notches, joists may be hung from girders with hangers as illustrated in Figure 6-7. These are strap iron stirrups which may be either single or double, rest on top of the girder, and carry the joists. Note that hangers allow as much shrinkage as if the joist rested directly on top of the girder.

The bearing surface of a joist on the sill should be at least 4 inches long. If the ends of joists are lapped at the girder, they should not project more than 1 inch beyond the sides of the girder. See Figure 6-8. If they are butted at this point, a scab should be nailed across the sides of the joists to hold them together.

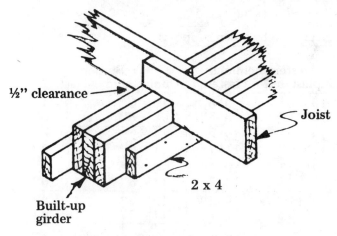

Framing joist over a girder
Figure 6-6

The full width of a joist should be supported at the sill and at the girder. If a joist is notched and supported as shown in Figure 6-9, the part of the joist carrying the load, in this case, is reduced to 5 inches in width, the lower part of the joist carrying very little load. A joist framed in this manner is likely to split eventually as shown.

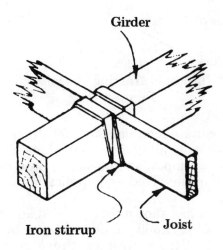

Joist supported by an iron stirrup
Figure 6-7

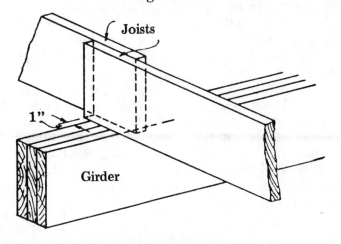

Lap of joists at girder
Figure 6-8

If joists are notched where they are framed at the girder, the sill supporting the other end of the joist should be doubled (Figure 6-10) and the full width of the joist (Figure 6-10,A) will be supported by the bottom member.

Table 6-11 is a simplified table for determining the size of joists for ordinary houses where the load is normal. Generally, it is not economical to use joists over 16 feet long.

In-Line Floor System
Floor joists positioned in-line (see Figure

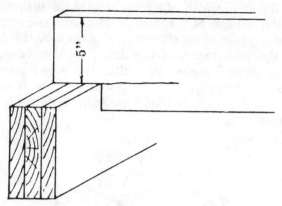

Improper framing of a joist
Figure 6-9

6-12,A) facilitate modular floor sheathing application. When holding floor sheathing to its 4 foot by 8 foot modular dimension, less cutting and waste are involved and application is much faster.

Research has shown that joints in adjacent 4 foot by 8 foot plywood panels can occur over the same joist. Staggering of these joints is unnecessary.

By keeping floor sheathing modular and floor joists in-line, the double joist at the ends of the house, traditionally used when joists are lapped, can be eliminated, for such double joists serve only as header joist fillers in cases where joists are lapped.

Some lumber yards supply preassembled in-line joists (see Figure 6-12,B). They are made up in lengths equal to the widths of the house. Preassembled in-line joists can be positioned much faster than lapped joists, resulting in substantial labor savings.

The availability of structural end-jointed lumber is another concept making in-line joists feasible. End-jointed lumber is produced in a variety of practical lengths. Such lumber is manufactured by gluing shorter lengths together to make longer lengths. This has to be done under a controlled manufacturing process in accordance with standards covering machining and gluing structural joints. When this is the case, a single, continuous piece of lumber can span the entire house width with bearing points at the ends and intermediate locations as required by the design. Structural end-jointed lumber should conform to requirements set forth in the United States Department of Housing and Urban Development publications "F.H. A. Use of Materials Bulletin" No. UM 51.

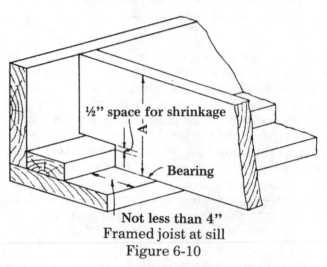

Framed joist at sill
Figure 6-10

Tables 6-13 and 6-14 present allowable spans for joists continuous over a central support, that is "two-span" conditions. The tables show the

Maximum spans for floor joists - Number 1 common
Live load for residential use of 40 pound per square feet
Uniformly distributed with ceiling as indicated

Nominal lumber size	Spacing on centers	Maximum clear span between supports					
		Southern pine and Douglas fir		Western hemlock		Spruce	
		Unplastered	Plastered	Unplastered	Plastered	Unplastered	Plastered
2"x 6"	12"	10'-11"	10'- 0"	10'- 5"	9'-6"	10'- 0"	9'- 1"
	16"	9'- 6"	9'- 1"	9'- 1"	8'-8"	8'- 8"	8'- 3"
2"x 8"	12"	14'- 5"	13'- 3"	13'-10"	12'-8"	13'- 2"	12'- 0"
	16"	12'- 7"	12'- 1"	12'- 0"	11'-7"	11'- 7"	11'- 0"
2"x10"	12"	18'- 2"	16'- 8"	17'- 4"	16'-0"	16'- 7"	15'- 2"
	16"	15'-10"	15'- 3"	15'- 2"	14'-7"	14'- 6"	13'-10"
2"x12"	12"	21'-11"	20'- 1"	20'-11"	19'-3"	19'-11"	18'- 3"
	16"	19'- 1"	18'- 5"	18'- 3"	17'-7"	17'- 5"	16'- 9"

Wood joist sizes
Table 6-11

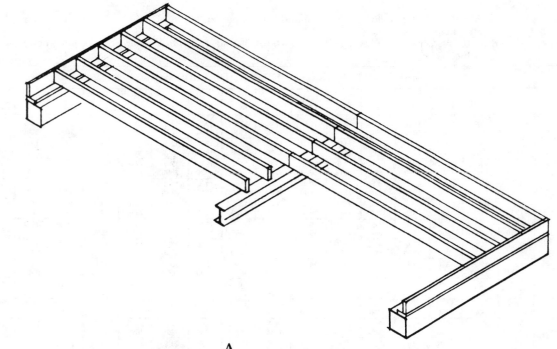

A

In-line joist system not preassembled

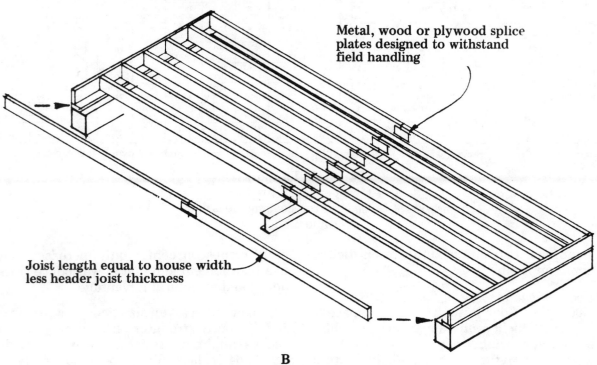

Metal, wood or plywood splice
plates designed to withstand
field handling

Joist length equal to house width
less header joist thickness

B

Preassembled in-line floor joists

In-line floor joist systems
Figure 6-12

allowable (single) clear span by joist size, spacing, and modulus of elasticity. The number given below the span refers to the required allowable fiber bending stress (Fb) of the joist.

Continuity over the center support increases floor stiffness significantly when all other variables are held constant or allows the span to be increased substantially for the same size joist. However, this also requires a considerably higher allowable fiber bending stress. Appendix

DESIGN CRITERIA:
Deflection - For 40 lbs. per sq. ft. live load on one span and 20 lbs. per sq. ft. on other. Limited to span in inches divided by 360.

Strength - Live load of 40 lbs. per sq. ft. plus dead load of 10 lbs. per sq. ft. on both spans determines the required fiber stress value.

JOIST SIZE (IN)	SPACING (IN)	Modulus of elasticity, "E", in 1,000,000 psi															
		0.8	0.9	1.0	1.1	1.2	1.3	1.4	1.5	1.6	1.7	1.8	1.9	2.0	2.2	2.4	2.6
2x6	12.0	10-3 / 1060	10-8 / 1150	11-1 / 1230	11-5 / 1310	11-9 / 1390	12-1 / 1460	12-5 / 1540	12-8 / 1610	13-0 / 1680	13-3 / 1750	13-6 / 1820	13-9 / 1880	14-0 / 1950	14-2 / 2010	14-5 / 2080	14-8 / 2140
	13.7	9-10 / 1110	10-3 / 1200	10-7 / 1280	10-11 / 1370	11-3 / 1450	11-7 / 1530	11-10 / 1610	12-2 / 1680	12-5 / 1760	12-8 / 1830	12-11 / 1900	13-1 / 1970	13-4 / 2040	13-7 / 2100	13-9 / 2170	14-0 / 2240
	16.0	9-4 / 1170	9-9 / 1260	10-1 / 1350	10-5 / 1440	10-8 / 1530	11-0 / 1610	11-3 / 1690	11-6 / 1770	11-9 / 1850	12-0 / 1920	12-3 / 2000	12-6 / 2070	12-8 / 2140	12-11 / 2210	13-2 / 2280	13-4 / 2350
	19.2	8-9 / 1240	9-2 / 1340	9-6 / 1440	9-9 / 1530	10-1 / 1620	10-4 / 1710	10-7 / 1800	10-10 / 1880	11-1 / 1960	11-4 / 2040	11-6 / 2120	11-9 / 2200	11-11 / 2280	12-2 / 2350	12-4 / 2430	12-6 / 2500
	24.0	8-2 / 1330	8-6 / 1440	8-9 / 1550	9-1 / 1650	9-4 / 1750	9-7 / 1840	9-10 / 1940	10-1 / 2030	10-3 / 2120	10-6 / 2200	10-8 / 2290	10-11 / 2370	11-1 / 2450	11-3 / 2530	11-5 / 2610	11-7 / 2690
2x8	12.0	13-7 / 1060	14-1 / 1150	14-7 / 1230	15-1 / 1310	15-6 / 1390	15-11 / 1460	16-4 / 1540	16-9 / 1610	17-1 / 1680	17-5 / 1750	17-9 / 1820	18-1 / 1880	18-5 / 1950	18-9 / 2010	19-0 / 2080	19-4 / 2140
	13.7	13-0 / 1110	13-6 / 1200	14-0 / 1290	14-5 / 1370	14-10 / 1450	15-3 / 1530	15-8 / 1610	16-0 / 1680	16-4 / 1760	16-8 / 1830	17-0 / 1900	17-4 / 1970	17-7 / 2040	17-11 / 2110	18-2 / 2180	18-6 / 2240
	16.0	12-4 / 1170	12-10 / 1260	13-3 / 1350	13-8 / 1440	14-1 / 1530	14-6 / 1610	14-10 / 1690	15-2 / 1770	15-6 / 1850	15-10 / 1930	16-2 / 2000	16-5 / 2070	16-9 / 2150	17-0 / 2220	17-3 / 2290	17-6 / 2360
	19.2	11-7 / 1240	12-1 / 1340	12-6 / 1440	12-11 / 1530	13-3 / 1620	13-8 / 1710	14-0 / 1800	14-4 / 1880	14-7 / 1970	14-11 / 2050	15-2 / 2130	15-6 / 2200	15-9 / 2280	16-0 / 2360	16-3 / 2430	16-6 / 2500
	24.0	10-9 / 1340	11-2 / 1440	11-7 / 1550	12-0 / 1650	12-4 / 1750	12-8 / 1840	13-0 / 1940	13-3 / 2030	13-7 / 2120	13-10 / 2200	14-1 / 2290	14-4 / 2370	14-7 / 2460	14-10 / 2540	15-1 / 2620	15-4 / 2700
2x10	12.0	17-4 / 1060	18-0 / 1150	18-8 / 1230	19-3 / 1310	19-10 / 1390	20-4 / 1460	20-10 / 1540	21-4 / 1610	21-10 / 1680	22-3 / 1750	22-8 / 1820	23-1 / 1880	23-6 / 1950	23-11 / 2010	24-3 / 2080	24-8 / 2140
	13.7	16-7 / 1110	17-3 / 1200	17-10 / 1290	18-5 / 1370	19-0 / 1450	19-6 / 1530	20-0 / 1610	20-5 / 1680	20-10 / 1760	21-4 / 1830	21-9 / 1900	22-1 / 1970	22-6 / 2040	22-10 / 2110	23-3 / 2170	23-7 / 2240
	16.0	15-9 / 1170	16-4 / 1260	16-11 / 1350	17-6 / 1440	18-0 / 1530	18-6 / 1610	19-0 / 1690	19-5 / 1770	19-10 / 1850	20-3 / 1930	20-7 / 2000	21-0 / 2070	21-4 / 2150	21-9 / 2220	22-1 / 2290	22-5 / 2360
	19.2	14-10 / 1240	15-5 / 1340	15-11 / 1440	16-6 / 1530	16-11 / 1620	17-5 / 1710	17-10 / 1800	18-3 / 1880	18-8 / 1970	19-0 / 2050	19-5 / 2130	19-9 / 2200	20-1 / 2280	20-5 / 2360	20-9 / 2430	21-1 / 2500
	24.0	13-9 / 1340	14-3 / 1440	14-9 / 1550	15-3 / 1650	15-9 / 1750	16-2 / 1840	16-7 / 1940	16-11 / 2030	17-4 / 2120	17-8 / 2200	18-0 / 2290	18-4 / 2370	18-8 / 2460	19-0 / 2540	19-3 / 2620	19-7 / 2700
2x12	12.0	21-1 / 1060	21-11 / 1150	22-8 / 1230	23-5 / 1310	24-1 / 1390	24-9 / 1460	25-5 / 1540	26-0 / 1610	26-7 / 1680	27-1 / 1750	27-7 / 1820	28-1 / 1880	28-7 / 1950	29-1 / 2010	29-6 / 2080	30-0 / 2140
	13.7	20-2 / 1110	20-11 / 1200	21-8 / 1290	22-5 / 1370	23-1 / 1450	23-8 / 1530	24-2 / 1610	24-10 / 1680	25-5 / 1760	25-11 / 1830	26-5 / 1900	26-11 / 1970	27-4 / 2040	27-9 / 2110	28-3 / 2170	28-8 / 2240
	16.0	19-2 / 1170	19-11 / 1260	20-7 / 1350	21-3 / 1440	21-11 / 1530	22-6 / 1610	23-1 / 1690	23-7 / 1770	24-1 / 1850	24-7 / 1930	25-1 / 2000	25-7 / 2070	26-0 / 2150	26-5 / 2220	26-10 / 2290	27-3 / 2360
	19.2	18-0 / 1240	18-9 / 1340	19-5 / 1440	20-0 / 1530	20-7 / 1620	21-2 / 1710	21-8 / 1800	22-3 / 1880	22-8 / 1970	23-2 / 2050	23-7 / 2130	24-0 / 2200	24-5 / 2280	24-10 / 2360	25-3 / 2430	25-7 / 2500
	24.0	16-9 / 1340	17-5 / 1440	18-0 / 1550	18-7 / 1650	19-2 / 1750	19-8 / 1840	20-2 / 1940	20-8 / 2030	21-1 / 2120	21-6 / 2200	21-11 / 2290	22-4 / 2370	22-8 / 2460	23-1 / 2540	23-5 / 2620	23-9 / 2700

Note: The required extreme fiber stress in bending, "F_b", in pounds per square inch is shown below each span.

Two-span floor joists
40 lbs. per sq. ft. live load
(All rooms except those used for sleeping areas and attic floors)
Table 6-13

B lists bending stress and modulus of elasticity for all common lumber.

Economical Spacing Of Floor Joists

Traditionally, floor joists have been spaced either 16 or 24 inches on center because this provides an even number of spaces with 8 foot floor sheathing lengths. Figure 6-15 illustrates three other joist spacings that are modular in 8 feet: 8 spaces at 12 inches; 7 spaces at 13.7 inches (13-23/32 inches), and 5 spaces at 19.2 inches (19-13/64 inches). These may be more economical joist spaces than either 16 or 24 inches. For example, use of 12 or 13.7 inch spacing may avoid the need to increase joist depth and be both cost and material use effective. Or, in another example, use of 2 x 10's instead of 2 x 8's increases material use 25 percent, but using 2 x 8's at 13.7 inches instead of 16 inches on center only increases material use 16.7 percent and that may produce sufficient additional structural advantage.

Likewise, a given size floor joist that is either insufficiently strong or stiff for a spacing of 24 inches might be satisfactory on a spacing of 19-13/64 inches. Similarly, if joists spaced 16 inches on center would be strong and stiff enough for a 19-13/64 inch spacing, it would be more economical to use the wider spacing. The 19.2 inch spacing also reduces joist labor, material handling and plywood fastening time. Since 5/8 inch thick plywood is frequently used with joists 16 inches on center, the 19.2 inch spacing has another advantage in that 5/8 inch Douglas fir or southern pine plywood may be used on joist spacings of 20 inches.

Engineering design allows selection of the

DESIGN CRITERIA:

Deflection - For 30 lbs. per sq. ft. live load on one span and 15 lbs. per sq. ft. on other. Limited to span in inches divided by 360.

Strength - Live load of 30 lbs. per sq. ft. plus dead load of 10 lbs. per sq. ft. on both spans determines the required fiber stress value.

JOIST SIZE (IN)	SPACING (IN)	Modulus of elasticity, "E", in 1,000,000 psi															
		0.8	0.9	1.0	1.1	1.2	1.3	1.4	1.5	1.6	1.7	1.8	1.9	2.0	2.2	2.4	2.6
2x6	12.0	11-4 / 1030	11-9 / 1110	12-2 / 1190	12-7 / 1270	13-0 / 1340	13-4 / 1420	13-8 / 1490	14-0 / 1560	14-3 / 1630	14-7 / 1700	14-10 / 1760	15-1 / 1830	15-5 / 1890	15-8 / 1950	15-10 / 2010	16-1 / 2070
	13.7	10-10 / 1070	11-3 / 1160	11-8 / 1250	12-0 / 1330	12-5 / 1410	12-9 / 1480	13-1 / 1560	13-4 / 1630	13-8 / 1700	13-11 / 1770	14-2 / 1840	14-5 / 1910	14-8 / 1970	14-11 / 2040	15-2 / 2100	15-5 / 2170
	16.0	10-3 / 1130	10-8 / 1220	11-1 / 1310	11-5 / 1400	11-9 / 1480	12-1 / 1560	12-5 / 1640	12-8 / 1720	13-0 / 1790	13-3 / 1870	13-6 / 1940	13-9 / 2010	14-0 / 2080	14-2 / 2150	14-5 / 2210	14-8 / 2280
	19.2	9-8 / 1200	10-1 / 1300	10-5 / 1390	10-9 / 1480	11-1 / 1570	11-5 / 1660	11-8 / 1740	11-11 / 1820	12-2 / 1900	12-5 / 1980	12-8 / 2060	12-11 / 2130	13-2 / 2210	13-4 / 2280	13-7 / 2350	13-9 / 2420
	24.0	9-0 / 1290	9-4 / 1400	9-8 / 1500	10-0 / 1600	10-3 / 1690	10-7 / 1790	10-10 / 1880	11-1 / 1960	11-4 / 2050	11-7 / 2130	11-9 / 2220	12-0 / 2300	12-2 / 2380	12-5 / 2460	12-7 / 2530	12-9 / 2610
2x8	12.0	14-11 / 1030	15-6 / 1110	16-1 / 1190	16-7 / 1270	17-1 / 1350	17-7 / 1420	18-0 / 1490	18-5 / 1560	18-10 / 1630	19-2 / 1700	19-7 / 1760	19-11 / 1830	20-3 / 1890	20-7 / 1950	20-11 / 2010	21-3 / 2070
	13.7	14-3 / 1080	14-10 / 1160	15-5 / 1250	15-11 / 1330	16-4 / 1410	16-10 / 1480	17-3 / 1560	17-7 / 1630	18-0 / 1700	18-4 / 1770	18-9 / 1840	19-1 / 1910	19-5 / 1980	19-9 / 2040	20-0 / 2110	20-4 / 2170
	16.0	13-7 / 1130	14-1 / 1220	14-7 / 1310	15-1 / 1400	15-6 / 1480	15-11 / 1560	16-4 / 1640	16-9 / 1720	17-1 / 1790	17-5 / 1870	17-9 / 1940	18-1 / 2010	18-5 / 2080	18-9 / 2150	19-0 / 2220	19-4 / 2280
	19.2	21-9 / 1200	13-3 / 1300	13-9 / 1390	14-2 / 1490	14-7 / 1570	15-0 / 1660	15-5 / 1740	15-9 / 1830	16-1 / 1910	16-5 / 1980	16-9 / 2060	17-0 / 2140	17-4 / 2210	17-7 / 2280	17-11 / 2350	18-2 / 2430
	24.0	11-10 / 1290	12-4 / 1400	12-9 / 1500	13-2 / 1600	13-7 / 1690	13-11 / 1790	14-3 / 1880	14-7 / 1970	14-11 / 2050	15-3 / 2140	15-6 / 2220	15-10 / 2300	16-1 / 2380	16-4 / 2460	16-7 / 2540	16-10 / 2610
2x10	12.0	19-1 / 1030	19-10 / 1110	20-6 / 1190	21-2 / 1270	21-10 / 1350	22-5 / 1420	23-0 / 1490	23-6 / 1560	24-0 / 1630	24-6 / 1700	25-0 / 1760	25-5 / 1830	25-11 / 1890	26-4 / 1950	26-9 / 2010	27-1 / 2070
	13.7	18-3 / 1080	19-0 / 1160	19-8 / 1250	20-3 / 1330	20-11 / 1410	21-5 / 1480	22-0 / 1560	22-6 / 1630	23-0 / 1700	23-5 / 1770	23-11 / 1840	24-4 / 1910	24-9 / 1980	25-2 / 2040	25-7 / 2110	25-11 / 2170
	16.0	17-4 / 1120	18-0 / 1220	18-8 / 1310	19-3 / 1400	19-10 / 1480	20-4 / 1560	20-11 / 1640	21-4 / 1720	21-10 / 1790	22-3 / 1870	22-8 / 1940	23-1 / 2010	23-6 / 2080	23-11 / 2150	24-3 / 2220	24-8 / 2280
	19.2	16-4 / 1200	16-11 / 1300	17-7 / 1390	18-1 / 1490	18-8 / 1570	19-2 / 1660	19-8 / 1740	20-1 / 1830	20-6 / 1910	20-11 / 1980	21-4 / 2060	21-9 / 2140	22-1 / 2210	22-6 / 2280	22-10 / 2350	23-2 / 2430
	24.0	15-1 / 1290	15-9 / 1400	16-4 / 1500	16-10 / 1600	17-4 / 1690	17-9 / 1790	18-3 / 1880	18-8 / 1970	19-1 / 1970	19-5 / 2140	19-10 / 2220	20-2 / 2300	20-6 / 2380	20-11 / 2460	21-2 / 2540	21-6 / 2610
2x12	12.0	23-2 / 1030	24-1 / 1110	25-0 / 1190	25-10 / 1270	26-7 / 1350	27-3 / 1420	27-11 / 1490	28-7 / 1560	29-3 / 1630	29-10 / 1700	30-5 / 1760	30-11 / 1830	31-6 / 1890	32-0 / 1950	32-6 / 2010	33-0 / 2070
	13.7	22-2 / 1080	23-1 / 1160	23-11 / 1250	24-8 / 1330	25-5 / 1410	26-1 / 1480	26-9 / 1560	27-4 / 1630	27-11 / 1700	28-6 / 1770	29-1 / 1840	29-7 / 1910	30-1 / 1980	30-7 / 2040	31-1 / 2110	31-7 / 2170
	16.0	21-1 / 1130	21-11 / 1220	22-8 / 1310	23-5 / 1400	24-1 / 1480	24-9 / 1560	25-5 / 1640	26-0 / 1720	26-7 / 1790	27-1 / 1870	27-7 / 1940	28-1 / 2010	28-7 / 2080	29-1 / 2150	29-6 / 2220	30-0 / 2280
	19.2	19-10 / 1200	20-7 / 1300	21-4 / 1390	22-1 / 1490	22-8 / 1570	23-4 / 1660	23-11 / 1740	24-5 / 1830	25-0 / 1910	25-6 / 1980	26-0 / 2060	26-6 / 2140	26-11 / 2210	27-4 / 2280	27-9 / 2350	28-2 / 2430
	24.0	18-5 / 1290	19-2 / 1400	19-10 / 1500	20-6 / 1600	21-1 / 1690	21-8 / 1790	22-2 / 1880	22-8 / 1970	23-2 / 2050	23-8 / 2140	24-2 / 2220	24-7 / 2300	25-0 / 2380	25-5 / 2460	25-10 / 2540	26-2 / 2610

Note: The required extreme fiber stress in bending, "F_b", in pounds per square inch is shown below each span.

Two-span floor joists
30 lbs. per sq. ft. live load
(All rooms used for sleeping and attic floors)
Table 6-14

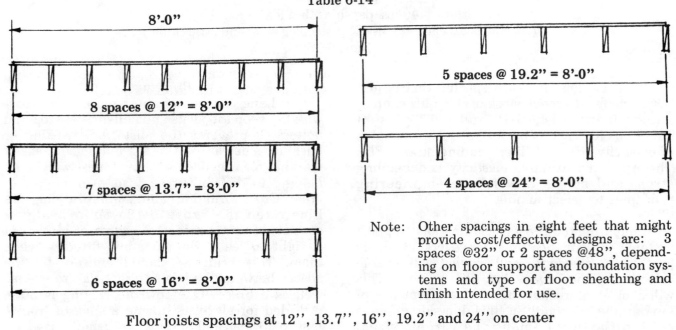

8'-0"
8 spaces @ 12" = 8'-0"
7 spaces @ 13.7" = 8'-0"
6 spaces @ 16" = 8'-0"
5 spaces @ 19.2" = 8'-0"
4 spaces @ 24" = 8'-0"

Note: Other spacings in eight feet that might provide cost/effective designs are: 3 spaces @32" or 2 spaces @48", depending on floor support and foundation systems and type of floor sheathing and finish intended for use.

Floor joists spacings at 12", 13.7", 16", 19.2" and 24" on center
Figure 6-15

DESIGN CRITERIA:
Deflection - For 40 lbs. per sq. ft. live load. Limited to span in inches divided by 360.
Strength - Live load of 40 lbs. per sq. ft. plus dead load of 10 lbs. per sq. ft. determines the required fiber stress value.

JOIST SIZE (IN)	SPACING (IN)	Modulus of elasticity, "E", in 1,000,000 psi																		
		0.4	0.5	0.6	0.7	0.8	0.9	1.0	1.1	1.2	1.3	1.4	1.5	1.6	1.7	1.8	1.9	2.0	2.2	2.4
2x6	12.0	6-9 / 450	7-3 / 520	7-9 / 590	8-2 / 660	8-6 / 720	8-10 / 780	9-2 / 830	9-6 / 890	9-9 / 940	10-0 / 990	10-3 / 1040	10-6 / 1090	10-9 / 1140	10-11 / 1190	11-2 / 1230	11-4 / 1280	11-7 / 1320	11-11 / 1410	12-3 / 1490
	13.7	6-6 / 470	7-0 / 550	7-5 / 620	7-9 / 690	8-2 / 750	8-6 / 810	8-9 / 870	9-1 / 930	9-4 / 980	9-7 / 1040	9-10 / 1090	10-1 / 1140	10-3 / 1190	10-6 / 1240	10-8 / 1290	10-10 / 1340	11-1 / 1380	11-5 / 1470	11-9 / 1560
	16.0	6-2 / 500	6-7 / 580	7-0 / 650	7-5 / 720	7-9 / 790	8-0 / 860	8-4 / 920	8-7 / 980	8-10 / 1040	9-1 / 1090	9-4 / 1150	9-6 / 1200	9-9 / 1250	9-11 / 1310	10-2 / 1360	10-4 / 1410	10-6 / 1460	10-10 / 1550	11-2 / 1640
	19.2	5-9 / 530	6-3 / 610	6-7 / 690	7-0 / 770	7-3 / 840	7-7 / 910	7-10 / 970	8-1 / 1040	8-4 / 1100	8-7 / 1160	8-9 / 1220	9-0 / 1280	9-2 / 1330	9-4 / 1390	9-6 / 1440	9-8 / 1500	9-10 / 1550	10-2 / 1650	10-6 / 1750
	24.0	5-4 / 570	5-9 / 660	6-2 / 750	6-6 / 830	6-9 / 900	7-0 / 980	7-3 / 1050	7-6 / 1120	7-9 / 1190	7-11 / 1250	8-2 / 1310	8-4 / 1380	8-6 / 1440	8-8 / 1500	8-10 / 1550	9-0 / 1610	9-2 / 1670	9-6 / 1780	9-9 / 1880
	32.0					6-2 / 1010	6-5 / 1090	6-7 / 1150	6-10 / 1230	7-0 / 1300	7-3 / 1390	7-5 / 1450	7-7 / 1520	7-9 / 1590	7-11 / 1660	8-0 / 1690	8-2 / 1760	8-4 / 1840	8-7 / 1950	8-10 / 2060
2x8	12.0	8-11 / 450	9-7 / 520	10-2 / 590	10-9 / 660	11-3 / 720	11-8 / 780	12-1 / 830	12-6 / 890	12-10 / 940	13-2 / 990	13-6 / 1040	13-10 / 1090	14-2 / 1140	14-5 / 1190	14-8 / 1230	15-0 / 1280	15-3 / 1320	15-9 / 1410	16-2 / 1490
	13.7	8-6 / 470	9-2 / 550	9-9 / 620	10-3 / 690	10-9 / 750	11-2 / 810	11-7 / 870	11-11 / 930	12-3 / 980	12-7 / 1040	12-11 / 1090	13-3 / 1140	13-6 / 1190	13-10 / 1240	14-1 / 1290	14-4 / 1340	14-7 / 1380	15-0 / 1470	15-6 / 1560
	16.0	8-1 / 500	8-9 / 580	9-3 / 650	9-9 / 720	10-2 / 790	10-7 / 850	11-0 / 920	11-4 / 980	11-8 / 1040	12-0 / 1090	12-3 / 1150	12-7 / 1200	12-10 / 1250	13-1 / 1310	13-4 / 1360	13-7 / 1410	13-10 / 1460	14-3 / 1550	14-8 / 1640
	19.2	7-7 / 530	8-2 / 610	8-9 / 690	9-2 / 770	9-7 / 840	10-0 / 910	10-4 / 970	10-8 / 1040	11-0 / 1100	11-3 / 1160	11-7 / 1220	11-10 / 1280	12-1 / 1330	12-4 / 1390	12-7 / 1440	12-10 / 1500	13-0 / 1550	13-5 / 1650	13-10 / 1750
	24.0	7-1 / 570	7-7 / 660	8-1 / 750	8-6 / 830	8-11 / 900	9-3 / 980	9-7 / 1050	9-11 / 1120	10-2 / 1190	10-6 / 1250	10-9 / 1310	11-0 / 1380	11-3 / 1440	11-5 / 1500	11-8 / 1550	11-11 / 1610	12-1 / 1670	12-6 / 1780	12-10 / 1880
	32.0					8-1 / 990	8-5 / 1080	8-9 / 1170	9-0 / 1230	9-3 / 1300	9-6 / 1370	9-9 / 1450	10-0 / 1520	10-2 / 1570	10-5 / 1650	10-7 / 1700	10-10 / 1790	11-0 / 1840	11-4 / 1950	11-6 / 2070
2x10	12.0	11-4 / 450	12-3 / 520	13-0 / 590	13-8 / 660	14-4 / 720	14-11 / 780	15-5 / 830	15-11 / 890	16-5 / 940	16-10 / 990	17-3 / 1040	17-8 / 1090	18-0 / 1140	18-5 / 1190	18-9 / 1230	19-1 / 1280	19-5 / 1320	20-1 / 1410	20-8 / 1490
	13.7	10-10 / 470	11-8 / 550	12-5 / 620	13-1 / 690	13-8 / 750	14-3 / 810	14-9 / 870	15-3 / 930	15-8 / 980	16-1 / 1040	16-6 / 1090	16-11 / 1140	17-3 / 1190	17-7 / 1240	17-11 / 1290	18-3 / 1340	18-7 / 1380	19-2 / 1470	19-9 / 1560
	16.0	10-4 / 500	11-1 / 580	11-10 / 650	12-5 / 720	13-0 / 790	13-6 / 850	14-0 / 920	14-6 / 980	14-11 / 1040	15-3 / 1090	15-8 / 1150	16-0 / 1200	16-5 / 1250	16-9 / 1310	17-0 / 1360	17-4 / 1410	17-8 / 1460	18-3 / 1550	18-9 / 1640
	19.2	9-9 / 530	10-6 / 610	11-1 / 690	11-8 / 770	12-3 / 840	12-9 / 910	13-2 / 970	13-7 / 1040	14-0 / 1100	14-5 / 1160	14-9 / 1220	15-1 / 1280	15-5 / 1330	15-9 / 1390	16-0 / 1440	16-4 / 1500	16-7 / 1550	17-2 / 1650	17-8 / 1750
	24.0	9-0 / 570	9-9 / 660	10-4 / 750	10-10 / 830	11-4 / 900	11-10 / 980	12-3 / 1050	12-8 / 1120	13-0 / 1190	13-4 / 1250	13-8 / 1310	14-0 / 1380	14-4 / 1440	14-7 / 1500	14-11 / 1550	15-2 / 1610	15-5 / 1670	15-11 / 1780	16-5 / 1880
	32.0					10-4 / 1000	10-9 / 1080	11-1 / 1150	11-6 / 1240	11-10 / 1310	12-2 / 1380	12-5 / 1440	12-9 / 1520	13-0 / 1580	13-3 / 1640	13-6 / 1700	13-9 / 1770	14-0 / 1830	14-6 / 1970	14-11 / 2080
2x12	12.0	13-10 / 450	14-11 / 520	15-10 / 590	16-8 / 660	17-5 / 720	18-1 / 780	18-9 / 830	19-4 / 890	19-11 / 940	20-6 / 990	21-0 / 1040	21-6 / 1090	21-11 / 1140	22-5 / 1190	22-10 / 1230	23-3 / 1280	23-7 / 1320	24-5 / 1410	25-1 / 1490
	13.7	13-3 / 470	14-3 / 550	15-2 / 620	15-11 / 690	16-8 / 750	17-4 / 810	17-11 / 870	18-6 / 930	19-1 / 980	19-7 / 1040	20-1 / 1090	20-6 / 1140	21-0 / 1190	21-5 / 1240	21-10 / 1290	22-3 / 1340	22-7 / 1380	23-4 / 1470	24-0 / 1560
	16.0	12-7 / 500	13-6 / 580	14-4 / 650	15-2 / 720	15-10 / 790	16-5 / 860	17-0 / 920	17-7 / 980	18-1 / 1040	18-7 / 1090	19-1 / 1150	19-6 / 1200	19-11 / 1250	20-4 / 1310	20-9 / 1360	21-1 / 1410	21-6 / 1460	22-2 / 1550	22-10 / 1640
	19.2	11-10 / 530	12-9 / 610	13-6 / 690	14-3 / 770	14-11 / 840	15-6 / 910	16-0 / 970	16-7 / 1040	17-0 / 1100	17-6 / 1160	17-11 / 1220	18-4 / 1280	18-9 / 1330	19-2 / 1390	19-6 / 1440	19-10 / 1500	20-2 / 1550	20-10 / 1650	21-6 / 1750
	24.0	11-0 / 570	11-10 / 660	12-7 / 750	13-3 / 830	13-10 / 900	14-4 / 980	14-11 / 1050	15-4 / 1120	15-10 / 1190	16-3 / 1250	16-8 / 1310	17-0 / 1380	17-5 / 1440	17-9 / 1500	18-1 / 1550	18-5 / 1610	18-9 / 1670	19-4 / 1780	19-11 / 1880
	32.0					12-7 / 1000	13-1 / 1080	13-6 / 1150	13-11 / 1220	14-4 / 1300	14-9 / 1380	15-2 / 1450	15-6 / 1520	15-10 / 1580	16-2 / 1650	16-5 / 1700	16-9 / 1770	17-0 / 1830	17-7 / 1950	18-1 / 2070

Note: The required extreme fiber stress in bending "Fb", in pounds per square inch is shown below each span.

Floor joists
40 lbs. per sq. ft. live load
(All rooms except those used for sleeping areas and attic floors)
Table 6-16

smallest size joist for each spacing that is just large enough to meet structural requirements. Tables 6-13, 6-14, 6-16 and 6-17 present numerous joist spans by size, spacing, modulus of elasticity (E), and fiber bending stress (Fb). The Appendix explains elasticity and bending stress and shows how to use engineering principles to select lumber.

Comparing total costs including joist support system and floor sheathing, for several available joist sizes, spacings, and grades will enable selection of the lowest cost floor system. This will usually represent the most efficient use of lumber and plywood, but this depends on local price differences among the various sizes, grades, and species.

Bridging

Bridging consists of short pieces set crosswise between the joists and nailed at the top and bottom. It prevents the joist from twisting or buckling sideways. Bridging may also make it possible for the floor to carry more weight by helping to distribute the load over a greater area. Cross bridging and solid bridging are illustrated in Figure 6-18. Solid bridging consists of single pieces of boards or blocks set at right angles to the joists and fitted between them. This bridging should entirely fill the spaces between the joists to act as a fire stop as well as a brace. One row of bridging is often provided for all joists having a span of from 6 feet to 10 feet and 2 rows are usually used for spans from 10 feet to 20 feet.

DESIGN CRITERIA:
Deflection - For 30 lbs. per sq. ft. live load. Limited to span in inches divided by 360.
Strength - Live load of 30 lbs. per sq. ft. plus dead load of 10 lbs. per sq. ft. determines the required fiber stress value.

JOIST SIZE (IN)	SPACING (IN)	Modulus of elasticity, "E", in 1,000,000 psi																		
		0.4	0.5	0.6	0.7	0.8	0.9	1.0	1.1	1.2	1.3	1.4	1.5	1.6	1.7	1.8	1.9	2.0	2.2	2.4
2x6	12.0	7-5 440	8-0 510	8-6 570	8-11 640	9-4 700	9-9 750	10-1 810	10-5 860	10-9 910	11-0 960	11-3 1010	11-7 1060	11-10 1100	12-0 1150	12-3 1200	12-6 1240	12-9 1280	13-1 1370	13-6 1450
	13.7	7-1 460	7-8 530	8-2 600	8-7 670	8-11 730	9-4 790	9-8 840	10-0 900	10-3 950	10-6 1010	10-10 1060	11-1 1110	11-3 1160	11-6 1200	11-9 1250	11-11 1300	12-2 1340	12-7 1430	12-11 1510
	16.0	6-9 480	7-3 560	7-9 630	8-2 700	8-6 770	8-10 830	9-2 890	9-6 950	9-9 1000	10-0 1060	10-3 1110	10-6 1160	10-9 1220	10-11 1270	11-2 1320	11-4 1360	11-7 1410	11-11 1500	12-3 1590
	19.2	6-4 510	6-10 600	7-3 670	7-8 740	8-0 810	8-4 880	8-8 940	8-11 1010	9-2 1070	9-5 1130	9-8 1180	9-10 1240	10-1 1290	10-4 1350	10-6 1400	10-8 1450	10-10 1500	11-3 1600	11-7 1690
	24.0	5-11 550	6-4 640	6-9 720	7-1 800	7-5 880	7-9 950	8-0 1020	8-3 1080	8-6 1150	8-9 1210	8-11 1270	9-2 1330	9-4 1390	9-7 1450	9-9 1510	9-11 1560	10-1 1620	10-5 1720	10-9 1820
	32.0					6-9 960	7-0 1040	7-3 1110	7-6 1190	7-9 1270	7-11 1330	8-2 1410	8-4 1470	8-6 1530	8-8 1590	8-10 1650	9-0 1710	9-2 1780	9-6 1910	9-9 2010
2x8	12.0	9-10 440	10-7 510	11-3 570	11-10 640	12-4 700	12-10 750	13-4 810	13-9 860	14-2 910	14-6 960	14-11 1010	15-3 1060	15-7 1100	15-10 1150	16-2 1200	16-6 1240	16-9 1280	17-4 1370	17-10 1450
	13.7	9-4 460	10-1 530	10-9 600	11-4 670	11-10 730	12-3 790	12-9 840	13-2 900	13-6 950	13-11 1010	14-3 1060	14-7 1110	14-11 1160	15-2 1200	15-6 1250	15-9 1300	16-0 1340	16-7 1430	17-0 1510
	16.0	8-11 480	9-7 560	10-2 630	10-9 700	11-3 770	11-8 830	12-1 890	12-6 950	12-10 1000	13-2 1060	13-6 1110	13-10 1160	14-2 1220	14-5 1270	14-8 1320	15-0 1360	15-3 1410	15-9 1500	16-2 1590
	19.2	8-5 510	9-0 600	9-7 670	10-1 740	10-7 810	11-0 880	11-4 940	11-9 1010	12-1 1070	12-5 1130	12-9 1180	13-0 1240	13-4 1290	13-7 1350	13-10 1400	14-1 1450	14-4 1500	14-9 1600	15-3 1690
	24.0	7-9 550	8-5 640	8-11 720	9-4 800	9-10 880	10-2 950	10-7 1020	10-11 1080	11-3 1150	11-6 1210	11-10 1270	12-1 1330	12-4 1390	12-7 1450	12-10 1510	13-1 1560	13-4 1620	13-9 1720	14-2 1820
	32.0					8-11 970	9-3 1040	9-7 1120	9-11 1200	10-2 1260	10-6 1340	10-9 1410	11-0 1470	11-3 1540	11-5 1590	11-8 1660	11-11 1730	12-1 1780	12-6 1900	12-10 2010
2x10	12.0	12-6 440	13-6 510	14-4 570	15-1 640	15-9 700	16-5 750	17-0 810	17-6 860	18-0 910	18-6 960	19-0 1010	19-5 1060	19-10 1100	20-3 1150	20-8 1200	21-0 1240	21-5 1280	22-1 1370	22-9 1450
	13.7	11-11 460	12-11 530	13-8 600	14-5 670	15-1 730	15-8 790	16-3 840	16-9 900	17-3 950	17-9 1010	18-2 1060	18-7 1110	19-0 1160	19-4 1200	19-9 1250	20-1 1300	20-5 1340	21-1 1430	21-9 1510
	16.0	11-4 480	12-3 560	13-0 630	13-8 700	14-4 770	14-11 830	15-5 890	15-11 950	16-5 1000	16-10 1060	17-3 1110	17-8 1160	18-0 1220	18-5 1270	18-9 1320	19-1 1360	19-5 1410	20-1 1500	20-8 1590
	19.2	10-8 510	11-6 600	12-3 670	12-11 740	13-6 810	14-0 880	14-6 940	15-0 1010	15-5 1070	15-10 1130	16-3 1180	16-7 1240	17-0 1290	17-4 1350	17-8 1400	18-0 1450	18-3 1500	18-10 1600	19-5 1690
	24.0	9-11 550	10-8 640	11-4 720	11-11 800	12-6 880	13-0 950	13-6 1020	13-11 1080	14-4 1150	14-8 1210	15-1 1270	15-5 1330	15-9 1390	16-1 1450	16-5 1510	16-8 1560	17-0 1620	17-6 1720	18-0 1820
	32.0					11-4 960	11-10 1050	12-3 1130	12-8 1200	13-0 1260	13-4 1330	13-8 1400	14-0 1460	14-4 1540	14-7 1590	14-11 1660	15-2 1720	15-5 1780	15-11 1890	16-5 2020
2x12	12.0	15-2 440	16-5 510	17-5 570	18-4 640	19-2 700	19-11 750	20-8 810	21-4 860	21-11 910	22-6 960	23-1 1010	23-7 1060	24-2 1100	24-8 1150	25-1 1200	25-7 1240	26-0 1280	26-10 1370	27-8 1450
	13.7	14-7 460	15-8 530	16-8 600	17-6 670	18-4 730	19-1 790	19-9 840	20-5 900	21-0 950	21-7 1010	22-1 1060	22-7 1110	23-1 1160	23-7 1200	24-0 1250	24-5 1300	24-10 1340	25-8 1430	26-5 1510
	16.0	13-10 480	14-11 560	15-10 630	16-8 700	17-5 770	18-1 830	18-9 890	19-4 950	19-11 1000	20-6 1060	21-0 1110	21-6 1160	21-11 1220	22-5 1270	22-10 1320	23-3 1360	23-7 1410	24-5 1500	25-1 1590
	19.2	13-0 510	14-0 600	14-11 670	15-8 740	16-5 810	17-0 880	17-8 940	18-3 1010	18-9 1070	19-3 1130	19-9 1180	20-2 1240	20-8 1290	21-1 1350	21-6 1400	21-10 1450	22-3 1500	22-11 1600	23-7 1690
	24.0	12-1 550	13-0 640	13-10 720	14-7 800	15-2 880	15-10 950	16-5 1020	16-11 1080	17-5 1150	17-11 1210	18-4 1270	18-9 1330	19-2 1390	19-7 1450	19-11 1510	20-3 1560	20-8 1620	21-4 1720	21-11 1820
	32.0					13-10 970	14-4 1040	14-11 1130	15-4 1190	15-10 1270	16-3 1340	16-8 1400	17-0 1460	17-5 1530	17-9 1590	18-1 1650	18-5 1720	18-9 1780	19-4 1890	19-11 2010

Note: The required extreme fiber stress in bending, "F_b", in pounds per square inch is shown below each span.

Floor joists
30 lbs. per sq. ft. live load
(All rooms used for sleeping areas and attic floors)
Table 6-17

Cross bridging may be made of material sized 1 x 3, 1 x 4, or 2 x 2. A Number 2 common grade of any softwood species is suitable. Figure 6-19 illustrates how cross bridging is first nailed at the top when placed in position. The lower end of each piece is left free until after the rough flooring has been nailed to the joists. Leave the lower ends loose until all the subfloor is nailed on to bring the tops of all the joists into alignment. As soon as the subfloor is in place, the bridging should be nailed at the bottom. To speed up the work of bridging, after the material has been cut for the bridging but before nailing begins, two nails are put on each end of each piece.

Metal joist bridges are more expensive than wood bridges but require considerably less labor

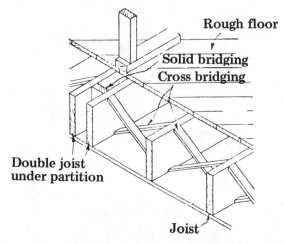

Solid and cross bridging
Figure 6-18

89

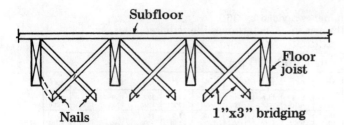

How bridging is nailed
Figure 6-19

54 x 2 = 108 total feet to be bridged
108 x 3 = 324 lineal feet of bridging stock required

Lumber Saving Techniques

Many modern building codes recognize that cross bridging is unnecessary between floor joists to obtain satisfactory structural perfor-

and no nails. Many contractors claim some total cost saving by using metal bridging.

Rows of bridging should be continuous from one side of the building to the other. If the rows are broken by a well or large chimney hole, bridging should be placed at the ends of the headers of the opening (Figure 6-20). Bridging is generally installed at intervals of not more than 8 feet along the joists.

To estimate the amount of bridging required for a floor, measure the length of the space in which bridging is to be placed and multiply this by 3. For example: In a house 24 feet wide and 54 feet long, the girder is placed 12 feet from each side wall. The span of the joists on each side is 12 feet, so each span would require one double row of bridging. Thus, each row of bridging would be 54 feet long.

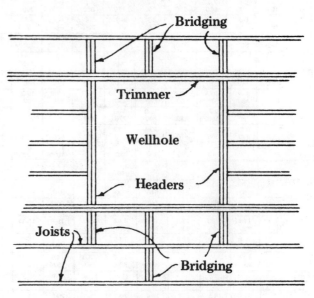

Bridging at well hole
Figure 6-20

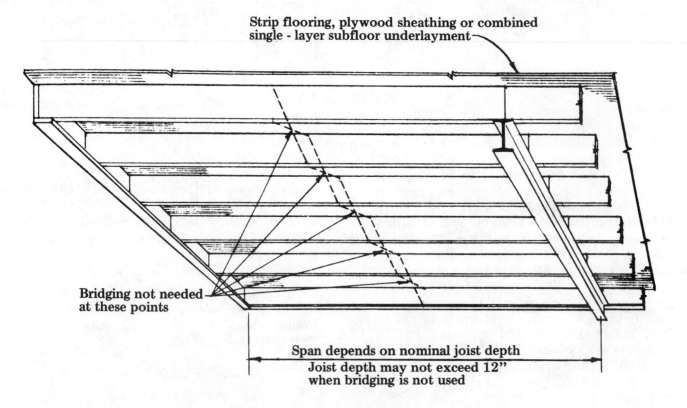

Eliminate bridging between floor joists
Figure 6-21

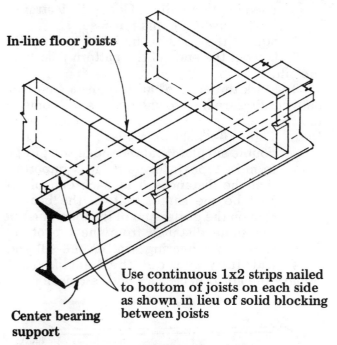

In-line floor joists

Use continuous 1x2 strips nailed to bottom of joists on each side as shown in lieu of solid blocking between joists

Center bearing support

Eliminate blocking through use of wood strips
Figure 6-22

mance (see Figure 6-21). Flooring and subfloor-ing attached to the joist perform the functions formerly attributed to bridging. Bridging can usually be eliminated between wood floor joists up to and including 2 x 12's.

Some building codes require solid blocking between floor joists over the center bearing support. According to some, blocking is required to keep the bottom of the joist from moving laterally. Figure 6-22 shows a method of using 1 x 2 wood strips nailed to the bottom of the wood joist on each side of the center beam that accomplishes the same function but uses less lumber. If a ceiling is placed on the under side of the joist, then neither the solid blocking nor 1 x 2 strips is required.

Joist Framing

The strength of joists may be seriously impaired if they are improperly framed and poorly secured at their bearing seats. In addition, the spacing and method of bridging has much to do with the strength of the floor. In the following paragraphs the common methods of framing joists to the sill and girder will be given. The reinforcement of joists and methods of framing around openings will be discussed later in this chapter.

How To Space Joists

Note: When marking the spacing of joists along the sill, it will save time if a strip of wood is laid along the sill and duplicate spacing is marked on the strip. This rod may be used to space the joists at the girder line and at the opposite side of the building. See Figure 6-23.

1. Measure in from the outside of the sill 3-5/8 inches and then 1-5/8 inch farther. This marks the inside edge of the first joist (Figure 6-23,A). Assume the joist spacing is to be 16 inches.

2. Place the body of the square along the edge of the sill with the figure 16 opposite the mark A on the sill. Mark along the outside edge of the tongue of the square. This marks the inside edge of the second joist (B). Mark an X on the left side of this mark to show where the joist is to be placed.

3. Move the square along the edge of the sill until the figure 16 on the body of the square is opposite the mark at B. Mark the line to show the position of the third joist. Continue in this manner until all the joists have been located.

4. If there is a well or chimney hole in the floor, mark the position of the headers wherever they come on the sill but also continue the regular spacing of the joists. This is shown at Figure 6-23, C.

5. Determine the location of the partitions running parallel to the joists. Lay out the location of double joists (Figure 6-24) directly

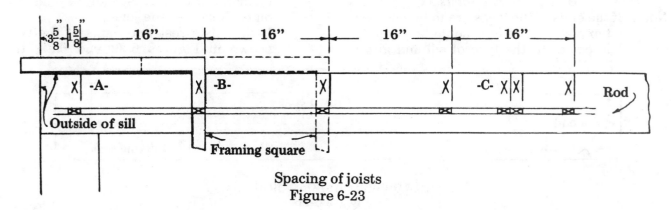

Spacing of joists
Figure 6-23

91

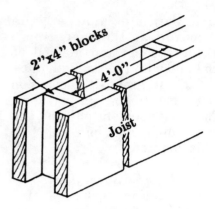

Double joint under partition
Figure 6-24

under such partitions. The two joists should be spaced apart the width of a 2 x 4.

Note: This will allow the partition studs to come directly over the space and will leave room for pipes and heating ducts to pass between the joists.

6. Mark the location of the joists on the sill on the opposite side of the building and on the girder in the same way, or use the rod if one has been made.

Note: If the joists are to be butted at the girder as shown in Figure 6-25, the spacing at the girder and at both sides of the building should be started from the same point on the sill. If, however, the joists are to be lapped at the girder as in Figure 6-8, the spacing for each joist of both sides should be marked on the girder so that the joists on both sides of the building may be kept in line. Also, the spacing on the sill at the opposite side of the building should be started 1-5/8 inch nearer the corner.

How To Frame Joists

Note: If the ends of the joists are to be notched, a template or pattern should be carefully laid out to fit the type of sill and girder

used in the building. Generally a straight joist is used for this purpose.

1. Square the ends of the joists to length.

2. Lay out the ends of the pattern joist to fit the sill.

3. Mark the ends of all the joists from this pattern. Be sure the crowned edges will come on top.

Note: Figure 6-26 shows a joist framed to fit a double sill plate and the top of the girder. In most types of framing it is important that the distances *A* at each end of the joist be the same, that is, the bearing seat on the girder end of the joist must be the same distance from the top of the joist as the bearing seat on the sill end.

4. Carefully saw out the bearing seat. Square the cut if necessary with a chisel.

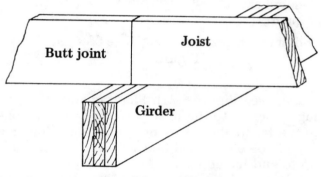

Butt joint at girder
Figure 6-25

How To Install Joists

1. Fasten two joists together as shown in Figure 6-24 and place them on the sill at the location of a bearing partition that runs parallel to the joists.

2. Nail all joists to the sill members and to the girder, being sure the joist rests in its proper place and on the spacing mark.

Note: If the joists rest upon a sill of the type shown in Figure 6-27, they should be toenailed with at least four 8d nails, two on each side of the joist.

If the joists rest upon sills of the types shown in Figures 6-28 and 6-29, they

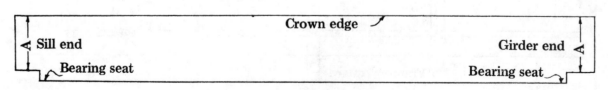

Framing on ends of joist
Figure 6-26

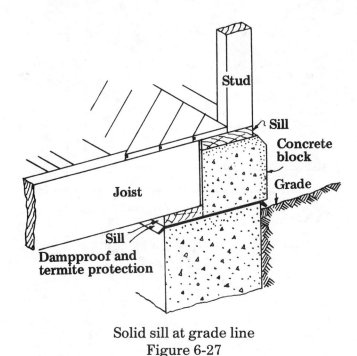

Solid sill at grade line
Figure 6-27

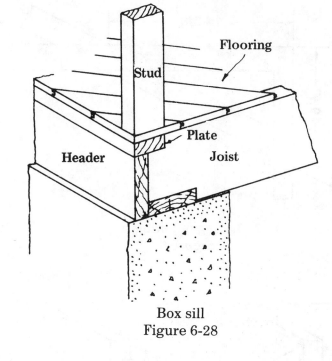

Box sill
Figure 6-28

should be toenailed into the sill plate. The header should be spiked to the ends of the joist, and the 2 x 4 plate should be fastened to the top of the joist with one spike.

3. Check across the tops of the joists with a straight edge. If they are not in a straight line, find the cause and remedy it.

How To Cut Bridging

1. Measure the actual distance between the joists. This distance, when the joists are 1-5/8 inch thick and spaced 16 inches on center, should be 14-3/8 inches.

2. Measure the actual height of the joist. If 2 inch by 10 inch joists are being used, for example, the actual height would be 9½ inches.

3. Lay the square on the edge of a piece of bridging stock as shown in Figure 6-30.

4. Hold the 9½ inch mark of the tongue of the square on the lower edge of the stock and the 14-3/8 inch mark of the body of the square on the upper edge of the stock (Figure 6-30).

5. Mark a line across the stock along the tongue of the square and mark a point on the top edge of the stock at the body of the square (Figure 6-30, A).

6. Reverse the square, keeping the same face up, so the 14-3/8 inch mark on the body is where the 9½ inch mark was originally, and so the 9½ inch mark on the tongue is where the 14-3/8 inch mark was originally (Figure 6-31).

7. Mark along the outside of the tongue. The two lines on the stock will be parallel.

8. Saw across the piece at these marks.

Note: Figures on the square that are about 3/8 inch less than the spacing and height of the joist may be used instead of the actual sizes. This will make the bridging piece shorter so the ends will not interfere with the floor or ceiling cover. See Figure 6-32. A miter box may be made to use in cutting the bridging.

9. To make the box shown in Figure 6-33, nail a piece of bridging stock to a 2 x 4 and lay out the cuts for the bridging as described in

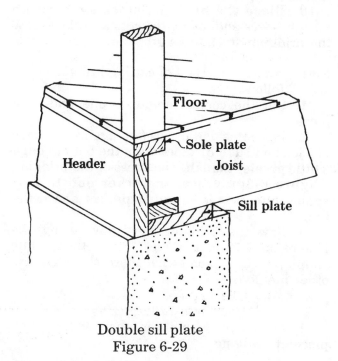

Double sill plate
Figure 6-29

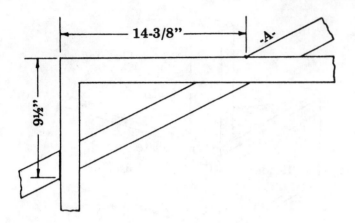

Laying out bridging
Figure 6-30

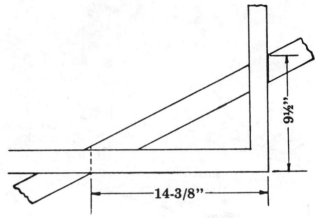

Laying out bridging
Figure 6-31

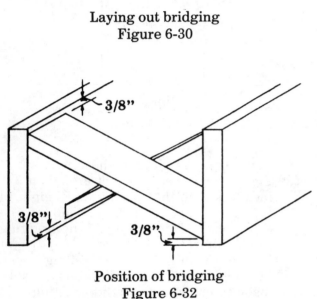

Position of bridging
Figure 6-32

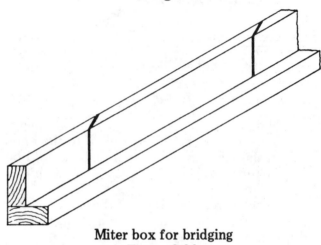

Miter box for bridging
Figure 6-33

steps 1 through 7. Cut along these lines down to the face of the 2 x 4.

10. Place the bridging stock against the upright piece and, using the cuts as guides, saw the bridging stock to length.

Note: Another method of cutting bridging is as follows:

1. Cut the correct angle on one end of a piece.

2. Place this end in position against the face of a joist and at a distance from the bottom edge a little greater than the thickness of the bridging (Figure 6-34). Allow the other end to rest against the next joist and to project above the top.

3. Saw the bridging vertically, using the face of the joist as a guide. This method, while simple, requires somewhat more skill than the other methods.

How To Install Bridging

1. Start two 8d nails in each end of each piece of bridging.

2. After having determined the location of the bridging, nail one end of a piece to the side of a joist near the top (Figure 6-35,A).

3. Nail another piece in the next opening to the same joist near the top (Figure 6-35,B).

4. Nail the rest of the bridging (C and D) in place at the top only.

Openings In Floor Joists

When large openings are made in floors, such as for stair wells and chimney holes, one or more main joists must be cut. The location of openings in the floor span has a direct bearing on the method of framing the joists. Openings for heating and plumbing fixtures that do not require the cutting of main joists to their full depth, but only the notching of the top or bottom of the joist, should be made by the carpenter who understands joist structure, rather than by the heating or plumbing contractor.

The engineering principles explained earlier in this chapter should be remembered in considering the framing around large openings in floors. However, the framing members

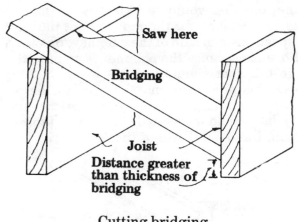

Cutting bridging
Figure 6-34

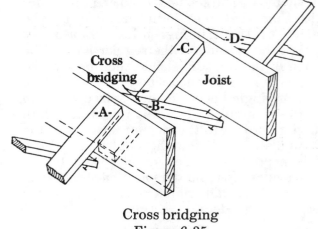

Cross bridging
Figure 6-35

around these openings are generally the same depth as the joists. Figure 6-36 shows a section of typical floor platform in which is located a stairwell and a chimney hole.

Headers are short joists at right angles to the regular joists on two sides of a floor opening. They support the ends of the joists that have been cut off, and are usually doubled (Figure 6-36,A) unless these cut off joists are quite short (Figure 6-36,B) in which case a single header may be used.

Trimmers at floor openings are joists that support the headers. They may be either regular joists or extra joists parallel to the regular joists (Figure 6-36). If the partition studs of a stair well support the trimmer joists, reinforcement of these joists is not necessary. However, in the case of the chimney hole (Figure 6-36), the trimmers joists support the headers and should be doubled.

Tail joists are joists that run from the headers to the supporting girder or side wall.

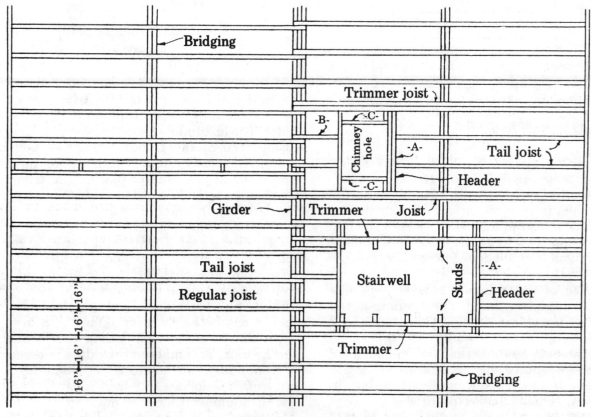

Floor plan of joists
Figure 6-36

The location of the trimmer joists (Figure 6-36) should be marked on the sill and the girder at the time the other joists are spaced. The location of the trimmer joists should not interrupt the spacing of the regular joists. The location of the headers should then be marked on the trimmers on each side of the well. The inside, or single header is then spiked between the side trimmers. The tail joists are then located according to the regular spacing of the joists and are spiked into the single header. If a double header is needed, it is spiked to the single header after the tail joists have been put in place. The small trimmers are then placed between the headers.

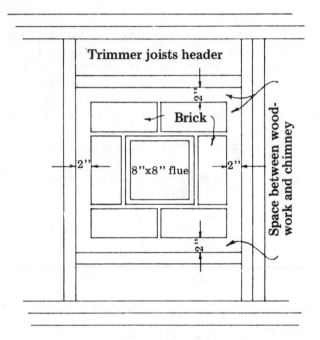

Chimney opening in joists
Figure 6-37

Framing A Chimney Hole

The headers, trimmers and studs should be kept 2 inches from each surface of the chimney to prevent the heat of the chimney from causing a fire. The space between the woodwork and the chimney should be filled with some noncombustible material.

The size of the chimney hole is generally figured from the flue lining. The opening in the floor joists for a chimney having one 8 inch by 8 inch flue (Figure 6-37) would be figured as follows: The outside dimensions of an 8 inch x 8 inch flue are approximately 9¼ inches x 9¼ inches. The brickwork surrounding the flue, in most cases, would consist of bricks 8 inches long. Therefore, two bricks would be required to cover one side of the flue, allowing ½ inch for

the joint between the bricks. The size of the chimney opening would be 16 inches for the two bricks plus ½ inch for the mortar joint plus a 2 inch space on each side of the chimney, or a total of 20½ inches. Since the chimney is square, the hole in the floor should be 20½ inches x 20½ inches. This procedure may be used for larger chimneys, but larger flues, more bricks and more joints would necessarily have to be figured to obtain the overall size of the brickwork.

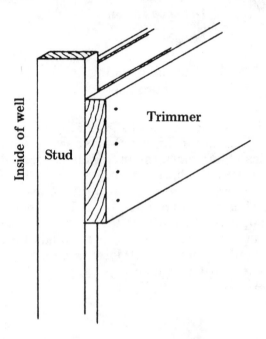

Detail of trimmer at wellhole
Figure 6-38

Framing A Stair Well

In framing a stair well, the trimmers should be supported by the studs of the walls of the well hole. These studs should project beyond the inside face of the trimmer (Figures 6-36 and 6-38). Allowance should be made for this stud projection at the time of placing the trimmers. The headers and tail joists should then be placed between the trimmer joists in the same manner as for the chimney hole.

The length of a stair well depends on the pitch of the stairs, the type of stairs, and the headroom required. The headroom is measured from the top side of the stairs to the ceiling line above the stairs and is determined by the location of the header. You can use an approximate method of finding the location of this header. This method provides for headroom of from 6 feet 8 inches to 7 feet when the risers are from 7½ inches to 8 inches each.

The number of risers must first be found. This number is then divided by two and the result is expressed in feet. The width of the joist

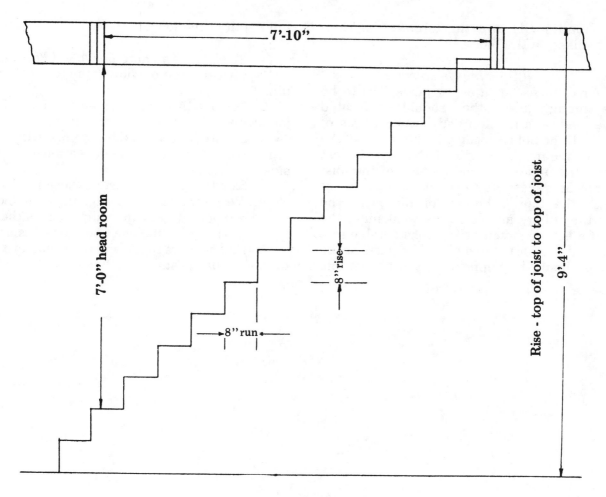

Finding headroom for stairwell
Figure 6-39

in inches is then added and the result is the length of the stair well. For example: Assume that the rise between the two floors is 9 feet 4 inches and that the joist is 10 inches (Figure 6-39). The number of risers at 8 inches each is 14. One half of this is 7. Therefore, the length of the stair well between the inside faces of the headers is 7 feet 10 inches.

The width of the stair well is dependent on the width of the stairs and the allowance made for the partition studs and wall finish on each side of the well.

How To Frame An Opening For A Chimney Or Stair Well

A definite procedure in cutting and assembling the various joist members is necessary in framing openings. This sequence of operations is about the same in most joist openings. Therefore, such typical openings as those for soil pipes, chimneys and stairs will serve as examples for all other openings in joists.

Note: A stair well and chimney hole are shown in Figure 6-36. The procedure in framing

both openings is similar.

1. Assuming that the opening is in the first floor, lay out the width of the well on the sill and on the girder.

2. Place the trimmers at these locations.

3. Mark the length of the well on both trimmers, allowing for the double headers if they are needed.

4. Cut the headers, taking the length at the sill or girder.

5. Spike the single headers between the trimmer joists at both ends of the well. Use three 20d spikes for 2 x 8 joists and four 20d spikes for 2 x 10's.

6. Mark the regular joist spacing on the single headers and spike the tail joists to the headers at these points.

7. Spike the double headers to the single headers.

Note: In Figure 6-36, short trimmers (C) are used at the chimney hole. In such a case, they are the last members to be placed and should be spiked to the single header. A block is spiked to the double header (*A*) and the trimmer is then spiked to the block.

Small Openings In Joists

If small holes, such as for pipes, are to be made through joists, they should be located halfway between the top and bottom edges of the joist. If the hole is located in the center of the joist, no material strength of the joist is lost because the upper and lower fibers of the joist have not been cut. However, if the hole is cut through the top or bottom of the joist, the supporting fibers are cut, thus weakening the joist. If a hole or notch of any considerable size is to be cut in a joist, the ends of the joists at the opening should be supported by a header the full depth of the joist.

To Cut An Opening For A Soil Pipe

1. Measure the outside diameter of the soil pipe.

2. Lay out this diameter on the face of the joist where the pipe is to cross it.

3. Square lines from these points to the edge of the joist, allowing ½ inch clearance on each side of the pipe.

4. Saw through the joist at these marks.

Note: While the joist is being cut, it should be supported by strips nailed across the top of this joist and the one on each side.

5. Nail headers to the end of the cut joist and to the regular joists on each side.

Chapter 7

Subflooring, Underlayment And Strip Flooring

Subflooring is used over the floor joists to form a working platform and base for finish flooring. It usually consists of (a) square-edge or tongued-and-grooved boards no wider than 8 inches and not less than ¾ inch thick or (b) plywood ½ to ¾ inch thick, depending on species, type of finish floor, and spacing of joists.

Subflooring may be applied either diagonally (most common) or at right angles to the joists. When subflooring is placed at right angles to the joists, the finish floor should be laid at right angles to the subflooring. Diagonal subflooring permits finish flooring to be laid either parallel or at right angles (most common) to the joists. End joints of the boards should always be made directly over the joists. Subfloor is nailed to each joist with two eightpenny nails for widths under 8 inches and three eightpenny nails for 8-inch widths.

The joist spacing should not exceed 16 inches on center when finish flooring is laid parallel to the joists, or where parquet finish flooring is used; nor should it exceed 24 inches on center when finish flooring at least 25/32 inch thick is at right angles to the joists.

Where balloon framing is used, blocking should be installed between ends of joists at the wall for nailing the ends of diagonal subfloor boards.

Laying the subfloor at right angles to the joists is used where economy of material and labor is the main consideration. It does not require the amount of cutting that a diagonal subfloor does, and consequently the labor and waste are less. With this method, it is necessary that the finished floor be laid parallel to the joists. If the finished floor were laid parallel to the subfloor, the shrinkage of the subfloor would tend to pull the finished floor boards apart.

The diagonal method of laying a subfloor causes more waste and requires more labor but it has advantages that should not be overlooked.

It braces the floor and the building more rigidly than the other method, particularly when the first and second floors are laid on opposite diagonals to the corners of the building. The joints of the subfloor run diagonally to the joints in the finished floor, thereby preventing the shrinkage of the subfloor from affecting the joints in the finished floor to any great extent. Diagonal subflooring also makes it possible to lay the finished floor in either direction.

In estimating quantities of subflooring needed, add a certain percentage to the total area to be covered. For 1 inch by 6 inch matched subflooring laid at right angles to the joists, add 20 percent to the total area. For 1 inch by 8 inch add 16 percent. Unless the floor openings are very large, they need not be deducted. If the flooring is laid diagonally, add 25 percent for 6 inch boards and 21 percent for 8 inch boards.

How To Lay A Board Subfloor At Right Angles To Joists

1. Locate the outside door openings, and provide a block on which to nail the ends of the

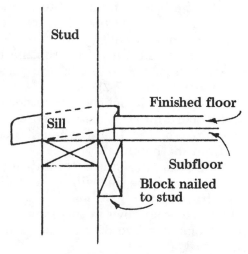

Block at door opening
Figure 7-1

99

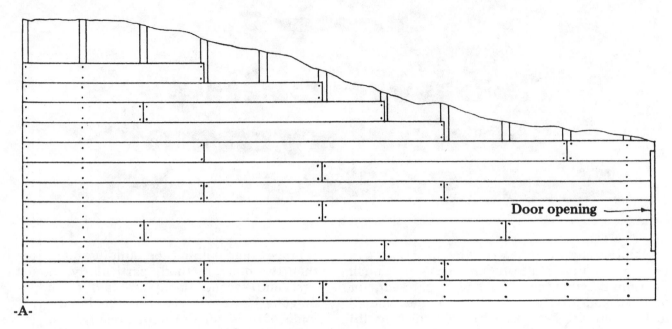

-A-

Subfloor laid at right angles to joists
Figure 7-2

subfloor. Keep the flooring back to allow for the door sill.

Note: The sill of an outside door frame projects about 1 inch inside the 2 x 4 studding (Figure 7-1).

2. Select several straight flooring boards with which to start.

3. Start at one side of the building and at the left end sill (Figure 7-2,A) and cut a board so the joint will come on the center of a joist.

4. Cut another board to butt against the end of the first board and so the other end comes on a joist. Continue in the same manner until the other end of the building is reached.

5. Nail the boards temporarily so the groove side is flush with the outside of the sill.

6. Align the boards from end to end.

7. Face nail this first row of boards to each joist with 8d common nails.

8. Lay the second row of boards the length of the floor. Be sure to break the joints on different joists (Figure 7-2).

9. If the boards do not come together easily, toenail the board into the joist. Be sure to clean off the tongue of the board when it becomes bruised from the hammer.

Note: Don't do any more toenailing than necessary since this will make it difficult to draw up the next board.

10. Lay five or six rows in this manner. Then face nail them with two 8d nails in each board into each joist.

11. Repeat these processes until the floor has been completed.

To Lay Subfloor Diagonally

1. Lay out a 45 degree angle across the joists by measuring equal distances along the sills (*AB* and *AC*, Figure 7-3).

2. Snap a chalk line from *B* to *C* on the top of the joists.

3. Cut a 45 degree angle on the end of the first piece (*D*). Use the combination square for laying out this angle.

4. Set this piece on the joists with the grooved edge at the chalk line and the bevel cut centered on that joist which will give the least amount of waste at the sill.

5. Nail this piece to each joist and sill with two 8d nails.

6. Cut a 45 degree angle on the piece *E* so it can butt against the bevel cut on the first piece as shown at *F* and nail this piece in place.

Note: Sometimes the joints over the joists are cut square with the edge of the board. This method is acceptable, providing the joint comes over a joist.

7. Cut the bevel on piece *G* using a shorter piece than *D* to break the joints. Nail this piece to the joist.

8. Fill in the rest of this strip in a similar manner.

9. Lay several more strips and then face nail the group as previously described. Break the joints so no joint will occur within several boards on the same joist.

Note: The overhang at each end may be then cut off or it may be all cut off at once after the entire floor has been laid. The

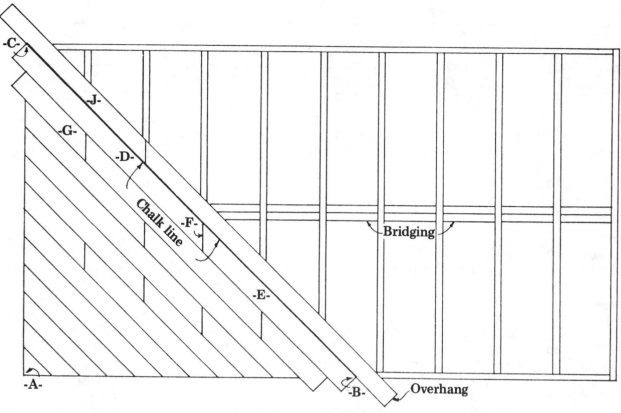

Subfloor laid diagonally to joists
Figure 7-3

overhang of subflooring at openings, such as chimney holes and stair wells, should be cut off. A portable electric saw is a great help in these operations.

10. Continue to lay the flooring until the corner A is reached.

Note: After this section of floor is laid to the corner, start to lay from the chalk line in the other direction.

11. Cut and nail the first piece (Figure 7-3,J) in place with the grooved edge to the grooved edge of the first section.

12. Fit in the remaining pieces of this strip. Break all the joints as before mentioned.

13. Complete laying the floor to the corner H in the same manner as the first section was laid.

Plywood

Plywood can be obtained in a number of grades designed to meet a broad range of end-use requirements. All interior type grades are also available with fully waterproof adhesive identical with those used in exterior plywood. This type is useful where a hazard of prolonged moisture exists, such as in underlayments or subfloors adjacent to plumbing fixtures and for roof sheathing which may be exposed for long periods during construction. Under normal conditions and for sheathing used on walls, standard sheathing grades are satisfactory.

Plywood suitable for subfloor, such as Standard sheathing, Structural I and II, and C-C Exterior grades, has a panel identification index marking on each sheet. These markings indicate the allowable spacing of rafters and floor joists for the various thicknesses when the plywood is used as roof sheathing or subfloor. For example, an index mark of 32/16 indicates that the plywood panel is suitable for a maximum spacing of 32 inches for rafters and 16 inches for floor joists. Thus, no problem of strength differences between species is involved as the correct identification is shown for each panel. Table 7-4 lists minimum plywood subfloor requirements.

Plywood should be installed with the grain direction of the outer plies at right angles to the joists. Plywood should be nailed to the joist at each bearing with eightpenny common or sevenpenny threaded nails for plywood ½ inch to ¾ inch thick. Space nails 6 inches apart along all edges and 10 inches along intermediate members. When plywood serves as both subfloor and underlayment, nails may be spaced 6 to 7 inches apart at all joists and blocking. Use

Panel Identification Index (1), (2), (3), and (4)	Plywood Thickness (inches)	Maximum Span (5) (inches)	Common Nail Size and Type	Nail Spacing (inches)	
				Panel Edges	Intermediate
30/12	5/8	12 (6)	8d	6	10
32/16	1/2	16 (7)	8d (8)	6	10
36/16	3/4	16 (7)	8d	6	10
42/20	5/8	20 (7)	8d	6	10
48/24	3/4	24	8d	6	10
1-1/8" groups 1&2	1-1/8	48	10d (9)	6	6
1-1/4" groups 3&4	1-1/4	48	10d (9)	6	6

Notes:

(1) These values apply for Structural I and II, Standard sheathing and C-C exterior grades only.

(2) Identification index appears on all panels except 1-1/8" and 1¼" panels.

(3) In some non-residential buildings, special conditions may impose heavy concentrated loads and heavy traffic requiring subfloor constructions in excess of these minimums.

(4) Edges shall be tongue and grooved or supported with blocking for square edge wood flooring, unless separate underlayment layer (¼" minimum thickness) is installed.

(5) Spans limited to values shown because of possible effect of concentrated loads. At indicated maximum spans, floor panels carrying identification index numbers will support uniform loads of more than 100 psf.

(6) May be 16" if 25/32" wood strip flooring is installed at right angles to joists.

(7) May be 24" if 25/32" wood strip flooring is installed at right angles to joists.

(8) 6d common nail permitted if plywood is ½".

(9) 8d deformed shank nails may be used.

Minimum thickness of plywood adequate
for subflooring
Table 7-4

eight or ninepenny common nails or seven or eightpenny threaded nails.

For the best performance, plywood should *not* be laid up with tight joints whether used on the interior or exterior. The spacings in Table 7-5 are recommendations by the American Plywood Association on the basis of field experience. Figure 7-6 illustrates the application method.

Feel free to use the minimum thickness plywood underlayment required whenever two-layer floors (see Figure 7-7) are specified. Table 7-8 provides information on the minimum thickness of plywood underlayment used over plywood subfloors.

Single-Layer Flooring (2-4-1 Plywood)

For many years, the practice has been to use two layers of floor sheathing in addition to the finish layer -- a subfloor and an underlayment floor. Single-layer floor sheathing using tongue and groove plywood that is designed as a combination subfloor and underlayment is more cost-effective than double-layer floors. Such

grades of plywood are readily available and may be used to reduce plywood cost wherever floorings such as tile, carpet, or sheet vinyl are used for the finish floor. Table 7-9 shows the various combinations of plywood grades used for subfloor-underlayment, their corresponding maximum span for various joist spacings, and the recommended nail spacing. Figure 7-10 shows details of single-layer plywood combination subfloor underlayment.

	Spacing	
	Edges (In.)	Ends (In.)
Underlayment or interior wall lining	1/32	1/32
Panel sidings and combination subfloor underlayment	1/16	1/16
Roof sheathing, subflooring, and wall sheathing (Under wet or humid conditions, spacing should be doubled.)	1/8	1/16

Plywood location and use
Table 7-5

102

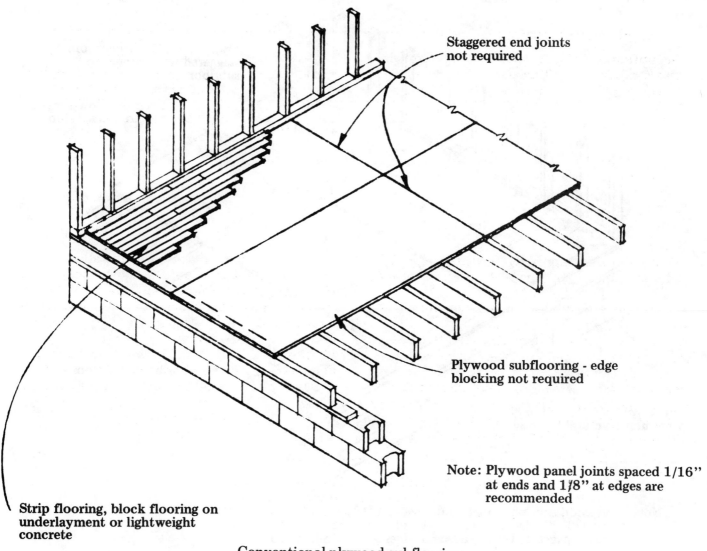

Staggered end joints not required

Plywood subflooring - edge blocking not required

Note: Plywood panel joints spaced 1/16" at ends and 1/8" at edges are recommended

Strip flooring, block flooring on underlayment or lightweight concrete

Conventional plywood subflooring
Figure 7-6

Plywood Glued To Floor Joists

When plywood is glued to floor joists, plywood flooring and wood joists act together as a single structural member. This single composite T-beam can span larger distances without the danger of excessive deflection or of being over-stressed than when plywood is fastened to joists only with nails. Since longer spans are possible with a given size floor joist when plywood is glued to the joist, floor joist framing lumber may be reduced for some spans when glued construction is used. Nail gluing of the floor sheathing to the joists also improves quality by reducing floor squeaking and increasing floor stiffness.

Single-Layer Hardwood Strip Flooring

Hardwood strip flooring 25/32 inch thick can be applied directly to floor joists for joist spacings up to 16 inches apart. Each piece of strip flooring must bear on at least two joists, and adjacent pieces cannot break or have end joints in the same span between joists.

When rain, snow, or heavy fog is not likely and before the house is closed in, lay hardwood strip flooring across joists to form a working platform (see Figure 7-11). If part of the floor surface is to be covered with resilient tile or carpeting, this area should be constructed with a single-layer combination subfloor underlayment plywood.

This method should be used when there is little chance of precipitation or heavy fog and when it is possible to rapidly erect exterior and interior bearing walls and roof trusses, and to dry-in the roof with building paper.

If rain, snow, or heavy fog is likely, erect exterior and interior bearing walls, complete the roof framing and dry-in the roof with building

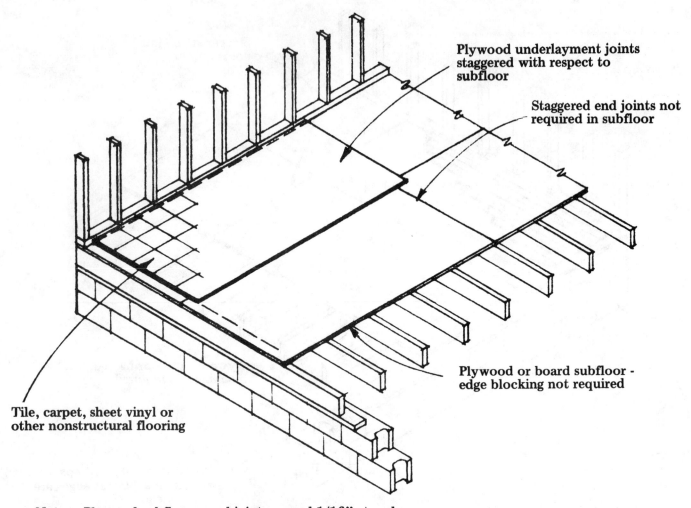

Plywood underlayment joints staggered with respect to subfloor

Staggered end joints not required in subfloor

Plywood or board subfloor - edge blocking not required

Tile, carpet, sheet vinyl or other nonstructural flooring

Note: Plywood subfloor panel joints spaced 1/16" at ends and 1/8" at edges and plywood underlayment panel joints spaced 1/32" at ends and edges are recommended

Minimum thickness plywood underlayment
under non-structural flooring
Figure 7-7

paper before laying hardwood strip flooring. Before erecting walls, place a filler block of hardwood strip flooring on top of the floor and header joists as shown in A and B of Figure 7-12. Then erect the exterior walls, load-bearing partitions, and roof. Place building paper over the roof to keep the flooring dry.

Strip flooring that runs parallel to the exterior wall can be installed as shown in Figure 7-12, A. Strip flooring running perpendicular to the exterior wall can be installed as shown in Figure 7-12, B. In this case, 2 x 2 blocking nailed to the header joist serves as a ledger for the ends of the flooring strips. Flooring can be face nailed into the 2 x 2 ledger.

C and D of Figure 7-12 show details for laying strip flooring up to interior load-bearing

Plywood Grades and Species Group (1) (7)	Minimum Plywood Thickness	Fastener Size (approx.) and Type (set nails 1/16") (5)	Fastener Spacing (inches)	
			Panel Edges	Inter-mediate
Groups 1, 2, 3, 4 Underlayment-Interior, Underlayment-Interior (with Exterior glue) or Underlayment-Exterior (C-C Plugged)	1/4" (2)	18 Ga. Staples (6) or 3d ring-shank nails	3	6 each way
	3/8" (3)	16 Ga. Staples (6)	3	6 each way
		3d ring-shank nails	6	8 each way
Same Grades as above, but Group 1 Only	1/4" (4)	18 Ga. Staples (6) or 3d ring-shank nails	3	6 each way

(1) For application of tile, carpeting, linoleum, or other non-structural flooring. Stagger all joints of underlayment with respect to subfloor joints. (For maximum stiffness, place face grain of underlayment across supports, and end joints over framing.) Fill any damaged or open areas, such as joints or splits wider than 1/16 inch. Do not fill nail holes. Lightly sand any surface roughness particularly at joints and around nails.
(2) Over plywood subfloor.
(3) Over lumber subfloor or other uneven

surfaces.
(4) Over lumber floor up to 4 inches wide. Face grain must be perpendicular to boards.
(5) Unless subfloor is thoroughly seasoned and dry material, counter-sink nail heads 1/16 inch below surface of the underlayment just prior to laying finish floors to avoid nail popping. Staples should be counter-sunk 1/32-inch.
(6) Crown width 3/8 inch for 16 gauge, 3/16 inch for 18 gauge staples; length sufficient to penetrate at least 5/8 inch or completely through subflooring.

Minimum thickness of plywood underlayment
required over plywood subfloors
Table 7-8

Plywood grade (3)	Plywood species group	16" o.c.		20" o.c.		24" o.c.		Nail Spacing (Inches)	
		Panel thickness	Deformed shank nail size (4)	Panel thickness	Deformed shank nail size (4)	Panel thickness	Deformed shank nail size (4)	Panel edges	Inter-mediate
C-C plugged exterior	1	1/2	6d	5/8"	6d	3/4"	6d	6	10
Underlayment with Ext. glue	2 & 3	5/8"	6d	3/4"	6d	7/8"	6d	6	10
Underlayment	4	3/4"	6d	7/8"	6d	1"	8d	6	10

Notes:
(1) Edges shall be tongue-and-grooved, or supported with framing.
(2) In some non-residential buildings, special conditions may impose heavy concentrated loads and heavy traffic requiring subfloor-underlayment constructions in excess of these minimums.
(3) For certain types of flooring such as hardwood strip, wood block, slate terrazzo, etc., sheathing grades of plywood may be used.
(4) Set nails 1/16" and lightly sand subfloor at joints if resilient flooring is to be applied.

Grades, thicknesses, and fastening details for single-layer plywood flooring
Table 7-9

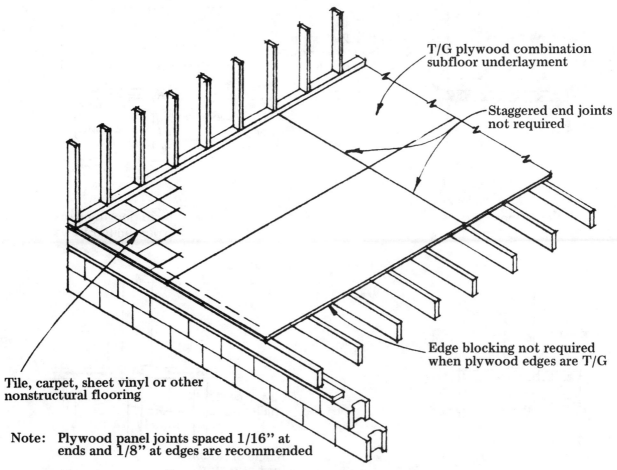

Note: Plywood panel joints spaced 1/16" at ends and 1/8" at edges are recommended

Single layer plywood combination subfloor underlayment
Figure 7-10

walls; *C*, for strip flooring layed parallel to walls; and *D*, for strip flooring layed perpendicular to walls.

It is suggested that hardwood strip flooring be layed from exterior walls toward interior load-bearing partitions. Then, non load-bearing partitions can be placed on top of the hardwood strip flooring.

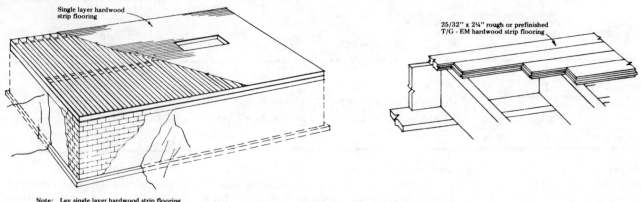

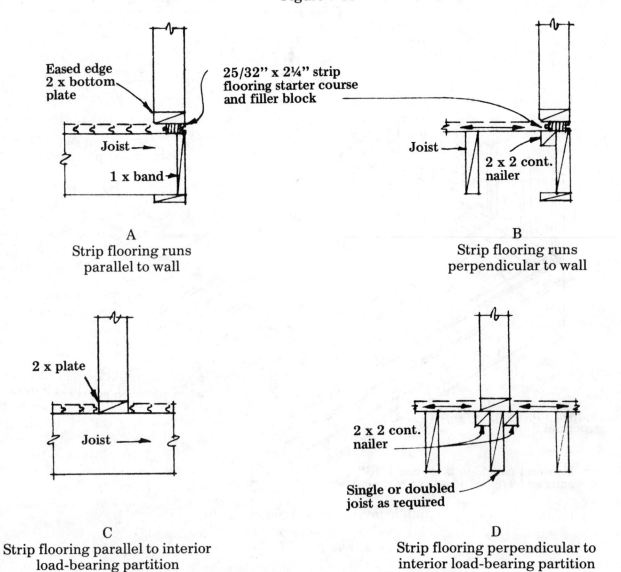

Single-layer hardwood strip flooring applied
directly to floor joists
Figure 7-11

A
Strip flooring runs
parallel to wall

B
Strip flooring runs
perpendicular to wall

C
Strip flooring parallel to interior
load-bearing partition

D
Strip flooring perpendicular to
interior load-bearing partition

Methods of laying hardwood strip flooring
after house is dried in
Figure 7-12

Chapter 8
Wall Framing

The floor framing with its subfloor covering has now been completed and provides a convenient working platform for construction of the wall framing. The term "wall framing" includes primarily the vertical studs and horizontal members (soleplates, top plates, and window and door headers) of exterior and interior walls that support ceilings, upper floors, and the roof. The wall framing also serves as a nailing base for wall covering materials.

The wall framing members used in conventional construction are generally nominal 2 x 4 studs. Depending on thickness of covering material, 24-inch spacing should be considered. Top plates and soleplates are also nominal 2 x 4's. Headers over doors or windows in load-bearing walls consist of doubled 2 x 6 inch and deeper members, depending on the span of the opening.

Requirements

The requirements for wall framing lumber are good stiffness, good nail-holding ability, freedom from warp, and ease of working. Species used may include Douglas fir, the hemlocks, southern pine, the spruces, pines, and white fir. The grades vary by species, but it is common practice to use the third grade for studs and plates and the second grade for headers over doors and windows.

All framing lumber for walls should be reasonably dry. Material at about 15 percent moisture content is desirable, with the maximum allowable considered to be 19 percent. When the higher moisture content material is used (as studs, plates, and headers), it is advisable to allow the moisture content to reach inservice conditions before applying interior trim.

Ceiling height for the first floor is 8 feet under most conditions. It is common practice to rough-frame the wall (subfloor to top of upper plate) to a height of 8 feet 1½ inch. In platform construction, precut studs are often supplied to a length of 7 feet 8-5/8 inches for plate thickness of 1-5/8 inch. When dimension material is 1½ inch thick, precut studs would be 7 feet 9 inches long. This height allows the use of 8-foot high wallboard sheets, or six courses of rock lath, and still provides clearance for floor and ceiling finish or for plaster grounds at the floor line.

Second-floor ceiling heights should not be less than 7 feet 6 inches in the clear, except that portion under sloping ceilings. One-half of the floor area, however, should have at least a 7-foot 6-inch clearance.

As with floor construction, two general types of wall framing are commonly used -- platform construction and balloon frame construction. The platform method is more often used because of its simplicity. Balloon framing is generally used where stucco or masonry is the exterior covering material in two-story houses.

Lumber Saving Techniques

The 2 x 4 stud framing in exterior walls of one-story houses can almost always be spaced 24 inches on center. There is no reason to space studs closer as long as facing materials on the exterior wall can span 24 inches. In fact, the commonly used 16 inch spacing may have arisen from when wood lath was used. Tradesmen may have found it difficult to plaster on lath when studs were spaced more than 16 inches apart. Today, many readily available sheet materials will span 24 inches.

Mid-height wall blocking is not necessary between studs and can usually be eliminated. Top and bottom plates of framed walls act as fire stops. Unless balloon framing is used, mid-height blocking is of little use as a fire block. Mineral wool or glass fiber insulation that fills the entire wall cavity acts as a fire stop. Mid-height wall fire blocking is not needed when this noncombustible insulation fills wall cavities.

One-half inch thick gypsum wallboard applied horizontally does not require blocking behind the horizontal taped joints when stud spacing is 24 inches on center or less. The taped joint is strong enough to resist usual occupancy loads placed on the board without wood back-up blocking.

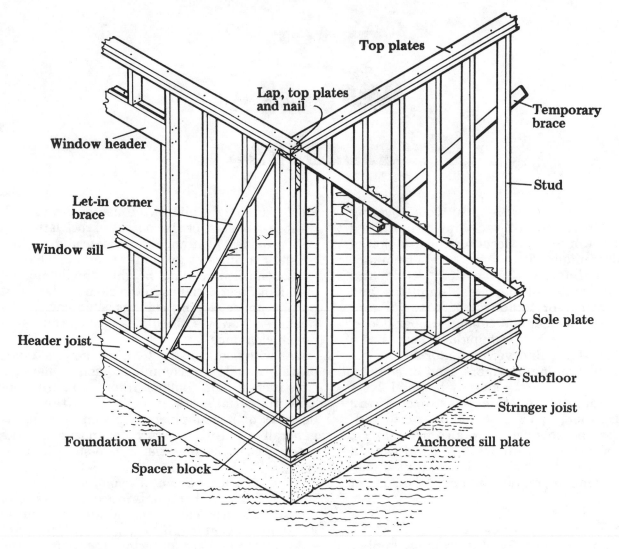

Wall framing used with platform construction
Figure 8-1

Platform Construction

The wall framing in platform construction is erected above the subfloor which extends to all edges of the building (Figure 8-1). A combination of platform construction for the first floor sidewalls and full-length studs for end walls extending to end rafters of the gable ends is commonly used in single-story houses.

One common method of framing is the horizontal assembly (on the subfloor) or "tilt-up" of wall sections. When a sufficient work crew is available, full-length wall sections are erected. Otherwise, shorter length sections easily handled by a smaller crew can be used. This system involves laying out precut studs, window and door headers, cripple studs (short-length studs), and window sills. Top and soleplates are then nailed to all vertical members and adjoining studs to headers and sills with sixteenpenny nails. Let-in corner bracing should be provided when required. The entire section is then erected, plumbed, and braced (Figure 8-1).

A variation of this system includes fastening the studs only at the top plate and, when the wall is erected, toenailing studs to the soleplates which have been previously nailed to the floor. Corner studs and headers are usually nailed together beforehand to form a single unit. Many contractors will also install sheathing before the wall is raised in place. Complete finished walls with windows and door units in place and most of the siding installed can also be fabricated in this manner.

When all exterior walls have been erected, plumbed, and braced, the remaining nailing is completed. Soleplates are nailed to the floor joists and headers or stringers (through the subfloor); corner braces, when used, are nailed to studs and plates; door and window headers are fastened to adjoining studs; and corner studs are nailed together. These and other

108

recommended nailing practices are shown in Figure 8-1.

In hurricane areas or areas with high winds, it is often advisable to fasten wall and floor framing to the anchored foundation sill when sheathing does not provide this tie. Figure 8-2 illustrates one system of anchoring the studs to the floor framing with steel straps.

Several arrangements of studs at outside corners can be used in framing the walls of a house. Figure 8-1 shows one method commonly used. Blocking between two corner studs is used to provide a nailing edge for interior finish (Figure 8-3, A). Figure 8-3,B and C show other methods of stud arrangement to provide the needed interior nailing surfaces as well as good corner support.

Interior walls should be well fastened to all exterior walls they intersect. This intersection

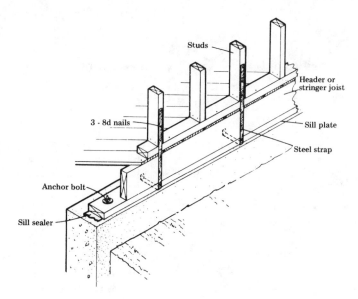

Anchoring wall to floor framing
Figure 8-2

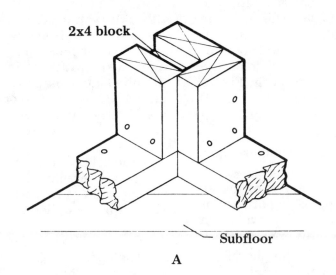

A

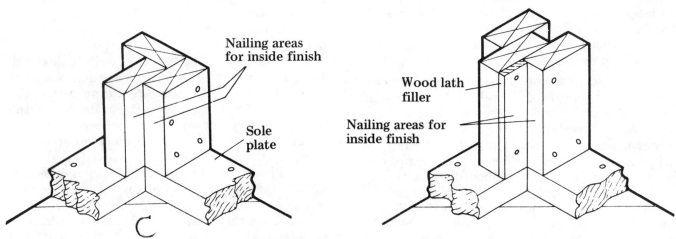

B

Examples of corner stud assembly: A, Standard outside corner; B, special corner with lath filler; C, special corner without lath filler
Figure 8-3

109

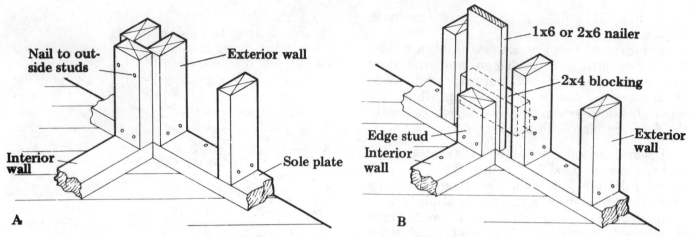

Intersection of interior wall with exterior wall:
A, With doubled studs on outside wall;
B, partition between outside studs
Figure 8-4

should also provide nailing surfaces for the plaster base or dry-wall finish. This may be accomplished by doubling the outside studs at the interior wall line (Figure 8-4, A). Another method used when the interior wall joins the exterior wall between studs is shown in Figure 8-4, B.

Short sections of 2 x 4 blocking are used between studs to support and provide backing for a 1 x 6 nailer. A 2 x 6 vertical member might also be used.

The same general arrangement of members is used at the intersection or crossing of interior walls. Nailing surfaces must be provided in some form or another at all interior corners.

The stud serving as a backer for the inside facing material can be eliminated by attaching back-up cleats to the corner stud as illustrated in Figure 8-5. The back-up cleats can be 3/8-inch thick plywood, 1-inch thick lumber, or specially designed metal clips.

Neither cleats nor back-up studs are required if framing of one wall is held back as shown in A of Figure 8-5. In this case, gypsum wallboard is inserted into the space between the corner studs of the two intersecting walls. The gypsum wallboard is not fastened to cleats. It is recommended that the wallboard sheet resting on the back-up cleats be installed first, then the wallboard sheet on the adjacent wall will lock into place the sheet resting on the back-up cleats.

After all walls are erected, a second top plate is added that laps the first at corners and wall intersections (Figure 8-1). This gives an additional tie to the framed walls. These top plates can also be partly fastened in place when the wall is in a horizontal position. Top plates are nailed together with sixteenpenny nails spaced 16 inches apart and with two nails at each wall intersection. Walls are normally plumbed and aligned before the top plate is added. By using 1 x 6 or 1 x 8 temporary braces on the studs between intersecting partitions, a straight wall is assured. These braces are nailed to the studs at the top of the wall and to a 2 x 4 inch block fastened to the subfloor or joists. The temporary bracing is left in place until the ceiling and the roof framing are completed and sheathing is applied to the outside walls.

How To Install The Sill

The operations of installing the sill plate are outlined in Chapter 3. To install the box sill, which is generally used in the platform frame, the following additional operations should be performed:

1. Toenail the headers to the sill plates with 16d nails, lapping the headers at each corner of the building. See Figure 8-6. Be sure that the lap of the sill plate is opposite to the lap of the header at each corner.

2. Spike the headers to each other at the corners of the building.

3. Place a short piece of 2 x 4 across the butt joint of the sill plate and against the inside face of the header as shown at A, Figure 8-6.

4. Spike the block to the header and to both sill plates.

How To Lay And Cut Plates To Length

1. After the subflooring has been laid, place a 2 x 4 around the outer edge of the building on the subfloor with a square end flush with the

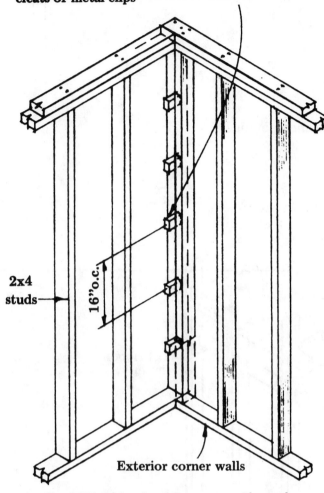

3/8" thick plywood, 1" thick lumber cleats or metal clips

16" o.c.

2x4 studs

Exterior corner walls

For 3/8" thick, plywood, use two 4d annular threaded nails in each cleat.
For 1" thick lumber, use two 5d annular threaded nails in each cleat.
For metal clips, use nails specified by manufacturers.

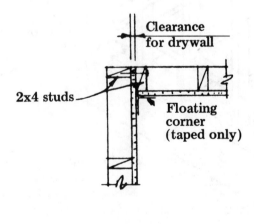

Clearance for drywall

2x4 studs

Floating corner (taped only)

A

Method of supporting interior facings at exterior wall corners
Figure 8-5

outside edge of the sill. This piece is for the bottom plate of the side wall.

2. Cut the other end of this plate off at the center of a stud location.

3. Continue with additional 2 x 4's until the other end of the building is reached. Cut the last 2 x 4 off flush with the outside of the sill.

4. Make another 2 x 4 plate for the top of the side wall.

5. Mark the upper and lower plates for the stud spacing.

Note: On the side walls, the stud markings should come directly over the joist markings because both joists and studs should have the same spacing.

6. Lay out the marking for the side studs of window and door openings, and also the locations of cross interior partitions on both the upper and the lower plates. See Figure 8-7.

7. Lay out the plates for the sides and ends of the building in a like manner, allowing for the insertion of the corner posts at the corners of the building.

How To Assemble, Erect And Brace Wall Sections

1. Provide a clear space on the subfloor to assemble the wall section.

2. Lay the shoe edgewise on the floor. Lay

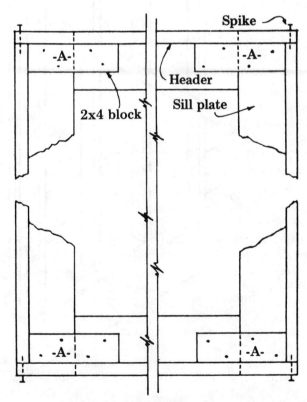

Spike

-A-

Header

2x4 block

Sill plate

-A-

-A-

-A-

Method of tying sill at corners
Figure 8-6

111

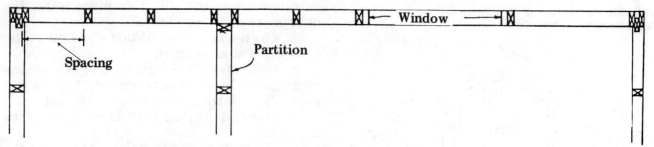

Spacing of studs at intersections and openings
Figure 8-7

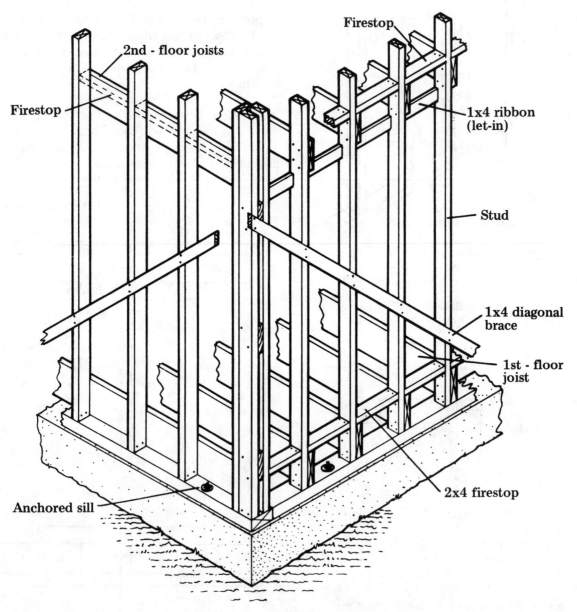

Wall framing used in balloon construction
Figure 8-8

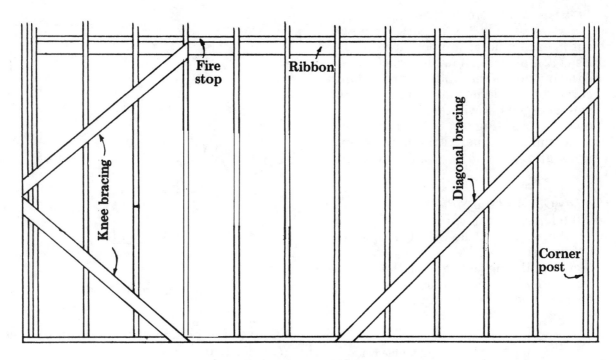

Bracing a wall section
Figure 8-9

one stud edgewise on the floor at every stud mark on the shoe.

3. Place the top plate at the upper ends of the studs.

4. Nail through each plate into the studs, holding the studs on their respective marks and flush with the edges of the plate. Use two 16d nails in each stud.

Note: Most carpenters prefer to frame in the openings for doors and windows before raising the wall sections. The openings may also be framed after the wall has been raised.

5. Raise the wall section into position, nail the shoe down and plumb and brace the section.

Note: The interior bearing partitions of all the floors are assembled and erected in the same manner.

6. Nail on the top piece of the double plate, breaking all joints by at least 2 feet.

7. Insert the diagonal or knee bracing into the studs as requred.

Balloon Construction

As described earlier, the main difference between platform and balloon framing is at the floor lines. The balloon wall studs extend from the sill of the first floor to the top plate or end rafter of the second floor, whereas the platform-framed wall is complete for each floor.

The balloon frame might also be called a skeleton frame since the joists, bearing parti-

tions, and outside wall sections are framed and erected before the sheathing or subfloor is applied. This form of construction had a particular advantage when long joists and studs were abundant and when length made no difference in the price per thousand feet of lumber. This type of construction permits rapid erection of the frame so the roof may be placed on the building as soon as possible. Then if the weather becomes bad, the mechanics can work inside.

In balloon frame construction, both the wall studs and the floor joists rest on the anchored sill (Figure 8-8). The studs and joists are toenailed to the sill with eightpenny nails and nailed to each other with at least three tenpenny nails.

The ends of the second-floor joists bear on a 1 x 4 inch ribbon that has been let into the studs. In addition, the joists are nailed with four tenpenny nails to the studs at these connections (Figure 8-8). The end joists parallel to the exterior on both the first and second floors are also nailed to each stud.

In most areas, building codes require that firestops be used in balloon framing to prevent the spread of fire through the open wall passages. These firestops are ordinarily of 2 x 4 inch blocking placed between the studs (Figure 8-8) or as required by local regulations.

Diagonal braces should be inserted in the corners of the side wall of each story after the side walls have been erected and plumbed. These braces are usually 1 x 4 stock. A notch must be cut in each stud at the point where the brace crosses it. This will allow the brace to set flush with the outside of the studs. If an opening near the corner of a building makes it impossible to use diagonal braces, it will then be necessary to use knee braces (Figure 8-9). These are installed in a similar manner. This same system of bracing is also used in the other types of frames.

How To Lay Out And Cut Side Wall Sections

To get the length of the stud and the proper location of the gain for the ribbon, make a pattern as follows:

1. Select a stud long enough for the height of the two stories of the frame. (Assume a non-standard ceiling height from the top of the

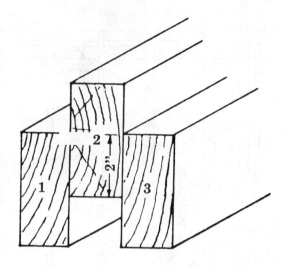

Corner posts
Figure 8-11

subfloor to the bottom of the joist on each story of 7 feet 7½ inches).

2. Lay off the width of the first floor joist (9½ inches in this case), starting at the bottom of the stud (Figure 8-10).

3. To this add ¾ inch for the thickness of the subfloor.

4. To this distance along the stud, add the height of the first story (7 feet 7½ inches). This point will be the top edge of the ribbon or the bottom edge of the second floor joist.

5. Measure up 9½ inches for the second floor joist and ¾ inch for the subfloor.

6. From this point, measure along the stud the height of the second story (7 feet 7½ inches). This point will be the top of the double plate upon which the ceiling joists of the second floor will rest.

7. Deduct the thickness of the double plate (3 inches) from the point representing the top of the plate. Cut off the stud at this point.

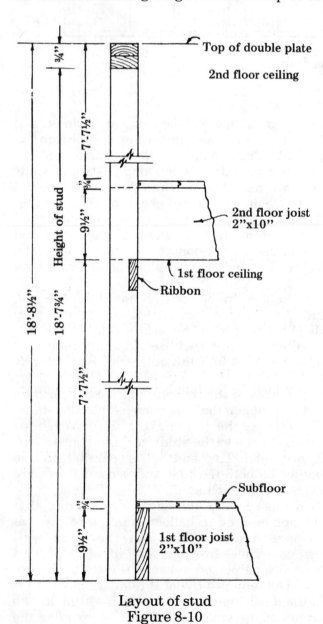

Layout of stud
Figure 8-10

```
        9½"  width of joist
         ¾"  thickness of subfloor
   7'- 7½"  height of first floor
        9½"  width of joist
         ¾"  thickness of subfloor
   7'- 7½"  height of second floor
  14'-35½"
        3"  thickness of double plate
  14'-32½" or
  16'- 8½"  total length of stud
```

8. Mark the stud for the location of the ribbon and make the gain so that the ribbon fits tightly and flush with the face of the stud.

9. Using this stud as a pattern, cut and notch the required number of studs for the side

114

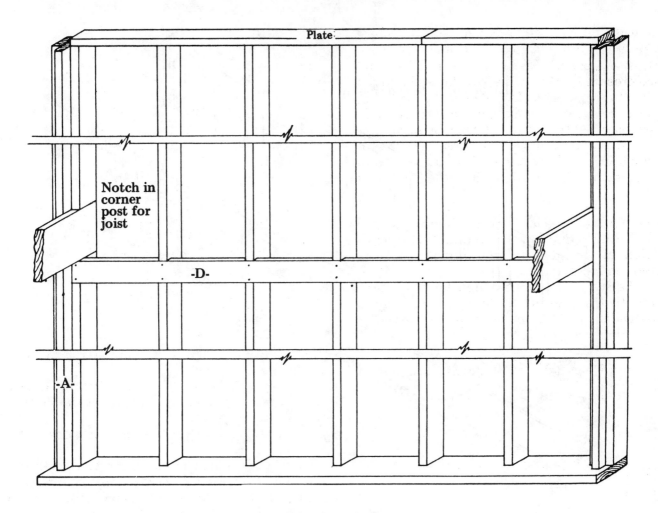

Balloon frame side wall section
Figure 8-12

walls, including one notched stud for each corner post.

Note: In cutting the studs, it is a good procedure to make a template on the saw the length of the studs, so that all the studs may be cut to the same dimensions as the pattern stud.

Caution: Be sure to mark the bottom end of each stud so that it will not be placed upside down in the wall.

How To Build Corner Posts

1. Select two straight studs without the ribbon notch, and one with the notch.

2. Gauge a pencil line along each face of one of the plain studs 2 inches from one edge and parallel to this edge. See stud marked 2 in Figure 8-11.

3. Place the plain stud marked *1* on this line and nail it about every 2 feet with 16d nails to the stud marked *2*. This will allow stud *2* to project 1½ inch beyond the edge of stud *1*.

4. Cut out a section of the notched stud to receive the width of the second floor joist. This is shown in Figure 8-12.

5. Make another similar corner post but with the notches for the joist and ribbon on the opposite face of the post. These two corner posts form a pair.

6. Make another pair of corner posts in the same manner. These two pairs of corner posts are provided for the four corners of the building.

How To Lay The Plates And Ribbon

1. Secure a sufficient number of straight 2 x 4's to form the single plate (Figure 8-12). It should be 6½ inches shorter than the outside length of the sill. The joints should be so arranged that they will come over the center of a stud.

2. Cut a 1 x 4 ribbon to exactly the same length. The joints in this member should also come on a stud.

3. Lay the plate and ribbon side by side (Figure 8-13) and lay out the spacing of the studs. This spacing should be the same as that

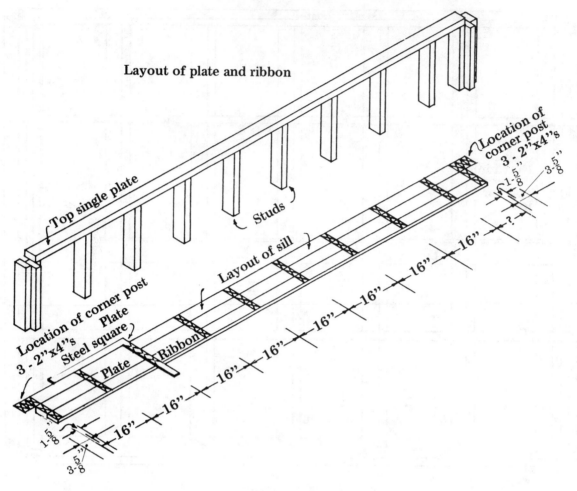

Layout of plate and ribbon
Figure 8-13

of the sill. The marking of the sill is shown in Figure 8-13 for comparison.

How To Assemble A Wall Section

1. Raise the corner post in its proper position so it can receive the ribbon. The notched stud should be toward the inside of the building.

2. Plumb and brace the post as shown in Figure 8-14.

3. Nail a stud at the end of the first piece of ribbon (Figure 8-14, A). The end of the ribbon should project only halfway on the stud and into the notch.

4. Erect this stud, toenail it in place and nail the opposite end of the ribbon into the notch in the corner post.

5. Plumb this stud and hold it in position with a brace as shown in Figure 8-14. The block on the subfloor should be nailed over a joist.

6. Erect studs *2, 3, 4, 5, 6* and *7* in a similar manner.

7. Erect the rest of the wall sections and the opposite corner posts in the same manner as the first section and corner post.

8. Tack a ¾ inch block on the outside of each corner post near the ribbon (Figure 8-15) and stretch a chalk line between these blocks.

9. Nail a brace temporarily in place about every fourth stud.

10. Adjust each stud until it is ¾ inch from the chalk line (Figure 8-15) and then fasten the brace firmly in place. A ¾ inch block may be slipped between the chalk line and each stud to find this ¾ inch distance.

11. Recheck the corner posts and side wall sections for plumbness.

12. Cut notches in the outside edges of the studs and nail the knee or diagonal braces in place.

How To Assemble And Erect The
Center Bearing Partition

1. Cut the required number of studs. The length of a stud is equal to the distance from the bottom of a side wall stud to the top of the ribbon minus 3¼ inches (the thickness of the plate).

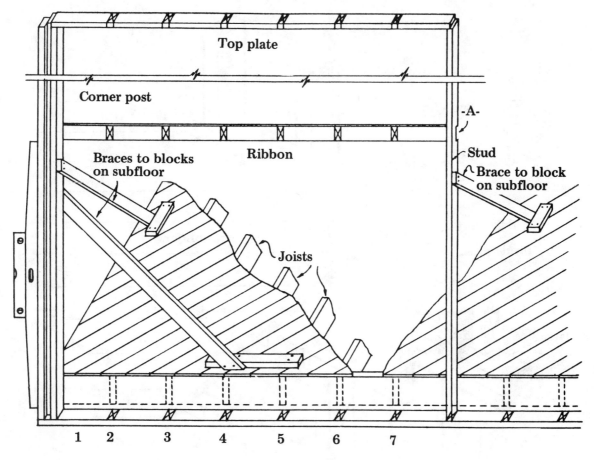

Bracing of wall section
Figure 8-14

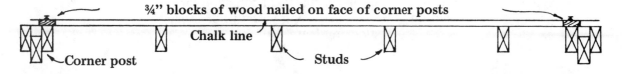

Chalk line stretched between posts
at ribbon line
Figure 8-15

2. Lay out a 2 x 4 plate spaced for studs 16 inches on center.

3. Spike the studs to the plate, using two 16d nails at each stud.

4. Raise the partition in sections and toenail the bottom of each stud to the girder.

5. Plumb and brace the partition.

6. Nail another 2 x 4 on top of the plate to form the double plate.

How To Place The Second Floor Joists

1. Place the second floor joists over the first floor joists, resting them on the center partition and on the ribbon.

2. Spike the joists to the studs on the outside wall sections.

3. Space and toenail the joists to the plate of the center partition.

4. Spike the joists together at the center partition if they are lapped. If they are butted, nail a piece of sheathing across the side of the joists to hold them together.

Caution: Do not nail the joists at the center partition unless the side walls are perfectly plumb on both sides of the building.

5. Plumb and brace the corner posts, as previously described, from the second floor line to the attic floor line.

6. Spike the single plate in place on top of

117

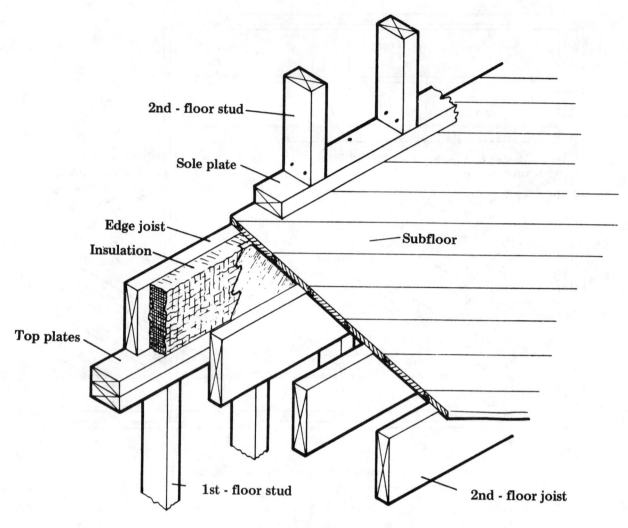

End-wall framing for platform construction
(junction of first-floor ceiling with upper-story
flooring framing)
Figure 8-16

the studs (Figure 8-14). Line up each stud with the spacing marks on the plate.

7. Spike the double plate over the single plate, lapping it over the single plate of the end wall section at the corners of the building.

8. Line up and brace the upper portions of the studs of the wall sections as described for the first floor wall sections.

Note: The center partition construction and erection is the same as described for the first floor. The attic floor joists are placed over the second floor joists. They are then toenailed to the top of the double plates of the side walls and center partition.

How To Erect And Assemble The End Wall Sections

The studs of the end walls are placed, erected and fastened to the sill and upper plates in about the same way as the studs of the side walls except that they are spiked to the side of the end joist at the second floor line. This joist acts on the end walls the same as the ribbon on the side walls.

The framing for the end walls in platform and balloon construction varies somewhat. Figure 8-16 shows a commonly used method of wall and ceiling framing for platform construction in 1½ or 2 story houses with finished rooms above the first floor. The edge floor joist is toenailed to the top wall plate with eightpenny nails spaced 16 inches on center. The subfloor, soleplate, and wall framing are then installed in the same manner used for the first floor.

In balloon framing, the studs continue through the first and second floors (Figure 8-17). The edge joist can be nailed to each stud with two or three tenpenny nails. As for the first floor, 2 x 4 firestops are cut between each stud.

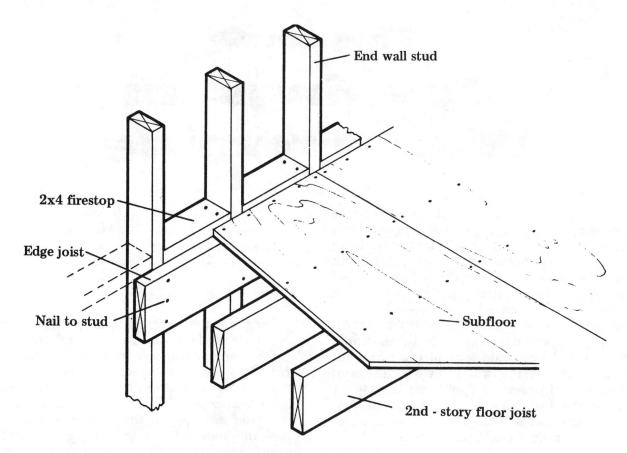

End-wall framing for balloon construction
(junction of first-floor ceiling and
upper-story floor framing)
Figure 8-17

Subfloor is applied in a normal manner.

Interior Walls

The interior walls in a house with conventional joist and rafter roof construction are normally located to serve as bearing walls for the ceiling joists as well as room dividers. Walls located parallel to the direction of the joists are commonly nonload-bearing. Studs are nominal 2 x 4's for load-bearing walls but can be 2 x 3's for nonload-bearing walls. However, most contractors use 2 x 4's throughout. Spacing of the studs is usually controlled by the thickness of the covering material. For example, 24 inch stud spacing will require ½ inch gypsum board for dry wall interior covering.

The interior walls are assembled and erected in the same manner as exterior walls, with a single bottom (sole) plate and double top plates. The upper top plate is used to tie intersecting and crossing walls to each other. A single framing stud can be used at each side of a door opening in nonload-bearing partitions. They must be doubled for load-bearing walls, however. When trussed rafters (roof trusses) are used, no load-bearing interior partitions are required. Thus, location of the walls and size and spacing of the studs are determined by the room size desired and type of interior covering selected. The bottom chords of the trusses are used to fasten and anchor crossing partitions. When partition walls are parallel to and located between trusses, they are fastened to 2 x 4 inch blocks which are nailed between the lower chords.

Chapter 9
Openings In Wall Framing

The members used to span over window and door openings are called headers or lintels (Figure 9-1). As the span of the opening increases, it is necessary to increase the depth of these members to support the ceiling and roof loads. A header is made up of two 2-inch members, usually spaced with 3/8-inch lath or wood strips, all of which are nailed together. They are supported at the ends by the inner studs of the double-stud joint at exterior walls and interior bearing walls. Two headers of species normally used for floor joists are usually appropriate for these openings in normal light-frame construction. The following sizes might be used as a guide for headers:

Maximum span Feet	Header size Inches
3½	2 x 6
5	2 x 8
6½	2 x 10
8	2 x 12

For other than normal light-frame construction, independent design may be necessary. Wider openings often require trussed headers, which may also need special design.

Two types of trussed openings are shown in Figure 9-2 and 9-3. The doubled headers may be turned on edge as in the narrower openings. These methods of trussing openings are used in interior bearing partitions, and over spans in porches and projecting bays. The technical information about the size and load bearing capacities of headers over openings is referred to later in this chapter.

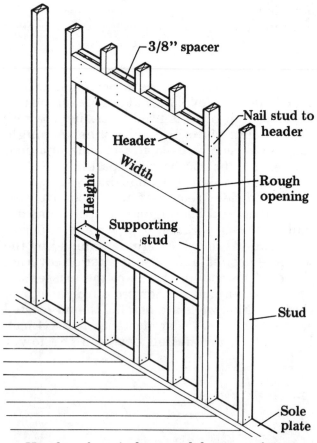

Headers for windows and door openings
Figure 9-1

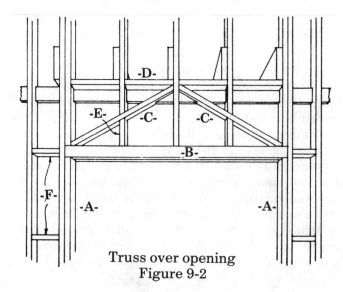

Truss over opening
Figure 9-2

Nonload-Bearing Walls

Lintels and headers can be eliminated from framing over doors and windows providing there are no floor or roof loads acting downward over the opening. End walls of most houses with gable end roofs are framed so roof trusses and floor joists bear on front and rear walls rather

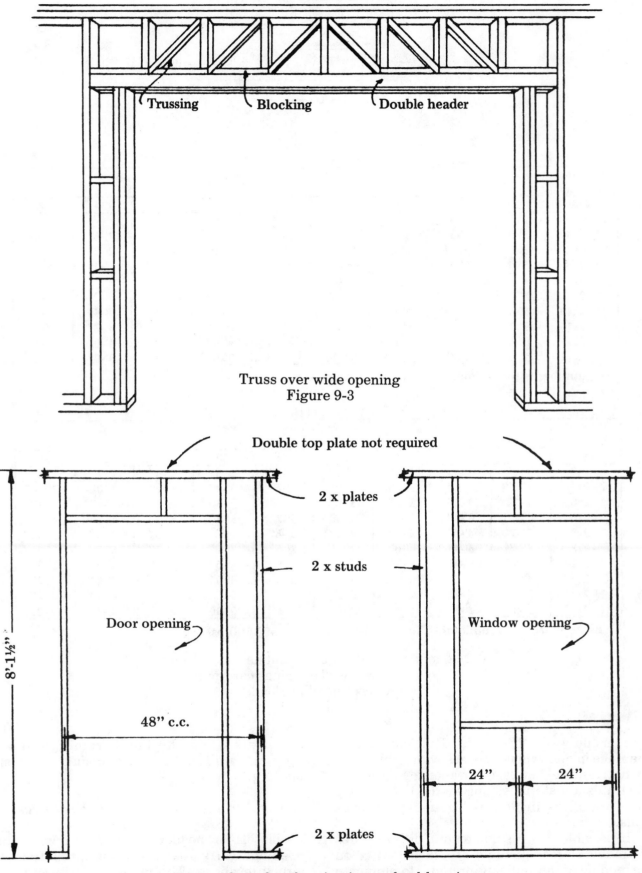

Truss over wide opening
Figure 9-3

Double top plate not required

2 x plates

2 x studs

Door opening

Window opening

8'-1½"

48" c.c.

24"   24"

2 x plates

Door and window framing in non-load-bearing
exterior walls
Figure 9-4

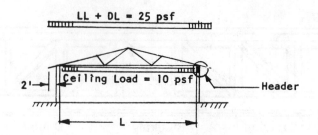

Lintel and Header Spans

Nominal Lumber Sizes	House Widths = L in Feet								
	20	22	24	26	28	30	32	34	36
2 - 2x3	2'-5"	2'-4"	2'-3"	2'-2"	2'-1"	2'-0"	—	—	—
2 - 2x4	3'-5"	3'-3"	3'-2"	3'-1"	2'-11"	2'-10"	2'-8"	2'-6"	2'-4"
1 - 2x6	3'-2"	2'-11"	2'-8"	2'-6"	2'-4"	2'-2"	2'-1"	2'-0"	—
2 - 2x6	5'-5"	5'-2"	5'-0"	4'-10"	4'-8"	4'-5"	4'-2"	3'-11"	3'-9"
1 - 2x8	4'-2"	3'-10"	3'-7"	3'-4"	3'-1"	2'-11"	2'-9"	2'-7"	2'-5"
2 - 2x8	7'-1"	6'-10"	6'-7"	6'-4"	6'-1"	5'-10"	5'-6"	5'-2"	4'-11"
1 - 2x10	5'-4"	4'-11"	4'-6"	4'-3"	3'-11"	3'-8"	3'-6"	3'-4"	3'-2"
2 - 2x10	9'-1"	8'-8"	8'-4"	8'-1"	7'-10"	7'-5"	7'-0"	6'-7"	6'-3"
1 - 2x12	6'-6"	5'-11"	5'-6"	5'-1"	4'-10"	4'-6"	4'-3"	4'-0"	3'-10"
2 - 2x12	11'-0"	10'-7"	10'-2"	9'-10"	9'-6"	9'-0"	8'-6"	8'-0"	7'-7"

Using lumber with a minimum allowable bending stress of 1000 psi

Linter and Header Spans

Nominal Lumber Sizes	House Widths = L in Feet								
	20	22	24	26	28	30	32	34	36
2 - 2x3	3'-0"	2'-10"	2'-9"	2'-8"	2'-7"	2'-5"	2'-3"	2'-2"	2'-0"
2 - 2x4	4'-2"	4'-0"	3'-10"	3'-9"	3'-7"	3'-4"	3'-2"	3'-0"	2'-10"
1 - 2x6	3'-10"	3'-6"	3'-3"	3'-0"	2'-10"	2'-8"	2'-6"	2'-4"	2'-3"
2 - 2x6	6'-7"	6'-4"	6'-1"	5'-10"	5'-7"	5'-3"	5'-0"	4'-8"	4'-6"
1 - 2x8	5'-0"	4'-7"	4'-3"	4'-0"	3'-8"	3'-6"	3'-3"	3'-1"	2'-11"
2 - 2x8	8'-8"	8'-4"	8'-0"	7'-9"	7'-5"	7'-0"	6'-7"	6'-2"	5'-11"
1 - 2x10	6'-5"	5'-10"	5'-5"	5'-1"	4'-9"	4'-5"	4'-2"	4'-0"	3'-9"
2 - 2x10	11'-1"	10'-8"	10'-3"	9'-10"	9'-5"	8'-11"	8'-4"	7'-11"	7'-7"
1 - 2x12	7'-9"	7'-2"	6'-7"	6'-2"	5'-9"	5'-5"	5'-1"	4'-10"	4'-7"
2 - 2x12	13'-6"	12'-11"	12'-5"	12'-0"	11'-6"	10'-10"	10'-2"	9'-8"	9'-2"

Using lumber with a minimum allowable bending stress of 1500 psi

Maximum span designs for lintels and headers
over exterior wall openings carrying
only roof-ceiling loads
Table 9-5

than on end walls. Therefore, door and window openings in end walls of houses with gable roofs do not require lintels or headers. Figure 9-4 illustrates a material saving method of framing such openings in nonload-bearing exterior walls.

If possible, jamb studs on one or both sides of window or door openings should be located on the module of the wall stud framing to eliminate extra studs required when openings are not so located. There is less waste in sheet facing

materials when the jamb framings of wall openings are located on the module of the stud spacing.

Door And Window Framing In Load-Bearing Exterior Walls

Lintels or headers over door and window openings carry roof and floor loads from the structure above. The minimum size and most economical grade of lumber in the lintel or header that will support the roof and floor loads

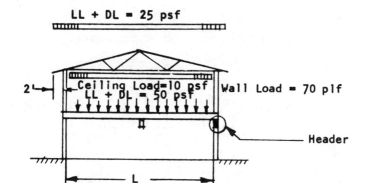

Lintel and Header Spans

Nominal Lumber Sizes	House Widths = L in Feet								
	20	22	24	26	28	30	32	34	36
2 - 2x4	2'-2"	2'-0"	—	—	—	—	—	—	—
2 - 2x6	3'-5"	3'-2"	2'-11"	2'-9"	2'-7"	2'-5"	2'-3"	2'-2"	2'-1"
1 - 2x8	2'-3"	2'-1"	—	—	—	—	—	—	—
2 - 2x8	4'-6"	4'-2"	3'-10"	3'-7"	3'-5"	3'-2"	3'-0"	2'-10"	2'-9"
1 - 2x10	2'-10"	2'-8"	2'-6"	2'-4"	2'-2"	2'-0"	—	—	—
2 - 2x10	5'-9"	5'-3"	4'-11"	4'-7"	4'-4"	4'-1"	3'-10"	3'-8"	3'-6"
1 - 2x12	3'-6"	3'-3"	3'-0"	2'-10"	2'-8"	2'-6"	2'-4"	2'-1"	2'-1"
2 - 2x12	6'-11"	6'-5"	6'-0"	5'-7"	5'-3"	4'-11"	4'-8"	4'-5"	4'-3"

Using lumber with a minimum allowable bending stress of 1000 psi

Lintel and Header Spans

Nominal Lumber Sizes	House Widths = L in Feet								
	20	22	24	26	28	30	32	34	36
2 - 2x4	2'-7"	2'-5"	2'-3"	2'-1"	2'-0"	—	—	—	—
2 - 2x6	4'-1"	3'-9"	3'-6"	3'-3"	3'-1"	2'-11"	2'-9"	2'-7"	2'-6"
1 - 2x8	2'-8"	2'-6"	2'-4"	2'-2"	2'-0"	—	—	—	—
2 - 2x8	5'-4"	5'-0"	4'-8"	4'-4"	4'-1"	3'-10"	3'-7"	3'-5"	3'-3"
1 - 2x10	3'-5"	3'-2"	2'-11"	2'-9"	2'-7"	2'-5"	2'-4"	2'-2"	2'-1"
2 - 2x10	6'-10"	6'-4"	5'-11"	5'-6"	5'-2"	4'-11"	4'-7"	4'-5"	4'-2"
1 - 2x12	4'-2"	3'-10"	3'-7"	3'-4"	3'-2"	3'-0"	2'-10"	2'-8"	2'-6"
2 - 2x12	8'-4"	7'-9"	7'-2"	6'-9"	6'-4"	5'-11"	4'-7"	5'-4"	5'-1"

Using lumber with a minimum allowable bending stress of 1500 psi

Maximum span designs for lintels and headers
over exterior wall openings carry roof-ceiling
plus one-story floor and wall loads
Table 9-6

from above should be selected. The practice of using over-size lintels or headers that are continuous along the top of the wall is wasteful.

Use of lintel and header design Tables 9-5 through 9-9 will help in the selection of a header size and grade of lumber that provides adequate strength and stiffness at a minimum cost. Some species and grades of lumber that have a minimum allowable bending stress of 1,000 pounds per square inch are: Douglas fir-larch Number 2, Douglas fir south Number 2, hem-fir Number 2, mountain hemlock Number 2, mountain hemlock-hem-fir Number 2, lodgepole pine Number 1, and southern pine Number 2. Some species and grades of lumber that have a minimum allowable bending stress of 1,500 pounds per square inch are: Douglas fir-larch Number 1, Douglas fir south select structural,

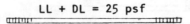

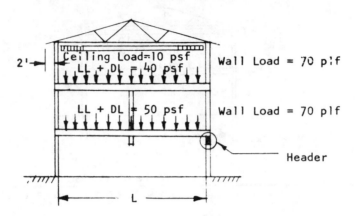

Nominal Lumber Sizes	Lintel and Header Spans House Widths = L in Feet								
	20	22	24	26	28	30	32	34	36
2 - 2x6	2'-5"	2'-3"	2'-1"	2'-0"	—	—	—	—	—
2 - 2x8	3'-3"	3'-0"	2'-9"	2'-7"	2'-6"	2'-4"	2'-2"	2'-1"	2'-0"
1 - 2x10	2'-1"	—	—	—	—	—	—	—	—
2 - 2x10	4'-1"	3'-10"	3'-7"	3'-4"	3'-2"	3'-0"	2'-10"	2'-8"	2'-6"
1 - 2x12	2'-6"	2'-4"	2'-2"	2'-0"	—	—	—	—	—
2 - 2x12	5'-0"	4'-8"	4'-4"	4'-1"	3'-10"	3'-7"	3'-5"	3'-3"	3'-1"

Using lumber with a minimum allowable bending stress of 1000 psi

Nominal Lumber Sizes	Lintel and Header Spans House Widths = L in Feet								
	20	22	24	26	28	30	32	34	36
2 - 2x6	2'-11"	2'-9"	2'-6"	2'-5"	2'-3"	2'-1"	2'-0"	—	—
2 - 2x8	3'-10"	3'-7"	3'-4"	3'-2"	2'-11"	2'-9"	2'-8"	2'-6"	2'-5"
1 - 2x10	2'-6"	2'-3"	2'-2"	2'-0"	—	—	—	—	—
2 - 2x10	4'-11"	4'-7"	4'-3"	4'-0"	3'-9"	3'-7"	3'-4"	3'-2"	3'-1"
1 - 2x12	3'-0"	2'-9"	2'-7"	2'-5"	2'-4"	2'-2"	2'-1"	—	—
2 - 2x12	6'-0"	5'-7"	5'-2"	4'-10"	4'-7"	4'-4"	4'-1"	3'-11"	3'-8"

Using lumber with a minimum allowable bending stress of 1500 psi

Maximum span designs for lintels and headers
over exterior wall openings carrying roof-
ceiling plus two-story floor and wall loads
Table 9-7

mountain hemlock select structural, and southern pine Number 1. Note that end splits may not exceed one times header depths. Linear interpolation within Tables 9-5 to 9-7 for house widths not given is permitted.

Tables 9-5 to 9-7 contain lintel and header designs for locations within houses having a front-to-back depth of from 20 to 36 feet. Table 9-5 applies to headers carrying roof truss loads; Table 9-6 to headers carrying roof truss loads

plus a one-story wall and floor loads; and Table 9-7 to headers carrying roof truss loads plus two-story wall and floor loads.

When window or door openings occur directly under an opening in the story or stories above, the header may be carrying floor loads only. When the lower opening width is equal to or smaller than the opening width above and the lintel or header is not loaded by the jamb studs from the opening above, a smaller lintel is

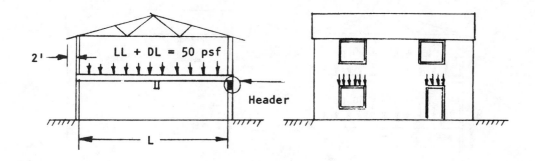

Lintel and Header Spans

Nominal Lumber Sizes	House Widths = L in Feet								
	20	22	24	26	28	30	32	34	36
2 - 2x3	2'-5"	2'-4"	2'-4"	2'-3"	2'-2"	2'-1"	2'-0"	—	—
2 - 2x4	3'-5"	3'-4"	3'-3"	3'-1"	3'-0"	2'-11"	2'-10"	2'-8"	2'-7"
1 - 2x6	3'-2"	3'-0"	2'-9"	2'-7"	2'-6"	2'-4"	2'-3"	2'-1"	2'-0"
2 - 2x6	5'-5"	5'-3"	5'-1"	4'-11"	4'-9"	4'-8"	4'-5"	4'-3"	4'-0"
1 - 2x8	4'-2"	3'-11"	3'-8"	3'-5"	3'-3"	3'-1"	2'-11"	2'-9"	2'-8"
2 - 2x8	7'-2"	6'-11"	6'-8"	6'-6"	6'-3"	6'-1"	5'-10"	5'-7"	5'-4"
1 - 2x10	5'-4"	5'-0"	4'-8"	4'-5"	4'-2"	3'-11"	3'-9"	3'-7"	3'-5"
2 - 2x10	9'-1"	8'-9"	8'-6"	8'-3"	8'-0"	7'-9"	7'-6"	7'-1"	6'-9"
1 - 2x12	6'-6"	6'-1"	5'-8"	5'-4"	5'-1"	4'-9"	4'-7"	4'-4"	4'-2"
2 - 2x12	11'-1"	10'-8"	10'-4"	10'-0"	9'-9"	9'-6"	9'-1"	8'-8"	8'-3"

Using lumber with a minimum allowable bending stress of 1000 psi

Lintel and Header Spans

Nominal Lumber Sizes	House Widths = L in Feet								
	20	22	24	26	28	30	32	34	36
2 - 2x3	3'-0"	2'-11"	2'-10"	2'-9"	2'-8"	2'-7"	2'-5"	2'-4"	2'-2"
2 - 2x4	4'-3"	4'-1"	3'-11"	3'-10"	3'-9"	3'-7"	3'-5"	3'-3"	3'-1"
1 - 2x6	3'-10"	3'-7"	3'-4"	3'-2"	3'-0"	2'-10"	2'-8"	2'-6"	2'-5"
2 - 2x6	6'-7"	6'-5"	6'-2"	6'-0"	5'-10"	5'-7"	5'-4"	5'-1"	4'-10"
1 - 2x8	5'-1"	4'-8"	4'-5"	4'-2"	3'-11"	3'-8"	3'-6"	3'-4"	3'-2"
2 - 2x8	8'-9"	8'-5"	8'-2"	7'-11"	7'-8"	7'-5"	7'-0"	6'-8"	6'-5"
1 - 2x10	6'-5"	6'-0"	5'-7"	5'-3"	5'-0"	4'-9"	4'-6"	4'-3"	3'-1"
2 - 2x10	11'-2"	10'-9"	10'-5"	10'-1"	9'-10"	9'-5"	9'-0"	8'-6"	8'-2"
1 - 2x12	7'-10"	7'-4"	6'-10"	6'-5"	6'-1"	5'-9"	5'-5"	5'-2"	4'-11"
2 - 2x12	13'-7"	13'-1"	12'-8"	12'-3"	11'-11"	11'-6"	10'-11"	10'-5"	9'-11"

Using lumber with a minimum allowable bending stress of 1500 psi

Maximum span designs for lintels and headers
over exterior wall openings carrying
only one floor load
Table 9-8

permissible (see Figure 9-11, A). Table 9-8 has lintel and header designs for locations within houses that carry one floor load only. Table 9-9 pertains to lintel and header designs that carry two floor loads and one wall load. When the lower opening lintel or header is loaded by jamb studs on the openings above (see Figure 9-11, B) or there are no openings in the story above (see Figure 9-11, C), smaller lintels or headers are *not* permissible, and lintels and headers must be selected from Tables 9-6 and 9-7.

When the header is kept directly under the field-applied top plate, it is most effective as a beam (see Figure 9-10). The double 2 x 4 top

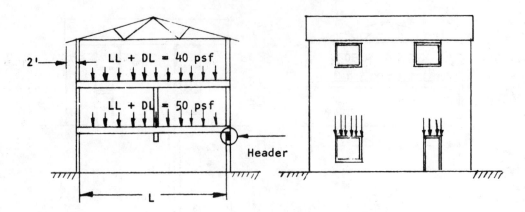

Nominal Lumber Sizes	Lintel and Header Spans House Widths = L in Feet								
	20	22	24	26	28	30	32	34	36
2 - 2x4	2'-8''	2'-6''	2'-4''	2'-2''	2'-1''	—	—	—	—
2 - 2x6	4'-2''	3'-11''	3'-8''	3'-5''	3'-3''	3'-1''	2'-11''	2'-9''	2'-8''
1 - 2x8	2'-3''	2'-2''	2'-0''	—	—	—	—	—	—
2 - 2x8	5'-6''	5'-1''	4'-10''	4'-6''	4'-3''	4'-0''	3'-10''	3'-8''	3'-6''
1 - 2x10	2'-11''	2'-9''	2'-7''	2'-5''	2'-3''	2'-2''	2'-1''	—	—
2 - 2x10	7'-0''	6'-6''	6'-1''	5'-9''	5'-5''	5'-2''	4'-11''	4'-8''	4'-6''
1 - 2x12	3'-7''	3'-4''	3'-1''	2'-1''	2'-9''	2'-7''	2'-6''	2'-4''	2'-3''
2 - 2x12	8'-6''	7'-11''	7'-5''	7'-0''	6'-8''	6'-3''	6'-0''	5'-8''	5'-5''

Using lumber with a minimum allowable bending stress of 1000 psi

Nominal Lumber Sizes	Lintel and Header Spans House Widths = L in Feet								
	20	22	24	26	28	30	32	34	36
2 - 2x4	2'-2''	2'-0''	—	—	—	—	—	—	—
2 - 2x6	3'-6''	3'-3''	3'-0''	2'-10''	2'-8''	2'-7''	2'-5''	2'-4''	2'-3''
1 - 2x8	2'-9''	2'-7''	2'-5''	2'-3''	2'-2''	2'-0''	—	—	—
2 - 2x8	4'-7''	4'-3''	4'-0''	3'-9''	3'-7''	3'-4''	3'-2''	3'-1''	2'-11''
1 - 2x10	3'-6''	3'-3''	3'-1''	2'-11''	2'-9''	2'-7''	2'-5''	2'-4''	2'-3''
2 - 2x10	5'-10''	5'-5''	5'-1''	4'-10''	4'-6''	4'-4''	4'-1''	3'-11''	3'-9''
1 - 2x12	4'-3''	4'-0''	3'-9''	3'-6''	3'-4''	3'-2''	3'-0''	2'-11''	2'-9''
2 - 2x12	7'-1''	6'-7''	6'-2''	5'-10''	5'-6''	5'-3''	5'-0''	4'-9''	4'-6''

Using lumber with a minimum allowable bending stress of 1500 psi

Maximum span designs for lintels and headers
over exterior wall openings carrying
two floors plus one wall load
Table 9-9

plate is not required over the header. Single 2 x 4's can serve as nailers at door and window heads and at sills for windows.

If possible, jamb studs on one or both sides of the window or door openings should be located on even 24 inch points of the wall stud framing, eliminating studs required when openings are not so located. There is less waste in sheet facing materials when jamb framings of wall openings are located to coincide with the stud spacing.

Nailed And Nail-Glued Plywood Headers
For Exterior Wall Openings
Performance tests have been made on two plywood header designs. These headers can be substituted for conventional lintels or headers in

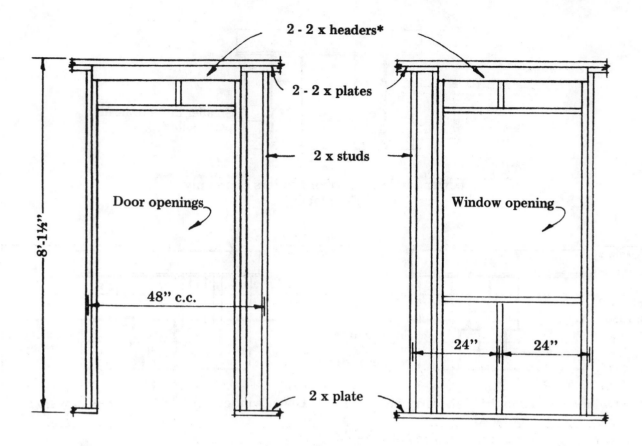

Load-bearing exterior wall openings

*See Tables 9-5 to 9-9, header designs for various loads and spans

Door and window framing in
load-bearing exterior walls
Figure 9-10

exterior wall openings in single-story houses or in the top story of multi-story houses up to 36 feet wide. These designs may be used whenever the wall is sheathed with ½-inch thick material. Roof and ceiling design loads are assumed to be transferred to the exterior walls through trusses or rafters.

Figure 9-12 shows construction details and material requirements for a nailed ½-inch thick plywood header. This design is suitable for rough opening widths up to nominal 4-foot openings.

Construction details and material requirements for nail-glued plywood headers spanning more than 4 feet and up to nominal 6-foot openings are shown in Figure 9-13.

Wall Opening Sizes

Location of the studs, headers, and sills around window openings should conform to the rough opening sizes recommended by the manufacturers of the millwork. The framing height to the bottom of the window and door headers should be based on the door heights, normally 6 feet 8 inches for the main floor. Thus to allow for the thickness and clearance of the head jambs of window and door frames and the finish floor, the bottoms of the headers are usually located 6 feet 10 inches to 6 feet 11 inches above the subfloor, depending on the type of finish floor used.

Rough opening sizes for exterior door and window frames might vary slightly between manufacturers, but the following allowances should be made for the stiles and rails, thickness of jambs, and thickness and slope of the sill:

Double-Hung Window (Single Unit)
Rough opening width = glass width plus 6 inches.

Rough opening height = total glass height plus 10 inches.

127

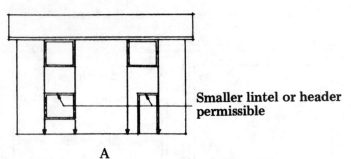

A

Lower window and door headers carry only
floor loads

Jamb stud
roof loads

Smaller lintel
or header
not permissible

B

Lower window and door headers carry floor and
roof loads

Smaller lintel
or header
not permissible

C

Lower window and door headers carry floor and
roof loads

Figure 9-11

½'' thick Standard Grade Plywood with exterior
glue conforming to U.S. Dept. of Commerce
Product Standard, PS1-66

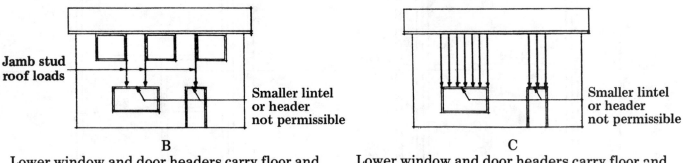

Plywood header nailed with 8d common wire nails
spaced 4'' o.c. along edges and intermediate members

Face grain

14½''
Minimum

Face grain

Typical
24''*

¾''

Up to nominal 4'-0''
Rough opening

24''*

48''

Door
opening

Window opening

½'' thick wall
sheathing

2 - 2x4 top
wall plates

½'' thick
plywood

2x4 cripples
spaced 24'' o.c.

2x4 door or
window head

Section X-X

*Applicable for studs spaced at 16'' o.c.

Exterior wall opening plywood nailed header
for spans up to nominal 4-feet
Figure 9-12

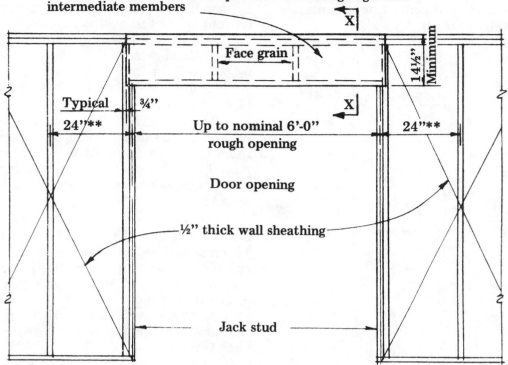

½'' thick Standard Grade Plywood with exterior
glue conforming to U.S.Dept. of Commerce
Product Standard, PS1-66

Plywood header nailed-glued with elastomeric adhesive &
with 8d common wire nails spaced 6" o.c. along edges and
intermediate members

X

Face grain

14½" Minimum

X

Typical   ¾"

24"**

Up to nominal 6'-0"
rough opening

Door opening

½" thick wall sheathing

24"**

Jack stud

*Any adhesive meeting requirements of American Plywood Association
Specification AFG-01

**Applicable for studs spaced 16" o.c.

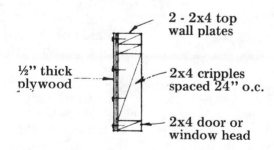

2 - 2x4 top
wall plates

½" thick
plywood

2x4 cripples
spaced 24" o.c.

2x4 door or
window head

Section X-X

Exterior wall opening plywood nailed-glued
header for spans up to nominal 6 feet
Figure 9-13

For example, the following tabulation illustrates several glass and rough opening sizes for double-hung windows:

Window glass size (each sash)

Window glass size (each sash)		Rough frame opening	
Width	Height	Width	Height
24" x	16"	30" x	42"
28" x	20"	34" x	50"
32" x	24"	38" x	58"
36" x	24"	42" x	58"

Casement Window (One Pair—Two Sash)

Rough opening width = total glass width plus 11¼ inches.

Rough opening height = total glass height plus 6-3/8 inches.

Doors

Rough opening width = door width plus 2½ inches.

Rough opening height = door height plus 3

inches.

The head jambs of all windows and doors should be at the same height. The openings for windows and doors may be marked out with a story pole or window rod made from a suitable strip of wood (Figure 9-14).

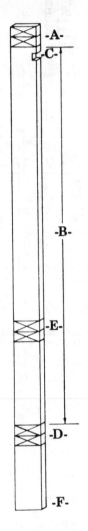

Layout of window rod
Figure 9-14

This rod can also be used when setting window frames and when framing the door headers on inside partitions. The window rod may be laid out as shown in Figure 9-14. The double header over the opening is shown at A. The height of the rough opening is shown at B, and the top jamb at C. A shallow notch should be cut here as shown. A clearance space of 1 inch is left between A and C. The double 2 x 4 at the bottom of the opening is shown at D. If more than one size window is to be laid out, additional double 2 x 4's may be marked as shown at E. The top of the subfloor is represented by F.

How To Lay Out A Rough Opening For A Window

Either of two methods may be used to frame window or door openings in side walls. In the balloon frame, the studs at the openings may be left out when the wall is erected. The headers and sills of the openings are then placed in position as shown by the broken lines in Figure 9-15. The short studs are next cut in and the side studs framed and placed according to the width of the opening.

Another method that may be used in the balloon frame and also in the other types of frames is shown in Figure 9-16. All the studs are placed in their proper positions, regardless of the openings. The openings are then cut out and framed.

It is assumed that a balloon frame is being erected and that studs have been placed every 16 inches. The opening is to be formed by cutting out parts of the studs (Figure 9-16). If the full length studs in the openings are omitted at the time of framing the wall, the procedure is very similar except that short studs above the header and below the sill will have to be inserted later (Figure 9-15).

1. Mark the width of the opening on the sill or plate. See the layout at the bottom of Figures 9-15 and 9-16. Notice that 1½ inch is allowed on each side of the opening for the doubled 2 x 4 upright E.

2. Place the window rod against a stud on each side of the opening and mark the location of the top and bottom headers (Figure 9-16).

How To Frame The Opening

1. Drive nails part way in at the points that show where the studs are to be cut off at the top and bottom of the opening.

2. Nail a supporting brace across four or more studs as at A, Figure 9-16.

3. Rest a straightedge on the nails that show the cutoff mark.

4. Mark along the under side of the straightedge across the studs that are to be cut out of the wall section.

5. Saw these studs off.

6. Mark and cut 4 pieces of 2 x 4 for the double header and sill of the opening as shown at B, Figure 9-15 and 9-16.

7. Nail the single sill and header to the studs on each side and to the cripples C, Figure 9-16.

8. Nail the other 2 x 4 to the single sill and header to form the double sill and header.

9. Cut and place the trimmer studs D, Figures 9-15 and 9-16 by toenailing them to the header and sill of the opening.

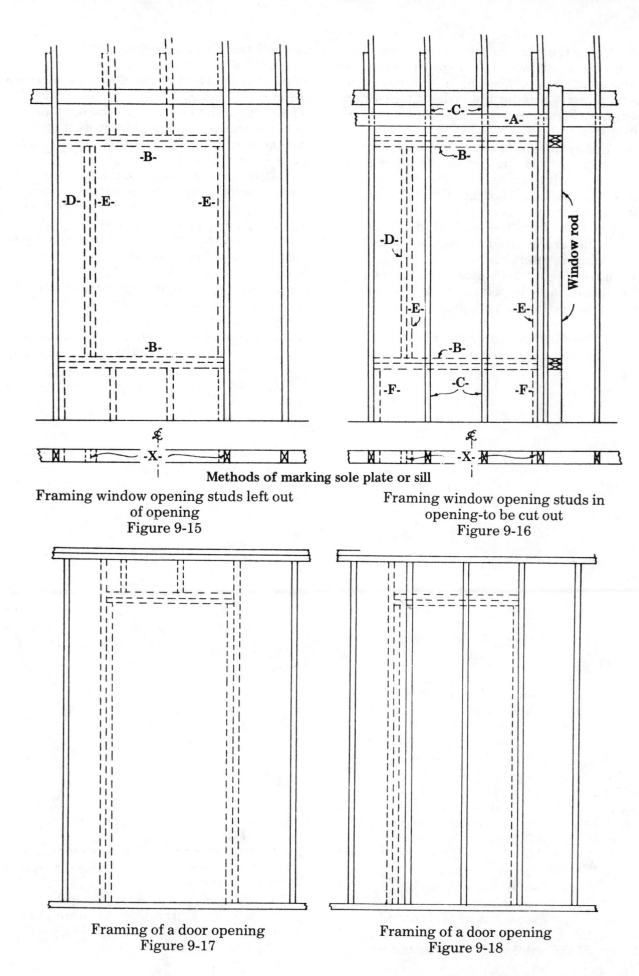

Methods of marking sole plate or sill

Framing window opening studs left out
of opening
Figure 9-15

Framing window opening studs in
opening-to be cut out
Figure 9-16

Framing of a door opening
Figure 9-17

Framing of a door opening
Figure 9-18

10. Double the trimmer studs on the inside of the opening as shown at *E*, Figures 9-15 and 9-16.

11. Nail 2 x 4's under the ends of the sill as shown at *F*, Figure 9-16.

How To Lay Out And Frame Door Openings

The process of laying out and framing door openings is similar to that explained for windows except that allowances must be made for the rough frame size.

Figure 9-17 shows the method in which the regular studs are omitted in the door opening. This opening is later framed as shown by the dotted lines. Figure 9-18 shows the method in which all studs are erected and then later cut as described under framing window openings.

How To Truss Large Openings

1. To truss an opening by the lintel method, cut the header to the proper size.

2. Spike the two members together with strips between them to space them to the width of a 2 x 4.

3. Frame the header into the opening.

4. In trusses such as shown in Figure 9-2, nail the double headers *B* between the double studs *A*.

5. Cut and nail the truss 2 x 4's *C* to the top of the header and to the center stud.

6. Cut and nail the cripple studs *E* on the regular 16 inch spacing.

7. Cut and nail stiffeners *F* between the side studs, every 2 feet from the top of the header down to the floor line.

Note: This type of trussing may be used with slight changes in the platform frame. Figure 9-3 shows another type of truss that may be used over a wide opening in the wall of a platform frame.

Chapter 10

Wall Sheathing

Wall sheathing is the outside covering used over the wall framework of studs, plates, and window and door headers. It forms a flat base upon which the exterior finish can be applied. Certain types of sheathing and methods of application can provide great rigidity to the house, eliminating the need for corner bracing. Sheathing serves to minimize air infiltration and, in certain forms, provides some insulation.

Some sheet materials serve both as sheathing and siding. Sheathing is sometimes eliminated from houses in the mild climates of the South and West. It is a versatile material and manufacturers produce it in many forms. Perhaps the most common types used in construction are boards, plywood, structural insulating board, and gypsum sheathing.

Wood Sheathing

Wood sheathing is usually of nominal 1-inch boards in a shiplap, a tongued-and-grooved, or a square-edge pattern. Resawn 11/16-inch boards are also allowed under certain conditions. The requirements for wood sheathing are easy working, easy nailing, and moderate shrinkage. It may be applied horizontally or diagonally (Figure 10-1, A). Sheathing is sometimes carried only to the subfloor (Figure 10-1, B), but when diagonal sheathing or sheet materials are placed as shown in Figure 10-1, C, greater strength and rigidity result. It is desirable to limit the wood moisture content to 15 percent to minimize openings between matched boards when shrinkage occurs.

Some manufacturers produce random-length side and end-matched boards for sheathing. Most softwood species, such as the spruces, Douglas fir, southern pine, hemlock, the soft pines, and others, are suitable for sheathing. Grades vary between species, but sheathing is commonly used in the third grade.

The minimum thickness of wood sheathing is generally ¾ inch. Widths commonly used are 6, 8, and 10 inches. The 6 and 8 inch widths will have less shrinkage than greater widths, so that smaller openings will occur between boards.

The boards should be nailed at each stud crossing with two nails for the 6 and 8 inch widths and three nails for the 10 and 12 inch widths. When diagonal sheathing is used, one more nail can be used at each stud; for example, three nails for 8 inch sheathing. Joints should be placed over the center of studs (Figure 10-1, A) unless end-matched (tongued - and - grooved) boards are used. End-matched tongued-and-grooved boards are applied continuously, either horizontally or diagonally, allowing end joints to fall where they may, even if between studs (Figure 10-1, A). However, when end-matched boards are used, no two adjoining boards should have end joints over the same stud space and each board should bear on at least two studs.

Two arrangements of floor framing and soleplate location may be used which affect wall sheathing application. The first method has the soleplate set in from the outside wall line so that the sheathing is flush with the floor framing (Figure 10-1, B). This does not provide a positive tie between wall and floor framing and in high wind areas should be supplemented with metal strapping placed over the sheathing. The second method has the sill plate located the thickness of the sheathing in from the edge of the foundation wall (Figure 10-1, C). When vertically applied plywood or diagonal wood sheathing is used, a good connection between the wall and floor framing is obtained. This method is usually preferred where good wall-to-floor-to-foundation connections are desirable.

Wood sheathing (Figure 10-1, A) is commonly applied horizontally because it is easy to apply and there is less lumber waste than with diagonal sheathing. Horizontal sheathing, however, requires diagonal corner bracing for wall framework.

Diagonal sheathing (Figure 10-1, A) should be applied at a 45 degree angle. This method of sheathing adds greatly to the rigidity of the wall and eliminates the need for corner bracing. There is more lumber waste than with horizontal sheathing because of angle cuts, and application

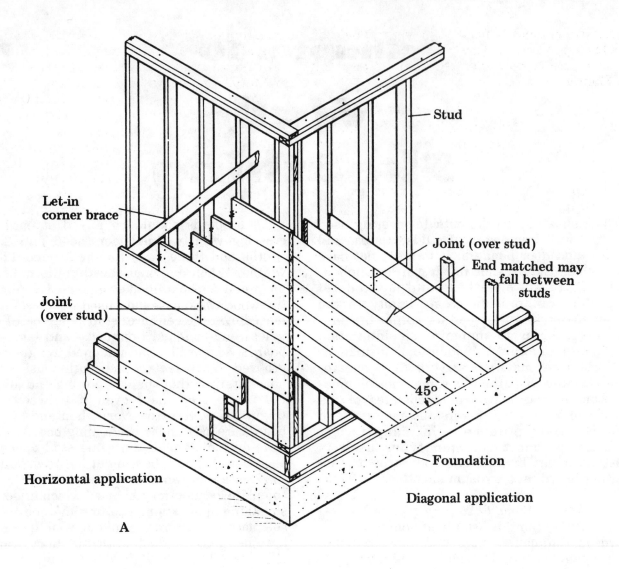

Stud

Let-in
corner brace

Joint (over stud)

End matched may
fall between
studs

Joint
(over stud)

45°

Foundation

Horizontal application

Diagonal application

A

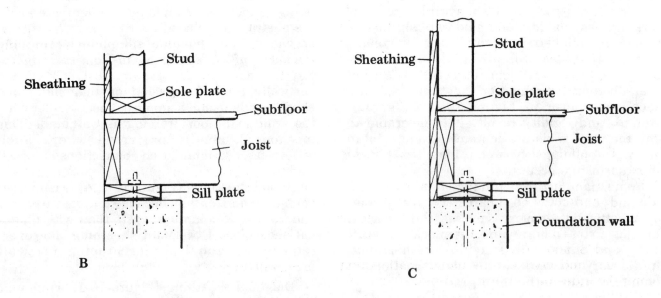

Stud

Sheathing

Sole plate

Subfloor

Joist

Sill plate

B

Sheathing

Stud

Sole plate

Subfloor

Joist

Sill plate

Foundation wall

C

Application of wood sheathing; A, Horizontal
and diagonal; B, started at subfloor;
C, started at foundation wall
Figure 10-1

is somewhat more difficult. End joints should be made over studs. This method is often specified in hurricane areas along the Atlantic Coast and in Florida.

How To Apply Sheathing Horizontally

Note: Refer to Chapter 7 for details of applying board sheathing since the operations are so similar to those of applying subflooring.

1. Line up the first course of sheathing with the bottom of the sill and nail it solidly with 8d nails. All joints must be made on studs.

2. Apply the next course. Cut the boards to such lengths that the joints will not come over those of the first course.

Note: If the waste piece cut off the end of the first course is more than 16 inches long, use it to start the second course so as to break the joints.

3. Continue to apply the sheathing up to the bottom of the first window opening or to 6 feet above the ground.

4. At this point, cut scaffold bracket holes in the sheathing every 10 feet. Cut these holes next to a stud, not halfway between two studs.

Note: There are many kinds of scaffold brackets on the market that are easily attached to the studs to support the scaffold planks. They should be well secured before workmen are allowed on the scaffold. The staging should be at least 12 inches wide and should be made of strong planks. Large amounts of lumber should not be piled on the scaffold brackets, but should be placed upright on the ground. Scaffolds should be placed along the side of the building no more than 7 feet above each other. Detailed information concerning scaffolds will be given in the next chapter.

How To Sheath Around Window And Door Openings

1. Cut off the sheathing flush with the inside of the rough window and door jambs.

Note: Some carpenters prefer to let the sheathing project beyond the rough jambs about 1 inch rather than to cut it flush with the jamb. This provides for better nailing of the window frame.

2. Cut and double face nail the sheathing along the sides of the rough jambs to the head and sill at the top and bottom of the opening.

Plywood Sheathing

Plywood is used extensively for sheathing of walls, applied vertically, normally in 4 by 8 foot and longer sheets (Figure 10-2). This method of sheathing eliminates the need for diagonal corner bracing; but, as with all sheathing materials, it should be well nailed.

Standard sheathing grade is commonly used for sheathing. For more severe exposures, this same plywood is furnished with an exterior glueline. While the minimum plywood thickness for 16 inch stud spacing is 5/16 inch, it is often desirable to use 3/8 inch and thicker, especially when the exterior finish must be nailed to the sheathing. The selection of plywood thickness is also influenced somewhat by standard jamb widths in window and exterior door frames. This may occasionally require sheathing of ½ inch or greater thickness. Some modification of jambs is required and readily accomplished when other plywood thicknesses are used.

Particleboard, hardboard, and other sheet materials may also be used as sheathing. However, their use is somewhat restricted because cost is usually substantially higher than the sheet materials previously mentioned.

Structural Insulating Board Sheathing

The three common types of insulating board (structural fiberboards) used for sheathing include regular density, intermediate density, and nail-base. Insulating board sheathings are coated or impregnated with asphalt or given other treatment to provide a water-resistant product. Occasional wetting and drying that occur during construction will not damage the sheathing materially.

Regular-density sheathing is manufactured in ½ and 25/32 inch thicknesses and in 2 by 8, 4 by 8, and 4 by 9 foot sizes. Intermediate-density and nail-base sheathing are denser products than regular-density. They are regularly manufactured only in ½ inch thickness and in 4 by 8 and 4 by 9 foot sizes. While 2 by 8 foot sheets with matched edges are used horizontally, 4 by 8 foot and longer sheets are usually installed with the long dimension vertical.

Corner bracing is required on horizontally applied sheets and usually on applications of ½ inch regular-density sheathing applied vertically. Additional corner bracing is usually not required for regular-density insulating board sheathing 25/32 inch thick or for intermediate-density and nail-base sheathing when properly applied (Figure 10-2) with long edges vertical. Naturally fastenings must be adequate around the perimeter and at intermediate studs (nails, staples, or other fastening system). Nail-base sheathing also permits the direct application of shingles as siding if fastened with special annular-grooved nails. Galvanized or other

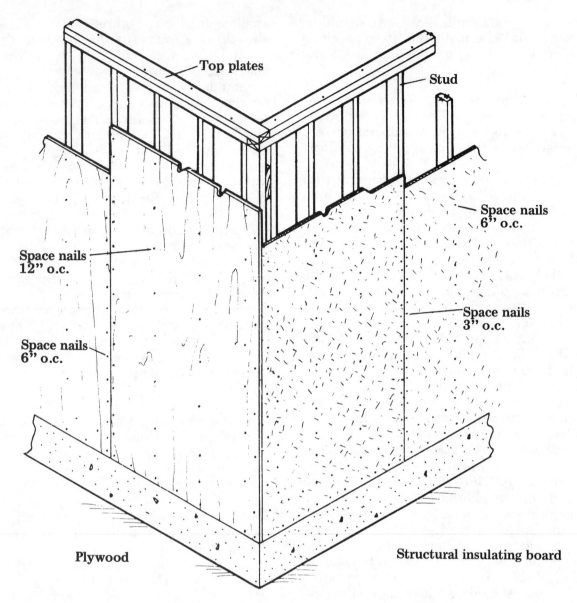

Vertical application of plywood or structural
insulating board sheathing
Figure 10-2

corrosion-resistant fasteners are recommended
for installation of insulating board sheathing.

Applying Structural Sheathing

Vertical application of structural insulating
board (Figure 10-2) in 4 by 8 foot sheets is
usually recommended by the manufacturer
because perimeter nailing is possible. Depending on local building regulations, spacing nails 3
inches on edges and 6 inches at intermediate
framing members usually eliminates the need
for corner bracing when 25/32 inch structural
insulating board sheathing or ½ inch medium-
density structural insulating board sheathing is
used. Use 1¾ inch galvanized roofing nails for
the 25/32 inch sheathing and 1½ inch nails for

the ½ inch sheathing. The manufacturers
usually recommend 1/8 inch spacing between
sheets. Joints are centered on framing members.

Plywood used for sheathing should be 4 by 8
feet or longer and applied vertically with
perimeter nailing to eliminate the need for
corner bracing (Figure 10-2). Sixpenny nails are
used for plywood 3/8 inch or less in thickness.
Use eightpenny nails for plywood ½ inch and
more in thickness. Spacing should be a
minimum of 6 inches at all edges and 12 inches
at intermediate framing members.

Plywood may also be applied horizontally,
but not being as efficient from the standpoint of
rigidity and strength, it normally requires

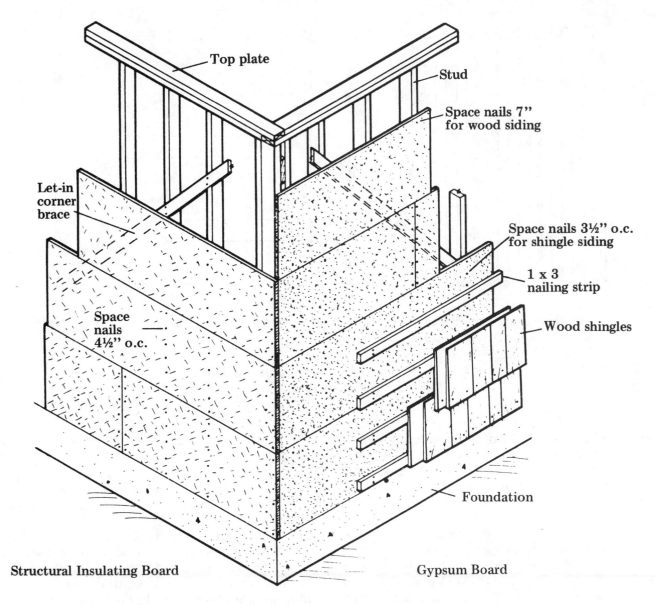

Top plate

Stud

Space nails 7"
for wood siding

Let-in
corner
brace

Space nails 3½" o.c.
for shingle siding

1 x 3
nailing strip

Space
nails
4½" o.c.

Wood shingles

Foundation

Structural Insulating Board

Gypsum Board

Horizontal application of 2-by 8-foot structural
insulating board or gypsum sheathing
Figure 10-3

diagonal bracing. However, blocking between studs to provide for horizontal edge nailing will improve the rigidity and usually eliminate the need for bracing. When shingles or similar exterior finishes are employed, it is necessary to use threaded nails for fastening when plywood is only 5/16 or 3/8 inch thick. Allow 1/8 inch edge spacing and 1/16 inch end spacing between plywood sheets when installing.

How To Apply Structural Sheathing

1. Start the first panel flush with the bottom of the sill and at the end of the wall. The other end must come on the center of a stud.

2. Continue with more panels until the end of the building is reached.

3. Erect the panels of the second row in the same manner.

Note: The piece that is cut off and left over from the first row of panels may be used to start the second row of panels, providing it is long enough to reach the center of a stud. If not, cut a full panel so that the first joint does not come directly over the first joint of the first row of panels.

Gypsum Sheathing

Gypsum sheathing is ½ inch thick, 2 by 8 feet in size, and is applied horizontally for stud spacing of 24 inches or less (Figure 10-3). It is composed of treated gypsum filler faced on two sides with water resistant paper, often having

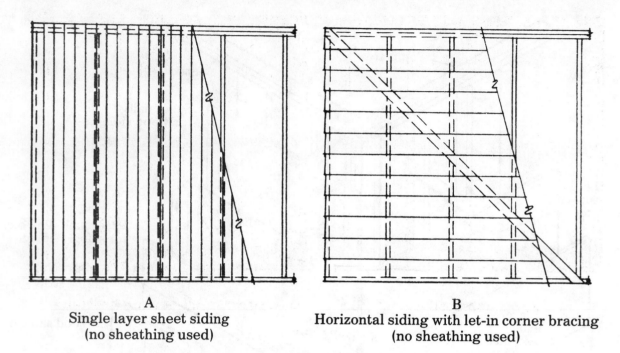

A	B
Single layer sheet siding (no sheathing used)	Horizontal siding with let-in corner bracing (no sheathing used)

Details of exterior walls without sheathing
Figure 10-4

one edge grooved, and the other with a matched V edge. This makes application easier, adds a small amount of tie between sheets, and provides some resistance to air and moisture penetration.

Applying 2' x 8' Sheathing

Gypsum and insulating board sheathing in 2 by 8 foot sheets applied horizontally require corner bracing (Figure 10-3). Vertical joints should be staggered. The 25/32 inch board should be nailed to each crossing stud with 1¾ inch galvanized roofing nails spaced about 4½ inches apart (six nails in the 2 foot height).

The ½ inch gypsum and insulating board sheathing should be nailed to the framing members with 1½ inch galvanized roofing nails spaced about 3½ inches apart (seven nails in the 2 foot height).

When wood bevel or similar sidings are used over plywood sheathing less than 5/8 inch thick, and over insulating board and gypsum board, nails must usually be located so as to contact the stud. When wood shingles and similar finishes are used over gypsum and regular density insulating board sheathing, the walls are stripped with 1 by 3 inch horizontal strips spaced to conform to the shingle exposure. The wood strips are nailed to each stud crossing with two eightpenny or tenpenny threaded nails, depending on the sheathing thickness (Figure 10-3). Nail-base sheathing board usually does

not require stripping when threaded nails are used.

Material Saving Techniques

On many residential jobs the wall sheathing can be eliminated entirely without loss of insulating value or wall strength. Figure 10-4, *A*, shows a low cost, sound wall construction that uses no sheathing, yet has high strength. Decorative 4 by 8 foot structural sheet siding of plywood, hardboard, or high-density fiberboard attached to 2 x 4 stud framing on 24 inch centers performs well. The sheet siding provides the structural racking strength required to resist horizontal wind and earthquake loads on the structure.

When siding provides low racking strength, such as with horizontal sidings, it may be necessary to use let-in 1 x 4 bracing at corners (see Figure 10-4, *B*). Each wall must contain at least one 8 foot or three 4 foot 1 x 4 braced wall sections. The 1 x 4 brace is fastened to each stud and each plate with two 8d common wire nails.

The above designs eliminate sheathing. If sheathing is used, however, careful consideration should be given to selecting materials having minimum cost such as low-density wood fiberboards and gypsum-board sheathing products.

Figure 10-5 shows two methods of using low cost sheathing for exterior walls. Figure 10-5, *A* illustrates use of a fiberboard or other

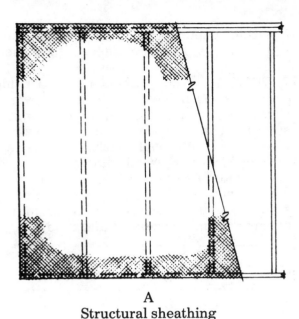

A
Structural sheathing

B
Non-structural sheathing with let-in
corner bracing

Details of exterior walls with low cost sheathing
Figure 10-5

sheet material that is classified as structural. Such materials must meet the requirements of FHA Technical Circular Number 12, "A Standard for Testing Sheathing Material for Resistance to Racking". If they do, no diagonal corner bracing is required. Figure 10-5, *B* illustrates use of materials classified as non-structural sheathing. These require the use of a 1 x 4 let-in corner bracing because they do not meet the requirements of FHA Technical Circular Number 12.

When plywood is used for sheathing or siding, it can be of a minimum thickness. Minimum thickness of plywood sheathing is 5/16 inch thick for studs spaced 16 inches on center and 3/8 inch thick for studs 24 inches on center.

Minimum thickness of plywood siding is ¼ inch if placed over sheathing. If no sheathing is used, the plywood siding must be 3/8 inch thick with studs spaced 16 inches on center or ½ inch thick with studs spaced 24 inches on center.

Often, sheathing or siding is used that is thicker than these minimums because the designer or carpenter believes the extra thickness gives added strength. The racking strength of a wall covered with plywood is governed mostly by the nailing or fastenings between plywood and framing. Increasing the thickness of plywood has little effect on improving racking strength of walls.

Corner Bracing
The purpose of corner bracing is to provide

rigidity to the structure and to resist the racking forces of wind. Corner bracing should be used at all external corners of houses where the type of sheathing used does not provide the bracing required (Figure 10-1, *A*). Types of sheathing that provide adequate bracing are: (a) wood sheathing, when applied diagonally; (b) plywood, when applied vertically in sheets 4 feet wide by 8 or more feet high and where attached with nails or staples spaced not more than 6 inches apart on all edges and not more than 12 inches at intermediate supports; and (c) structural insulating board sheathing 4 feet wide by 8 feet or longer (25/32 inch thick regular grade and ½ inch thick intermediate-density or nail-base grade) applied with long edges vertical with nails or staples spaced 3 inches along all edges and 6 inches at intermediate studs.

Another method of providing the required rigidity and strength for wall framing consists of a ½ inch plywood panel at each side of each outside corner and ½ inch regular-density fiberboard at intermediate areas. The plywood must be in 4 foot wide sheets and applied vertically with full perimeter and intermediate stud nailing.

Where corner bracing is required, use 1 by 4 inch or wider members let into the outside face of the studs, and set at an angle of 45 degrees from the bottom of the soleplate to the top of the wallplate or corner stud. Where window openings near the corner interfere with 45 degree braces, the angle should be increased but the full-length brace should cover at least three stud spaces. Tests conducted at the Forest Products Laboratory showed a full-length brace to be much more effective than a knee brace, even though the angle was greater than that of a 45 degree brace.

Sheathing Paper

Sheathing paper should be water-resistant but not vapor-resistant. It is often called "breathing" paper as it allows the movement of water vapor but resists entry of direct moisture. Materials such as 15-pound asphalt felt, rosin, and similar papers are considered satisfactory. Sheathing paper should have a "perm" value of 6.0 or more. It also serves to resist air infiltration.

Sheathing paper should be used behind a stucco or masonry veneer finish and over wood sheathing. It should be installed horizontally starting at the bottom of the wall. Succeeding layers should lap about 4 inches. Ordinarily, it is not used over plywood, fiberboard, or other sheet materials that are water-resistant. However, 8 inch or wider strips of sheathing paper should be used around window and door openings to minimize air infiltration.

Chapter 11
Staging And Scaffolding

It is difficult and unsafe for you to work on the walls of a building unless you can reach the work comfortably. Therefore, some arrangement must be made whereby you can work under safe conditions as the building increases in height. This arrangement is called staging or scaffolding. The height between lifts of a staging will vary with different kinds of work. Most carpenters can work comfortably up to a reach of six feet. The staging must be carefully planned to avoid accidents.

The most widely used type of scaffolding is galvanized steel tubing with a male fitting on one end and a female fitting on the other to form interlocking units. See Figure 11-1. These units are then coupled together with a patented coupling device and can be adapted to a variety of uses. They consist of three parts: interlocking steel tubes of various sizes, a base plate or caster, and an adjustable or right angle coupler. All parts are galvanized steel or aluminum to give permanent protection against rust.

On many smaller jobs it is not practical to rent or buy sectional steel scaffolding, and job built wood scaffolds must be constructed. The selection of job built staging material is as important as the method of erecting it. Material that is rather brittle, such as hemlock or white pine, should not be used. Spruce or fir, which are strong and tough are very satisfactory.

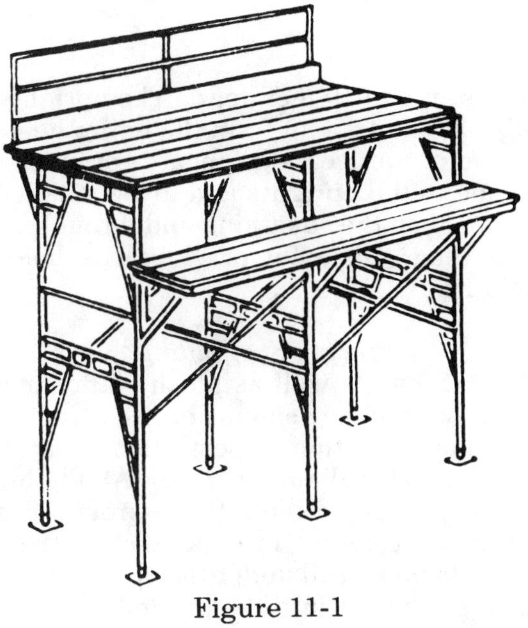

Good staging is important to the safety and efficiency of the workmen so its erection should be supervised by a competent carpenter. The use of scaffold nails is a safety precaution, since they may be driven home and later may be withdrawn easily. This easy removal encourages the workmen to pull the nails out of the scaffolding when it is taken down. If the lumber is then placed on the ground, there will be no nails protruding for the workmen to step on.

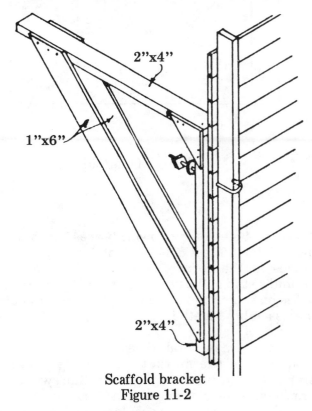

2"x4"

1"x6"

2"x4"

Scaffold bracket
Figure 11-2

Staging Brackets

The scaffolding or staging bracket shown in Figure 11-2 is simple to make and is an excellent means of supporting a staging on a frame building. This type of bracket consists of two 2 x

Figure 11-1

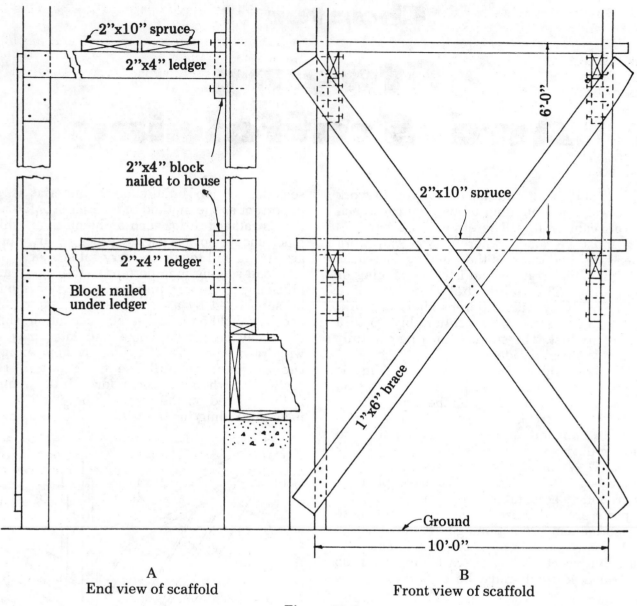

A
End view of scaffold

B
Front view of scaffold

Figure 11-3

4's about 4 feet long, two long braces 1 inch by 6 inches about 5 feet 8 inches long, and two short braces at the corner. A hooked bolt fastens around a stud and passes through the bracket. A nut is placed on the outer end of the bolt to hold the assembly together.

There are several types of steel scaffold brackets on the market. These are usually slightly smaller and lighter than wooden brackets. Some of them are made so they can be folded into a small space for convenience in storing.

The Single Post Scaffold

The single post scaffold (Figure 11-3, A) has an outside post only. These uprights are usually spaced from 8 to 10 feet apart. The uprights are usually 2 x 4 or 4 x 4 and run the entire height of the scaffold. Spruce ledger boards or outriggers are nailed to the uprights and to a 2 x 4 cleat or block nailed on the sheathing and through it into the studding. Diagonal or X braces hold the scaffold rigid.

Bracing Scaffolding

The braces, as well as all the other lumber used for scaffolding, should be free from large or loose knots or other imperfections that might cause them to break under strain. All the braces should be properly nailed. If the ground is soft, blocks that are larger in cross section than the post should be placed under the uprights to keep them from sinking into the ground.

How To Erect And Secure Scaffold Brackets On Wood Sheathing

1. Determine the height of the scaffold and mark this distance on a wall stud.

2. From this point measure down the distance between the top of the bracket and the center of the bolt. See Figure 11-2.

3. At this point bore a hole 1/16 inch larger than the bolt, through the sheathing and along side the stud.

4. Push the bolt through the hole from the inside so the hooked end will be around the stud.

5. Drive an 8d nail part way into the 2 inch edge of the stud and along side the hook of the bolt. Bend the nail over and around the bolt to hold it in place.

6. Place the bracket over the bolt on the outside of the building.

7. Put the nut and washer on the bolt and tighten the nut. After the brackets have been put in position on one side of the building, place the planks on the brackets and fasten them in place.

Note: When composition sheathing is used, it is necessary to put two boards between the bracket and the sheathing. These boards should run horizontally and should cover at least two studs. They must be nailed to keep them from falling. The top piece should have a hole bored through it to receive the bolt. These pieces will then carry the weight of the bracket, which would otherwise be forced down through the soft sheathing material.

How To Erect Single Post Scaffolds

Assume that the building is 40 feet long and 20 feet high from the ground to the cornice. The scaffold will require 5 uprights spaced at 10 foot intervals. Use 18 foot long 2 x 4's for the uprights.

1. Measure up 6 feet on the 2 x 4 upright.

2. At this point, nail on a 2 inch by 4 inch by 4 foot spruce ledger board with four 10d nails. Make sure that the ledger board is at right angles to the upright.

3. Nail a piece about 12 inches long on the upright directly under the ledger board (Figure 11-3).

4. Measure 6 feet from the ground up along a stud on the building.

5. Using this as a center point, nail a piece of 2 inch by 4 inch by 18 inch stock on the building with three 20d nails. The nails should go through the sheathing into the stud.

6. Stand the upright in place and approximately plumb.

7. Nail the ledger board to the side of the 2 x 4 on the building.

8. Prepare and erect the other four uprights in the same manner.

9. Nail on the diagonal braces as shown in Figure 11-3, B.

10. Put the planks in place.

11. After the building has been sheathed up another 6 feet, erect and brace the next lift or height of staging in the same manner.

Double Post Scaffold

Some types of work require the scaffolding supports to be free from the building. An example of this is in brick veneer work. This makes a double post scaffold necessary. This type is similar to the single post scaffold with the addition of an inside post of the same size as the outside post. The ledger boards are nailed to the posts and should have cleats nailed under them to help carry the load. The inside post should be braced diagonally the same as the outside posts. See Figure 11-3. The distance between the inside and outside posts should be at least 4 feet.

How To Erect Double Post Scaffolding

Assume that the building is 40 feet long and 20 feet high from the ground to the cornice. Five sets of uprights will be required. Each set will consist of two 2 x 4's, each 18 feet long.

1. Place one 18 foot upright not less than 6 inches from the building and the other one 4 feet outside the first one.

2. Measure 6 feet up from the bottom on each post.

3. Nail the ledger board on at this point with four 10d nails in each post.

4. Nail a 4 foot horizontal brace across the top of the two uprights. This is to keep the upright posts parallel.

5. Nail a block on each post directly under the ledger board.

6. Make another set of uprights in the same manner.

7. Stand the two sets of uprights in place.

8. Brace this scaffold in the same manner as the single post scaffold with the addition of another set of the same type of braces for the inside posts.

9. Build the rest of the uprights in the same manner.

10. Place the planks in the same way as for single post scaffolding.

11. Build each succeeding lift or height of scaffolding in a similar manner.

12. Brace the end uprights to the ends of the building.

Chapter 12
Roof Trusses

Mill made roof trusses are used widely in residential construction. Significant savings result from using engineered wood roof trusses because less material is required and job site roof framing labor can be reduced substantially. Trusses can be manufactured in a wide variety of sizes and types to fit the job requirements. Structural designs for roof trusses are not provided in this book because the design depends upon the type of truss connectors used as truss joints. Connectors at truss joints are proprietary fasteners and designs of trusses using them must be supplied by fastener manufacturers. However, as a carpenter or builder you should satisfy yourself that the truss specified has been designed for the minimum required roof loads for the area.

You probably will have to install factory made roof trusses and occasionally you may be called on to build a truss or assemble a truss from pre-cut materials. Often, as a carpenter or builder you can build a simple roof truss on the site when framing a roof. You should be familiar with truss types and how trusses can be used to good advantage in house framing.

The simple truss or trussed rafter is an assembly of members forming a rigid framework of triangular shapes capable of supporting loads over long spans without intermediate support. It has been greatly refined during its development over the years, and the gusset and other preassembled types of wood trusses are being used extensively in the housing field. They save material, can be erected quickly, and the house can be enclosed in a short time.

Trusses are usually designed to span from one exterior wall to the other with lengths from 20 to 32 feet or more. Because no interior bearing walls are required, the entire house becomes one large workroom. This allows increased flexibility for interior planning, as partitions can be placed without regard to structural requirements.

Wood trusses most commonly used for houses include the W-type truss, the King-post, and the scissors (Figure 12-1). These and similar trusses are most adaptable to houses with rectangular plans so that the constant width requires only one type of truss. However, trusses can also be used for L plans and for hip roofs as special hip trusses can be provided for each end and valley area.

Trusses are commonly designed for 2 foot spacing, which requires somewhat thicker interior and exterior sheathing or finish material than is needed for conventional joist and rafter construction using 16 inch spacing. Truss designs, lumber grades, and construction details are available from several sources including the American Plywood Association at 1119 A Street, Tacoma, Washington 98401 and The Small Homes Council at 1 E. St. Mary's Road, Champaign, Illinois 61820.

W-Type Truss

The W-type truss (Figure 12-1, *A*) is perhaps the most popular and extensively used of the light wood trusses. Its design includes the use of three more members than the King-post truss, but distances between connections are less. This usually allows the use of lower grade lumber and somewhat greater spans for the same member size.

King-Post Truss

The King-post truss is the simplest form of truss used for houses, as it is composed only of upper and lower chords and a center vertical post (Figure 12-1, *B*). Allowable spans are somewhat less than for the W-truss when the same size members are used, because of the unsupported length of the upper chord. For short and medium spans, it is probably more economical than other types because it has fewer pieces and can be fabricated faster. For example, under the same conditions, a plywood gusset King-post truss with 4 in 12 pitch and 2-foot spacing is limited to about a 26 foot span for 2 by 4 inch members, while the W-type truss with the same size members and spacing could be used for a 32 foot span. Furthermore, the grades of lumber used for the two types might

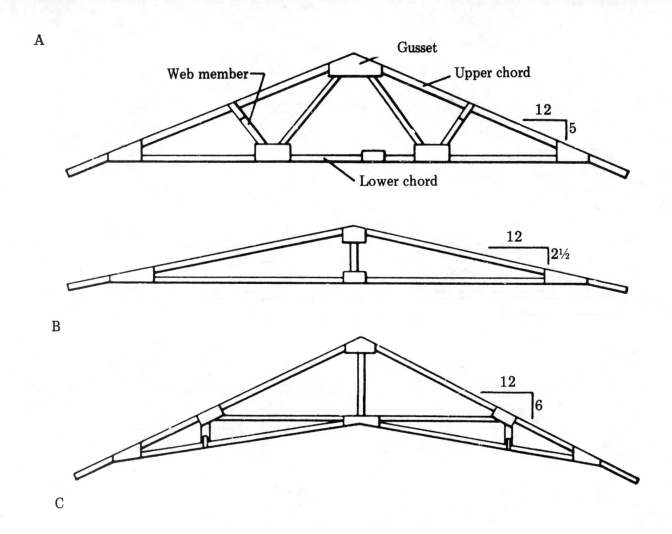

Light wood trusses: A, W-type; B, King post;
C, scissors
Figure 12-1

also vary.

Local prices and design load requirements (for snow, wind, etc.) as well as the span should likely govern the type of truss to be used.

Scissors Truss

The scissors truss (Figure 12-1, C) is a special type used for houses in which a sloping living room ceiling is desired. Somewhat more complicated than the W-type truss, it provides good roof construction for a "cathedral" ceiling with a saving in materials over conventional framing methods.

Design And Fabrication

The design of a truss not only includes snow and windload considerations but the weight of the roof itself. Design also takes into account the slope of the roof. Generally, the flatter the slope, the greater the stresses. This results not only in the need for larger members but also in

stronger connections. Consequently, all conditions must be considered before the type of truss is selected and designed.

A great majority of the trusses used are fabricated with gussets of plywood (nailed, glued, or bolted in place) or with metal gusset plates. Others are assembled with split-ring connectors. Designs for standard W-type and King-post trusses with plywood gussets are usually available through a local lumber dealer. Information on metal plate connectors for wood trusses is also available. Many lumber dealers are able to provide builders with completed trusses ready for erection.

To illustrate the design and construction of a typical wood W-truss more clearly, the following example is given: The span for the nail-glued gusset truss (Figure 12-2) is 26 feet, the slope 4 in 12, and the spacing 24 inches. Total roof load is 40 pounds per square foot, which is usually

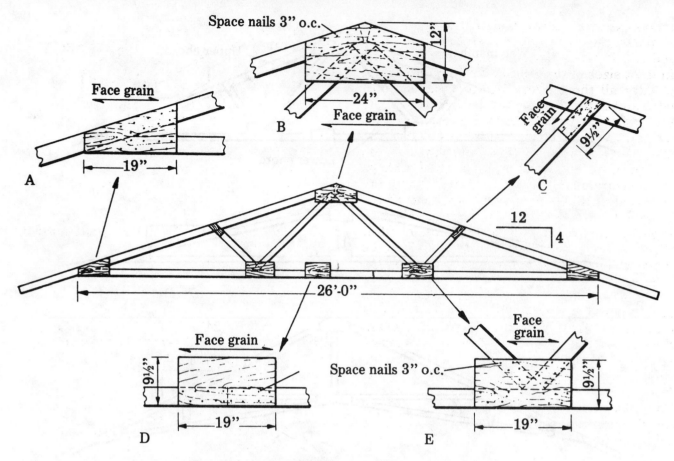

Construction of a 26-foot W truss: A, Bevel-heel
gusset; B, peak gusset; C, upper chord
intermediate gusset; D, splice of lower chord;
E, lower chord intermediate gusset
Figure 12-2

sufficient for moderate to heavy snow belt areas. Examination of truss tables and charts might show that the upper and lower chords can be 2 by 4 inches in size, the upper chord requiring a slightly higher grade of material. It is often desirable to use dimension material with a moisture content of about 15 percent with a maximum of 19 percent.

Plywood gussets can be made from 3/8 or ½ inch standard plywood with exterior glueline or exterior sheathing grade plywood. The cutout size of the gussets and the general nailing pattern for nail-gluing are shown in Figure 12-2. More specifically, fourpenny nails should be used for plywood gussets up to 3/8 inch thick and sixpenny for plywood 1/2 to 7/8 inch thick. Three inch spacing should be used when plywood is no more than 3/8 inch thick, and 4 inches for thicker plywood. When wood truss members are nominal 4 inches wide, use two rows of nails with a ¾ inch edge distance. Use three rows of nails when truss members are 6 inches wide. Gussets are used on both sides of the truss.

For normal conditions and where relative humidities in the attic area are inclined to be high, such as might occur in the southern and southeastern States, a resorcinol glue should be used for the gussets. In dry and arid areas where conditions are more favorable, a casein or similar glue might be considered.

Glue should be spread on the clean surfaces of the gusset and truss members. Either nails or staples might be used to supply pressure until the glue has set, although only nails are recommended for plywood ½ inch and thicker. Use the nail spacing previously outlined. Closer or intermediate spacing may be used to insure "squeezeout" at all visible edges. Gluing should be done under closely controlled temperature conditions. This is especially true if using the resorcinol adhesives. Follow the assembly temperatures recommended by the manufacturer.

Layout And Assembly
In laying out a truss, first get the material to a level spot of ground where workbenches will

be approximately level. Obtain from the plans the measurement of all pieces that are to be used in the truss. Lay out the lengths on the different sizes of timber and cut them accurately. After all the lengths of difficerent sizes of material for the truss have been cut, lay the pieces in their correct position to form a truss and nail them together temporarily. After the truss is assembled in this way, lay out the location of all holes to be bored, then recheck the measurements to be sure that they are correct; after this is done, bore the holes to the size called for on the plan. They should be bored perpendicular to the face of the timber. After the holes have been bored, the truss can be dismantled if final assembly is to be done on the job site.

The assembling of a truss after it has been cut and bored is simple. In most cases, timber connectors are used where the different members of the truss join. The truss is again assembled as it was for boring holes, with the timber connectors in place. The bolts are then placed in the holes and tightened, a washer being placed at the head and nut ends of each bolt. Straight and sound timber should be used in trusses to avoid weak places.

Handling

In handling and storage of completed trusses, avoid placing unusual stresses on them. They were designed to carry roofloads in a vertical position, and it is important that they be lifted and stored in an upright position. If they must be handled in a flat position, enough men or supports should be used along their length to minimize bending deflections. Never support them only at the center or only at each end when in a flat position.

Placing The Truss

After the trusses have been assembled, they must be placed on the building. The first set of rafters may be assembled in the end section of the building or at the center as shown in Figure 12-3. The rafter trusses are raised by hand into position and nailed into place with sixteenpenny nails. These trusses are temporarily braced to the end section of the building until the sheathing is applied. Temporary benches may be erected for the crew to stand on while erecting these trusses. Knee braces are not used on every truss unless needed. The trusses are installed as follows:

1. Mark proper positions of all truss assemblies on the top plate. The marks must show the exact position on the face of all rafters (south or north, etc.).

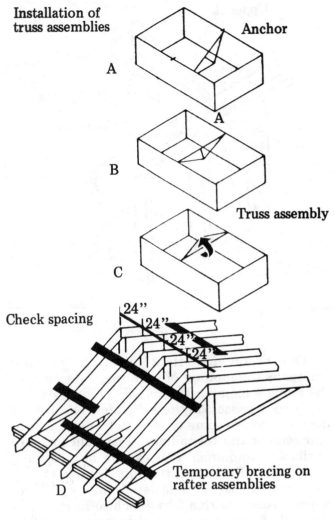

Erection of trusses
Figure 12-3

2. Rest one end of a truss assembly, peak down, on an appropriate mark on the top plate on one side of the structure (Figure 12-3, *A*).

3. Rest the other end of the truss on the opposing mark on the top plate on the other side of the structure (Figure 12-3, *B*).

4. Rotate the assembly into position by means of a pole or rope (Figure 12-3, *C*).

5. Line up the faces flush against the marks and secure them.

6. Raise and nail three assemblies into position. Nail temporary 1 by 6 inch braces across these three assemblies (Figure 12-3, *D*) and other assemblies as they are brought into position. Check rafter spacing at peaks as braces are nailed on.

7. Braces may be used as a platform when raising those trusses for which there is too little room to permit rotation.

Note: A small truck boom lift will make erection much easier and reduce the crew required to two men.

147

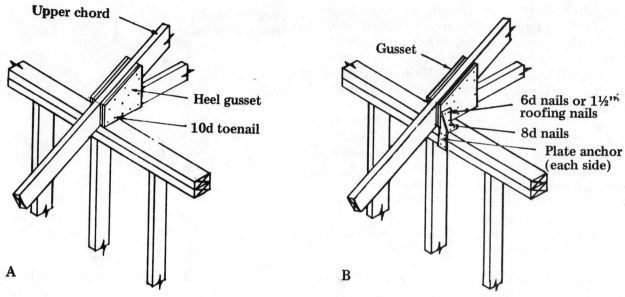

Upper chord

Heel gusset

10d toenail

A

Gusset

6d nails or 1½'' roofing nails

8d nails

Plate anchor (each side)

B

Fastening trusses to wallplate
Figure 12-14

One of the important details in erecting trusses is the method of anchoring. Because of their single member thickness and the presence of plywood gussets at the wallplates, it is usually desirable to use some type of metal connector to supplement the toenailings. Plate anchors are available commercially or can be formed from sheet metal. Resistance to uplift stresses as well as thrust must be considered. Many dealers supply trusses with a 2 by 4 inch soffit return at the end of each upper chord to provide nailing areas for the soffit.

Trusses can be fastened to the top wallplates by toenailing, but this is not always the most satisfactory method. The heel gusset in a plywood gusset truss is located at the wallplate and makes toenailing difficult. However, two tenpenny nails at each side of the truss can be used in nailing the lower chord to the plate (view A, Figure 12-4). Predrilling may be necessary to prevent splitting. A better system involves the use of a simple metal connector or plate anchor (available commercially or can be formed from sheet metal), shown in view B of Figure 12-4. Plate anchors should be nailed to the wallplates at sides and top with eightpenny nails and to the lower chords of the truss with sixpenny or 1½-inch roofing nails.

Chapter 13
Ceiling And Roof Framing

After exterior and interior walls are plumbed, braced, and top plates added, ceiling joists can be positioned and nailed in place. They are normally placed across the width of the house, as are the rafters. The partitions of the house are usually located so that ceiling joists of even lengths (10, 12, 14, and 16 feet or longer) can be used without waste to span from exterior walls to load-bearing interior walls. The sizes of the joists depend on the span, wood species, spacing between joists, and the load on the second floor or attic. The correct sizes for various conditions are designated by local building requirements or can be found in joist tables. When preassembled roof trusses are used, the lower chord acts as the ceiling joist. The truss also eliminates the need for load-bearing partitions.

Second grades of the various species are commonly used for ceiling joists and rafters. It is also desirable, particularly in two-story houses and when material is available, to limit the moisture content of the second-floor joists to no more than 15 percent. This applies as well to other lumber used throughout the house. Maximum moisture content for dimension material should be 19 percent.

Ceiling joists are used to support ceiling finishes. They often act as floor joists for second and attic floors and as ties between exterior walls and interior partitions. Since ceiling joists also serve as tension members to resist the thrust of the rafters of pitched roofs, they must be securely nailed to the plate at outer and inner walls. They are also nailed together, directly or with wood or metal cleats, where they cross or join at the load-bearing partition (Figure 13-1, A) and to the rafter at the exterior walls (Figure 13-1, B). Toenail at each roof.

In areas of severe windstorms, the use of metal strapping or other systems of anchoring ceiling and roof framing to the wall is good practice. When ceiling joists are perpendicular to rafters, collar beams and cross ties should be used to resist thrust. The in-line joist system described in Chapter 6 can also be adapted to ceiling or second floor joists.

Flush Ceiling Framing
In many house designs, the living room and the dining or family room form an open "L." A wide, continuous ceiling area between the two rooms is often desirable. This can be created with a flush beam, which replaces the load-bearing partitions used in the remainder of the house. A nail-laminated beam, designed to carry the ceiling load, supports the ends of the joists. Joists are toenailed into the beam and supported by metal joist hangers (Figure 13-2, A) or wood hangers (Figure 13-2, B). To resist the thrust of the rafters for longer spans, it is

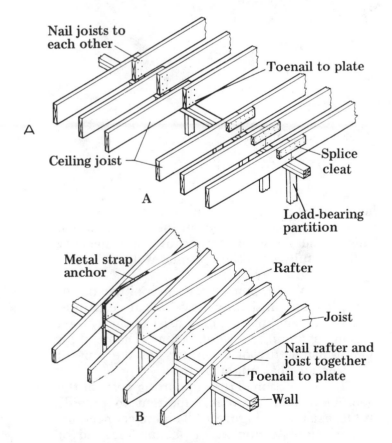

Ceiling joist connections: *A*, At center partition with joists lapped or butted; *B*, at outside wall

Figure 13-1

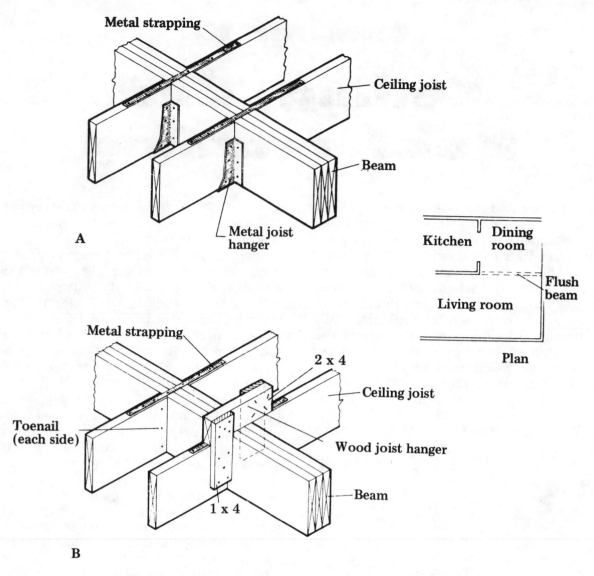

Flush ceiling framing: *A*, Metal joist hanger
B, wood hanger
Figure 13-2

often desirable to provide added resistance by using metal strapping. Strapping should be nailed to each opposite joist with three or four eightpenny nails.

Post And Beam Framing

In contemporary houses, exposed beams are often a part of the interior design and may also replace interior and exterior load-bearing walls. With post and beam construction, exterior walls can become fully glazed panels between posts, requiring no other support. Areas below interior beams within the house can remain open or can be closed in with wardrobes, cabinets, or light curtain walls.

This type of construction, while not adaptable to many styles of architecture, is simple and straightforward. However, design of the house should take into account the need for shear or racking resistance of the exterior walls. This is usually accomplished by solid masonry walls or fully sheathed frame walls between open glazed areas.

Roofs of such houses are often either flat or low-pitched, and may have a conventional rafter-joist combination or consist of thick wood decking spanning between beams. The need for a well-insulated roof often dictates the type of construction that might be used.

The connection of the supporting posts at the floor plate and beam is important to provide uplift resistance. Figure 13-3 shows connections at the soleplate and at the beam for solid or spaced members. The solid post and beam are fastened together with metal angles nailed to the top plate and to the soleplate as well as the

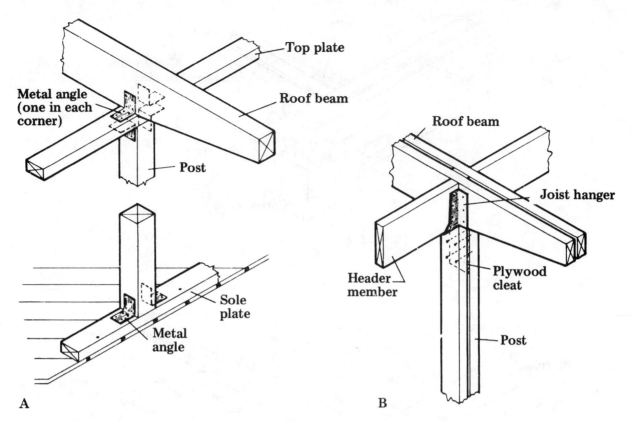

Post and beam connections: *A*, Solid post
and beam; *B*, spaced post and beam
Figure 13-3

roof beam (Figure 13-3, *A*). The spaced beam and post are fastened together with a 3/8 inch or thicker plywood cleat extending between and nailed to the spaced members (Figure 13-3, *B*). A wall header member between beams can be fastened with joist hangers.

Continuous headers are often used with spaced posts in the construction of framed walls or porches requiring large glazed openings. The beams should be well fastened and reinforced at the corners with lag screws or metal straps. Figure 13-4, *A* illustrates one connection method using metal strapping.

In low-pitch or flat roof construction for a post and beam system, wood or fiberboard decking is often used. Wood decking, depending on thickness, is frequently used for beam spacings up to 10 or more feet. However, for the longer spans, special application instructions are required. Depending on the type, 2 to 3 inch thick fiberboard decking normally is limited to a beam or purlin spacing of 4 feet.

Tongued-and-grooved solid wood decking, 3 by 6 and 4 by 6 inches in size, should be toenailed and face-nailed directly to the beams and edge-nailed to each other with long nails used in predrilled holes (Figure 13-4, *B*). Thinner decking is usually only face-nailed to the beams. Decking is usually square end-trimmed to provide a good fit. If additional insulation is required for the roof, fiberboard or an expanded foamed plastic in sheet form is fastened to the decking before the built-up or similar type of roof is installed. The moisture content of the decking should be near its in-service condition to prevent joints opening later as the wood dries.

Roof Slopes

The architectural style of a house often determines the type of roof and roof slope which are best suited. A contemporary design may have a flat or slightly pitched roof, a rambler or ranch type an intermediate slope, and a Cape Cod cottage a steep slope. Generally, however, the two basic types may be called flat or pitched. Flat roofs are flat or slightly pitched roofs in which roof and ceiling supports are furnished by one type of member. In pitched roofs both ceiling joists and rafters or trusses are required.

The pitch of the roof is generally expressed as the number of inches of vertical rise in 12

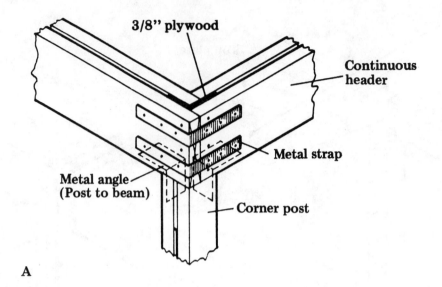

3/8" plywood

Continuous header

Metal strap

Metal angle (Post to beam)

Corner post

A

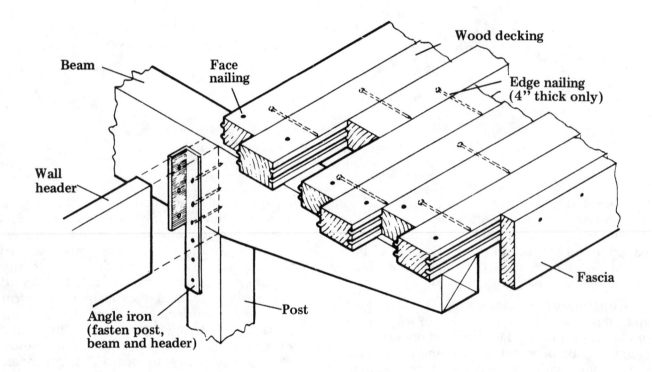

Beam

Face nailing

Wood decking

Edge nailing (4" thick only)

Wall header

Fascia

Angle iron (fasten post, beam and header)

Post

B

Post and beam details: *A*, Corner connection
with continuous header; *B*, with roof decking
Figure 13-4

inches of horizontal run. The rise is given first, for example, 4 in 12.

In terms of proportion, the pitch is the ratio of the rise of the rafter to the width of the building. For example, assume that the span of a roof is 24 feet and the rise is 12 feet. The ratio of the rise to the span is 12 to 24 or 12/24. This equals 1/2 pitch. If the rise were 8 feet the pitch would be 8/24 or 1/3. If the rise were 4 feet the pitch would be 4/24 or 1/6.

The pitch is often given in inches. For example, 8½ inch rise means 8½ inches of rise for every foot of run. If the span were 20 feet, the total run of a rafter would be ½ of 20 = 10 feet. The total rise of the rafter would be 10 feet x 8½ inches or 85 inches rise. If the rise were 10½ inches per foot of run, the total rise of the rafter would be 10 x 10½ inches or 105 inches rise.

If the total rise is given to the top of the

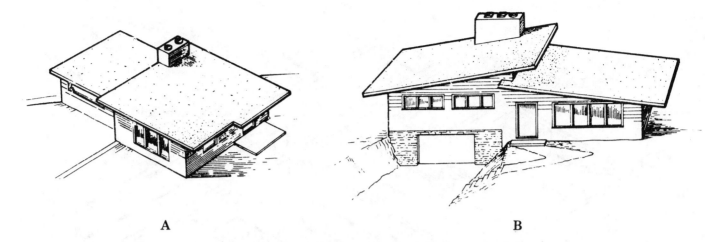

A   B

Roofs using single roof construction:
A, Flat roof; *B*, low-pitched roof
Figure 13-5

ridge, as in porch or bay window roofs, this rise is divided by the run of the rafter in feet to get the rise per foot. For example, assume that the rise of the rafter is 32 inches and the run is 36 inches. 36 inches = 3 feet. 32 ÷ 3 = $10\frac{2}{3}$ or approximately 10-5/8 inches rise per foot of run.

A further consideration in choosing a roof slope is the type of roofing to be used. However, modern methods and roofing materials provide a great deal of leeway in this. For example, a built-up roof is usually specified for flat or very low-pitched roofs, but with different types of asphalt or coal-tar pitch and aggregate surfacing materials, slopes of up to 2 in 12 are sometimes used. Also, in sloped roofs where wood or asphalt shingles might be selected, doubling the underlay and decreasing the exposure distance of the shingles will allow slopes of 4 in 12 and less.

Second grades of the various wood species are normally used for rafters. Most species of softwood framing lumber are acceptable for roof framing, subject to maximum allowable spans for the particular species, grade, and use. Because all species are not equal in strength properties, larger sizes, as determined from the design, must be used for weaker species for a given span.

All framing lumber should be well seasoned. Lumber 2 inches thick and less should have a moisture content not over 19 percent. When obtainable, lumber at about 15 percent is more desirable because less shrinkage will occur when moisture equilibrium is reached.

Flat Roofs

Flat or low-pitched roofs, sometimes known as shed roofs, can take a number of forms, two of which are shown in Figure 13-5. Roof joists for flat roofs are commonly laid level or with a slight pitch, with roof sheathing and roofing on top and with the underside utilized to support the ceiling. Sometimes a slight roof slope may be provided for roof drainage by tapering the joist or adding a cant strip to the top.

The house design usually includes an overhang of the roof beyond the wall. Insulation is sometimes used in a manner to provide for airways just under the roof sheathing to minimize condensation problems in winter. Flat or low-pitched roofs of this type, require larger sized members than steeper pitched roofs because they carry both roof and ceiling loads.

The use of solid wood decking often eliminates the need for joists. Roof decking used between beams serves as: (a) supporting members, (b) interior finish, and (c) roof sheathing. It also provides a moderate amount of insulation. In cold climates, rigid insulating materials are used over the decking to further reduce heat loss.

When overhang is involved on all sides of the flat roof, lookout rafters are ordinarily used (Figure 13-6). Lookout rafters are nailed to a doubled header and toenailed to the wallplate. The distance from the doubled header to the wall line is usually twice the overhang. Rafter ends may be finished with a nailing header which serves for fastening soffit and fascia boards. Care should be taken to provide some type of ventilation at such areas.

Gable Roof

Perhaps the simplest form of the pitched roof, where both rafters and ceiling joists are required because of the attic space formed, is

153

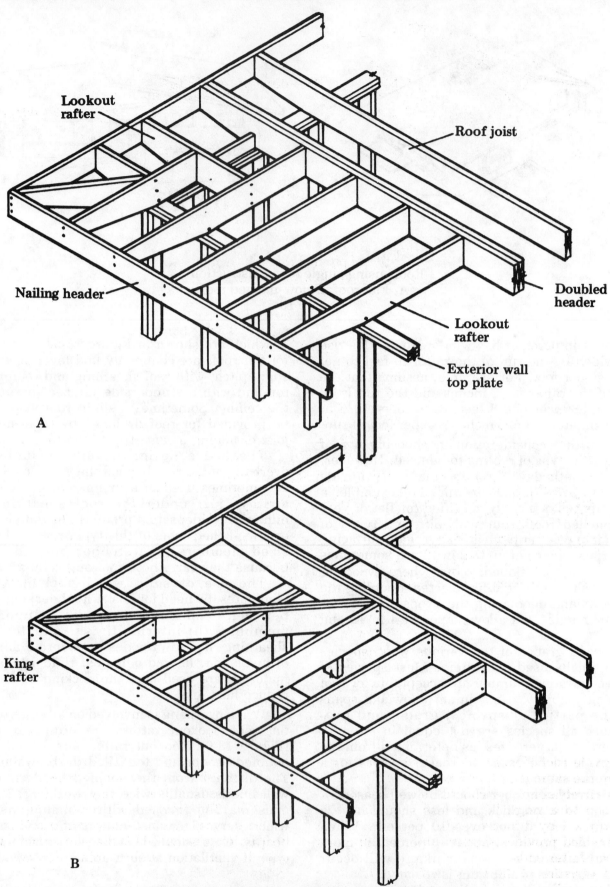

Typical construction of flat or low-pitched roof
with side and end overhang of: *A*, Less than 3
feet: *B*, more than 3 feet
Figure 13-6

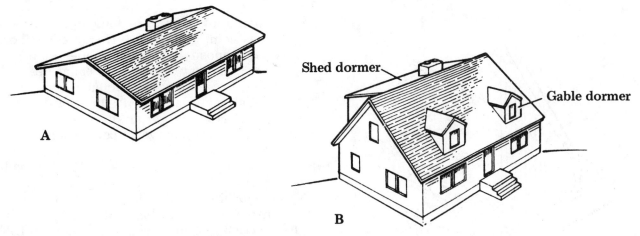

Types of pitched roofs: *A*, Gable;
B, gable with dormers
Figure 13-7

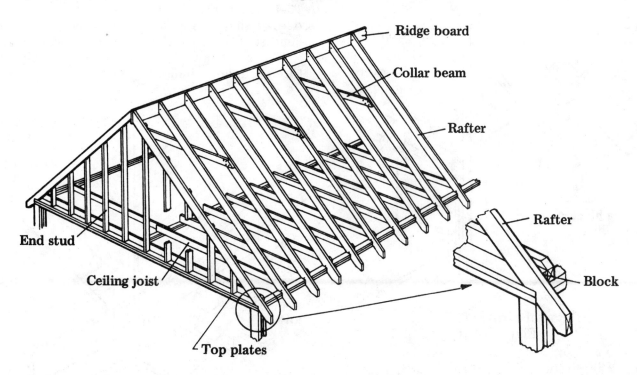

Ceiling and roof framing: Overall view
of gable roof framing
Figure 13-8

the gable roof (Figure 13-7, *A*). All rafters are cut to the same length and pattern and erection is relatively simple, each pair being fastened at the top to a ridge board. The ridge board is usually a 1 by 8 inch member for 2 by 6 inch rafters and provides support and a nailing area for the rafter ends.

A variation of the gable roof, used for Cape Cod or similar styles, includes the use of shed and gable dormers (Figure 13-7, *B*). Basically, this is a one-story house because the majority of the rafters rest on the first floor plate. Space

and light are provided on the second floor by the shed and gable dormers for bedrooms and bath. Roof slopes for this style may vary from 9 in 12 to 12 in 12 to provide the needed headroom.

In normal pitched-roof construction, the ceiling joists are nailed in place after the interior and the exterior wall framing are complete. Rafters should not be erected until ceiling joists are fastened in place, as the thrust of the rafters will otherwise tend to push out the exterior walls.

Rafters are usually precut to length with

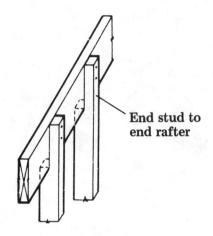

Connection of gable end studs to rafter
Figure 13-9

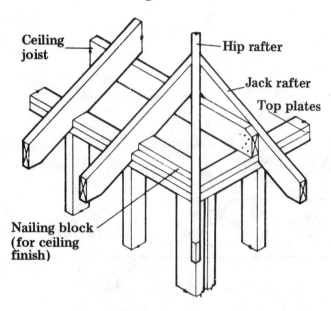

Detail of corner of hip roof
Figure 13-10

proper angle cut at the ridge and eave, and with notches provided for the top plates (Figure 13-8). Rafters are erected in pairs. Studs for gable end walls are cut to fit and nailed to the end rafter and the top-plate of the end wall sole-plate (Figure 13-9). With a gable (rake) overhang, a fly rafter is used beyond the end rafter and is fastened with blocking and by the sheathing.

Hip Roof

Hip roofs are framed the same as a gable roof at the center section of a rectangular house. The ends are framed with hip rafters which extend from each outside corner of the wall to the ridge board at a 45 degree angle. Jack rafters extend from the top plates to the hip rafters (Figure 13-10).

When roof spans are long and slopes are flat, it is common practice to use collar beams between opposing rafters. Steeper slopes and shorter spans may also require collar beams but only on every third rafter. Collar beams may be 1 by 6 inch material. In 1½ story houses, 2 by 4 inch members or larger are used at each pair of rafters which also serve as ceiling joists for the finished rooms.

Good practices to be followed in the nailing of rafters, ceiling joists, and end studs are shown in Figures 13-8 through 13-10.

The valley is the internal angle formed by the junction of two sloping sides of a roof. The key member of valley construction is the valley rafter. In the intersection of two equal size roof sections, the valley rafter is doubled (Figure 13-11) to carry the roof load, and is 2 inches deeper than the common rafter to provide full contact with jack rafters. Jack rafters are nailed to the ridge and toenailed to the valley with three tenpenny nails.

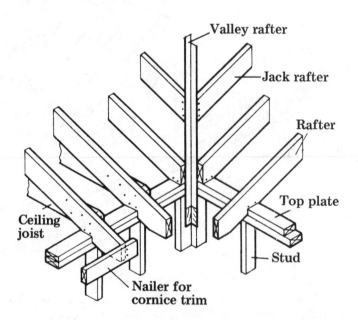

Framing at a valley
Figure 13-11

Other Roof Types

Flat, gable and hip roofs are used most frequently in residential construction. Occasionally you will see other roof types:

The single pitch or shed roof is perhaps the simplest roof, as it sheds water only one way. This type is used chiefly on garages or over entrances which are to be shielded from the weather (Figure 13-12).

The gambrel roof (Figure 13-13) is similar to the gable roof but has two different slopes on each of the two long sides of the roof. This type

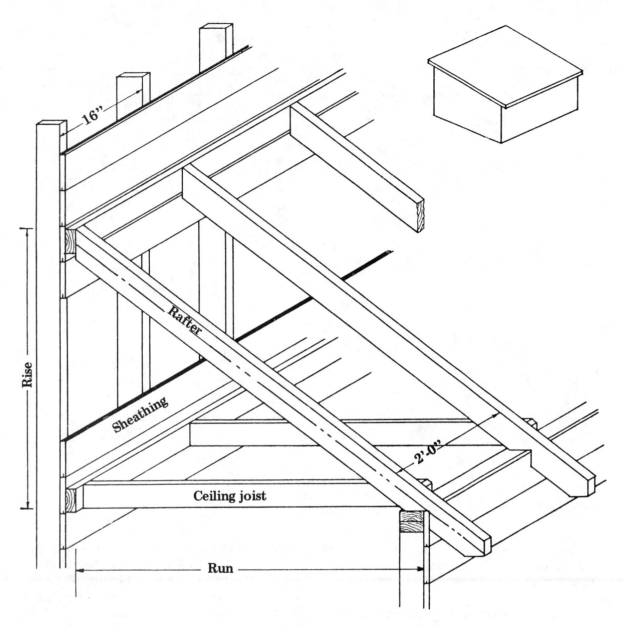

Shed roof over entrance
Figure 13-12

of roof is used in a modified form in the Dutch Colonial house.

The only difference in framing a gambrel roof rather than an ordinary gable or hip roof is at the point where the rafters of the two slopes join each other. This point may be framed in one of two ways, either with or without a purlin plate. The use of a purlin, as shown in Figure 13-13, is desirable in structures where it can be supported by partitions, as in a house supported with partition walls.

The chief advantage of the gambrel roof over the gable roof is the additional space the gambrel provides at no increase in the height of the ridge. This allows greater headroom.

The mansard roof (Figure 13-14) is similar to the gambrel roof except that each of the four sides has the double slope. When the upper part of this type of roof is flat, the roof is called a deck roof.

Overhangs

In two-story houses, the design often involves a projection or overhang of the second floor for the purpose of architectural effect, to accommodate brick veneer on the first floor, or for other reasons. This overhang may vary from 2 to 15 inches or more. The overhang should ordinarily extend on that side of the house where joist extensions can support the wall framing (Figure 13-15). This extension should be provided with insulation and a vapor barrier.

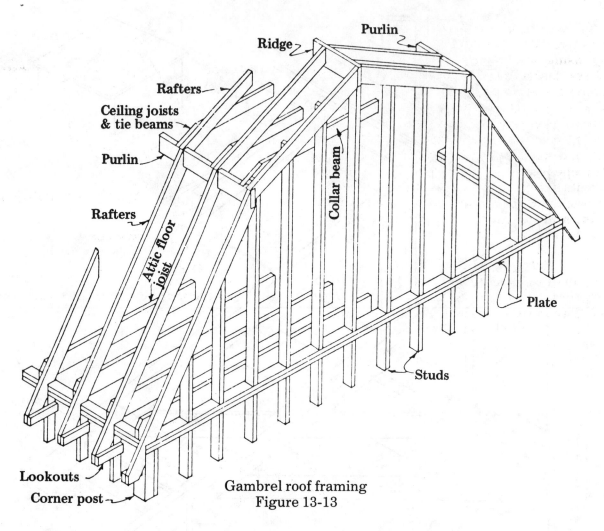

Gambrel roof framing
Figure 13-13

When the overhang parallels the second floor joists, a doubled joist should be located back from the wall at a distance about twice the overhang.

By eliminating eave and gable-end overhangs, substantial material savings can be realized. The typical eave overhang -- or cornice -- utilizes the rafter or truss projections as nailing surfaces for fascia boards and for 2 x 4 framing members which serve as nailing surfaces and supports for soffits. A frieze board is usually installed at the wall-soffit intersection along with moulding to cover the frieze-soffit intersection.

If there were no eave overhang, the 2 x 4 framing, plywood or board soffits, frieze boards, and moulding would be eliminated (see Figure 13-16, A).

Where the gable end and eave both have some amount of overhang, a box return is required. This box would be eliminated if no overhang were used (see Figure 13-16, B). The gable-end overhang is made with 2 x 4 lookout blocks, fascia board, soffit, frieze board, and bed moulding. If the overhang were removed,

all but the fascia board would be eliminated.

The wide-box eave overhang usually requires 2 x 4 framing members to serve as nailing surfaces and support for the soffit. Framing is nailed to the rafter or truss ends and to either the exterior wall studs or 2 x 4 blocking at the

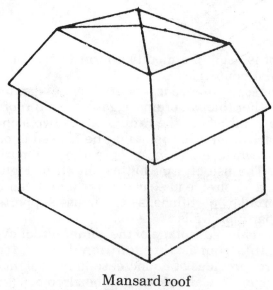

Mansard roof
Figure 13-14

wall. Generally, plywood or board soffits are nailed to the 2 x 4 framing.

By using the open soffit, 2 x 4 framing members, fascia board, soffit, frieze board, and moulding are eliminated on the eave overhang.

The typical gable-end overhang consists of 2 x 4 lookouts nailed to the gable end top chord 16 inches on center. A fascia board, soffit, frieze board, and moulding make up the overhang.

The open-gable overhang eliminates the 2 x 4 lookouts except at the eave-gable intersection and possibly at the gable ridge. Soffits, frieze boards, and mouldings are eliminated (see Figure 13-17).

Ridge Beam Roof Details

In low-slope roof designs, the style of architecture often dictates the use of a ridge beam. These solid, glue-laminated, or nail-

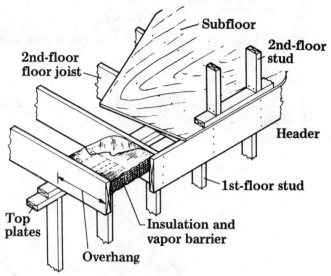

Construction of overhang at second floor
Figure 13-15

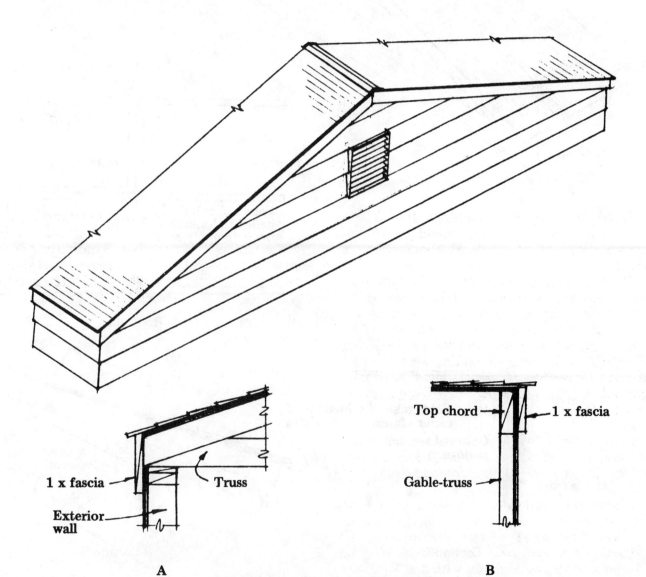

A

B

Soffit and gable closed overhang
Figure 13-16

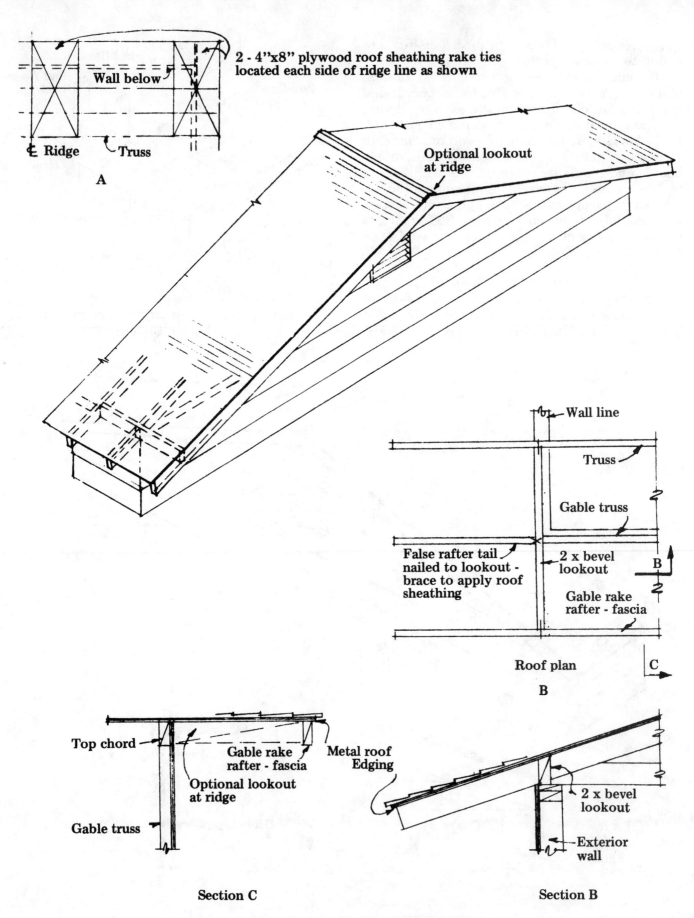

2 - 4"x8" plywood roof sheathing rake ties
located each side of ridge line as shown

Wall below

Ridge   Truss

A

Optional lookout
at ridge

Wall line

Truss

Gable truss

False rafter tail
nailed to lookout -
brace to apply roof
sheathing

2 x bevel
lookout

B

Gable rake
rafter - fascia

Roof plan

C

B

Top chord

Gable rake
rafter - fascia

Optional lookout
at ridge

Gable truss

Metal roof
Edging

2 x bevel
lookout

Exterior
wall

Section C

Section B

Open soffit and gable overhangs
Figure 13-17

160

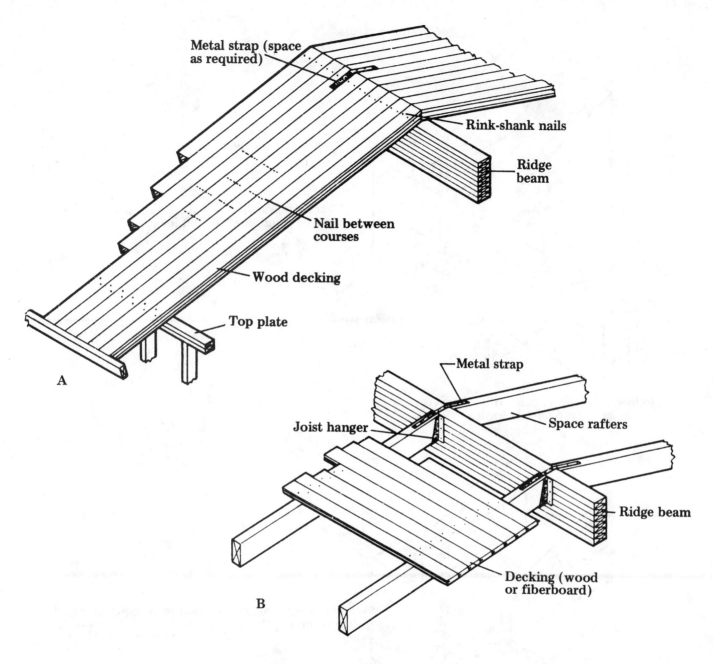

Ridge beam for roof: *A*, With wood decking;
B, with rafters and decking
Figure 13-18

laminated beams span the open area and are usually supported by an exterior wall at one end and an interior partition wall or a post at the other. The beam must be designed to support the roof load for the span sheathing. Spaced rafters placed over the ridge beam or hung on metal joist hangers serve as alternate framing methods. When a ridge beam and wood decking are used (Figure 13-18, *A*), good anchoring methods are needed at the ridge and outer wall. Long ringshank nails and supplemental metal strapping or angle irons can be used at both bearing areas.

A combination of large spaced rafters (purlin rafters) which serve as beams for longitudinal wood or structural fiberboard decking is another system which might be used with a ridge beam. Rafters can be supported by metal hangers at the ridge beam (Figure 13-18, *B*) and extend beyond the outer walls to form an overhang. Fastenings should be supplemented by strapping or metal angles.

Saving Time And Material
When you plan the roof, remember that time and material can be saved if the roof dimensions

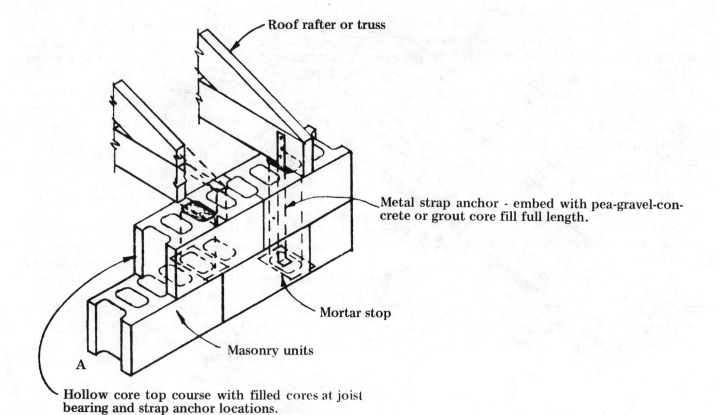

Roof rafter or truss

Metal strap anchor - embed with pea-gravel-concrete or grout core fill full length.

Mortar stop

Masonry units

A

Hollow core top course with filled cores at joist bearing and strap anchor locations.

Metal strap anchor - embed into concrete bond beam full depth.

Concrete bond beam

Mortar stop

Masonry units

B

Methods of eliminating wood sill plates on
top of masonry block walls
Figure 13-19

are multiples of 48 or 24 inches. Particular attention should be given to the length of roof slope from the peak to the overhang. If this length is a multiple of 48 inches or 24 inches, there will be no waste in the width (4 foot) direction of the sheathing. Similarly, if the roof length is a multiple of 48 inches or 24 inches, there will be no waste in the length (8 foot) direction of the sheathing. Observations and structural research have shown that sheathing butt joints need not be staggered for roofs, walls, and floors. This is particularly helpful in saving sheathing material for roofs with hips or valleys.

If possible, place roof openings for chimneys between truss spacings, preventing waste or an additional roof truss or roof framing.

Prepare a roof materials schedule from drawings of the roof framing. The schedule should list the exact quantity, description, and grade required for each item. Particular emphasis should be placed on using lowest cost lumber and plywood substitutes for soffits or fascia, and on eliminating overhang lumber or plywood.

Wood sill plates often are used as a bearing and nailer plate on top of masonry block walls. If a roof truss or a rafter bears on webs or on grout or pea-gravel concrete-filled cores of masonry block, a sill plate is not required for bearing and may be eliminated as shown in Figure 13-19, *A*. A truss or rafter can be anchored directly to the masonry wall in this case rather than to a wood sill plate.

Figure 13-19, *B* illustrates the same concept when used in conjunction with cast-in-place bond beams or cast-in-place concrete walls.

Design Of Ceiling Joists And Rafters

The principles governing size of ceiling joists and rafters are the same as for girders. The first consideration, of course, is that there shall be ample strength. Many times, however, a ceiling joist or rafter which is strong enough may still be limber enough to permit a noticeable bending or vibration from walking. This not only is annoying but may be sufficient to crack the roof surface or ceiling below. Stiffness, therefore, must also be considered as a factor. Tables A-12 to A-21 in Appendix A give the spans for various depths and grades of lumber.

Chapter 14

The Gable Roof

There are several methods of finding the lengths and cuts of roof rafters. However, the basic principle of all methods is geometric construction. Each one of the commonly used methods, the graphic, the rafter table and the step-off, is used where it can be applied most conveniently. The architectural draftsman might find the graphic method most convenient because the work may be laid out by using drafting instruments. The carpenter finds the step-off method or the rafter table method convenient because the work is laid out with a framing or steel square.

The step-off method of laying out rafters is considered the most logical for the carpenter to use. This chapter will describe that method. The rafter tables will also be described so the two methods may be used by the carpenter as a check against inaccurate work.

To avoid confusion, you should study only one system of rafter layout at a time until you are thoroughly familiar with that system. These chapters on roof framing contain only the fundamental principles that are necessary for laying out the type of rafter under discussion.

Rafter layout is a somewhat complicated procedure to learn. Therefore, it is important that some definite habits be formed as to the order in which the various steps of the layout work are performed. In laying out rafters, they should all be placed in the same relative position. The crowned edge, which should be considered the top edge, should be toward you as you are laying out the rafter. The tongue of the steel square should be held in the left hand and the body of the square in the right. If the square is pictured as greatly enlarged, the tongue should form the vertical or top cut of the rafter, and the body, the level or seat cut (Figure 14-1).

The safety and convenience of the workmen must be considered in the erection of roof rafters. The scaffolds along the sides and ends of the building should be about 3 feet below the attic floor line, and should be made secure and wide enough for safe walking. The attic floor should be laid so that scaffolds needed for erection of ridgeboards and rafters may be erected.

The spacing of rafters depends on the weight of the roofing material to be used and the span of the rafter. In most cases, the spacing is 24 inches on centers for light frame construction.

Finding The Rough Length
Of A Common Rafter

The approximate length of a common rafter may be found by representing the rise of the rafter in feet on the tongue of the square and the run in feet on the blade. The length of the diagonal between these two points is then measured. This measurement, expressed in feet, is the rough length of the rafter. If an overhang for a cornice is needed, this length should be added. For example, assume that the total rise of a rafter is 9 feet and the run is 12 feet. Locate 9 and 12 on the square (Figure 14-2) and measure the diagonal. This will be found to be 15. Therefore, the rough length of the rafter will be 15 feet plus the overhang of 1 foot. Stock 16 feet long would have to be used.

Locating The Measuring Line On A Rafter

The stock of which the first rafter is to be made should be straight and of the correct rough length. The width and thickness may be determined as described in Appendix A. This piece is laid flat across two sawhorses and the square is placed near the right hand end (Figure 14-3).

The inch mark on the outside edge of the body corresponding to the run of the roof (12 inches in all cases) and the inch mark on the outside edge of the tongue corresponding to the rise of the roof, should both come at the edge of the rafter. This is to be the top edge. The line AB, Figure 14-3 is then drawn on the plank to represent the level of the top of the wall plate. A distance of 3-5/8 inches is measured along this line from B to C to locate the outside top corner of the plate. This should be far enough from the right hand end of the rafter to allow for the tail.

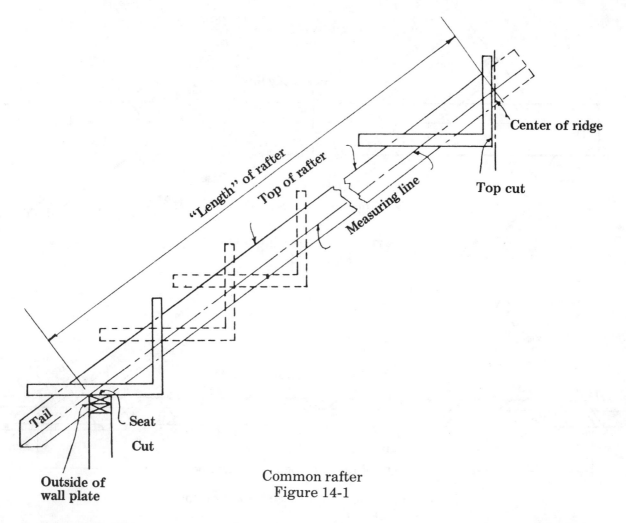

Common rafter
Figure 14-1

The measuring line (CD) is then gauged parallel to the edge of the rafter.

Laying Out A Common Rafter
(Step-Off Method)

A wooden fence may be made and clamped at the proper points on the tongue and body of the square to aid in laying out a rafter. Small metal clamps may also be used for this purpose.

Assume that a building is 24 feet wide and the pitch of the rafter is to be 3/8, or 9 inches rise on 12 inches run. It will then be necessary to find the exact length of the rafter. The square is laid on the stock so the 12 inch mark on the outside of the body is at point C, Figure 14-4, and the 9 inch mark on the outside edge of the tongue is on the measuring line at E. The square is kept in this position and the fence is adjusted so its edge lies against the top edge of the rafter (Figure 14-4). The fence is then tightened on the square. It is important that the 12 inch mark on the body of the square and the 9 inch mark on the tongue be exactly on the measuring line when the fence is against the edge of the rafter.

Place the fence against the rafter stock as shown at the extreme right of Figure 14-4. Mark along the outside edge of the body and tongue of the square. This locates point E on the measuring line. Slide the square to the left until the 12 inch mark is over E (position 2) and again mark along the tongue and body. Continue this operation as many times as there are feet of run

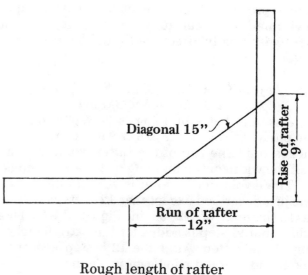

Rough length of rafter
Figure 14-2

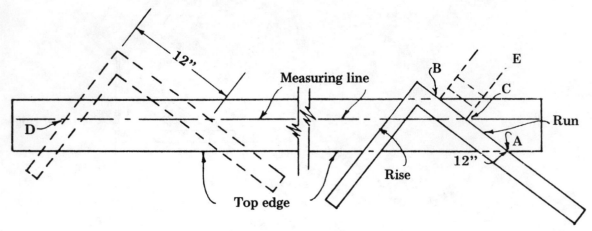

Laying out measuring line
Figure 14-3

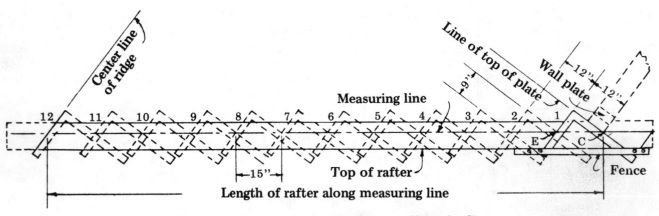

Laying out a common rafter (step-off method)
Figure 14-4

of the rafter (12 in this case). The successive positions of the square are shown by the figures 1 through 12 on Figure 14-4. When the last position has been reached, a line drawn along the tongue across the rafter will indicate the center line of the ridgeboard. Each step in this whole process should be clearly marked and performed very carefully since even a slight error will greatly affect the fit and length of the rafter.

Layout Of A Rafter When The Span Is An Odd Number Of Feet

Assume that a building is 25 feet wide and the pitch of the rafter is 3/8 or 9 inches rise to 12 inches run. The run of the rafter would be one half the width of the building or 12 feet 6 inches.

The layout of the rafter is very similar to the one just described except that 12½ steps will be taken instead of 12 as in Figure 14-4. This additional ½ step is added for the extra 6 inches run of the rafter. After the 12th step has been taken and marked as shown in Figure 14-5, the square is placed on the top edge of the rafter

with the figures 9 and 12 coinciding with the edge as shown. The square is then moved until the figure 6 on the outside of the body is directly over the 12th step plumb line A, Figure 14-5. A line is marked along the outside edge of the tongue to indicate the center line of the ridgeboard.

Allowance For Ridgeboard

The last line marked on the rafter shows where the rafter would be cut off if there were no ridgeboard and the rafters were to be butted against each other. Since a ridgeboard is used in most cases, the rafter as previously laid out will be a little too long. It will be necessary to cut a piece off the end of the rafter equal to half the thickness of the ridgeboard (Figure 14-6).

To lay out this line, slide the square back away from the last line (position 12 in Figure 14-4), keeping the fence tight against the upper edge of the rafter. When the square has been moved back half the thickness of the ridgeboard from this last line, mark the plumb cut along the edge of the tongue. This line should be inside of,

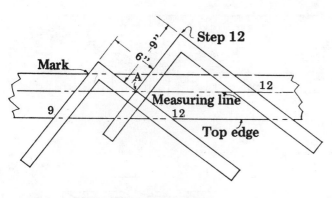

Additional half step
Figure 14-5

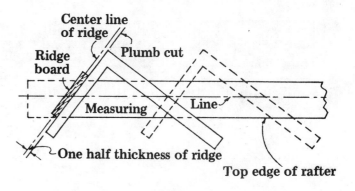

Allowance for ridgeboard
Figure 14-6

and parallel to, the original line marking the end of the rafter. The measurement should be taken from this original line and at right angles to it (Figure 14-6).

Bottom Or Seat Cut

The bottom or seat cut of the rafter is a combination of the level and plumb cuts. The level cut (BC, Figure 14-3) rests on the top face of the side wall plate, and the plumb cut (CE) fits against the outside edge of the wall plate. The plumb cut is laid out by squaring a line from line AB, Figure 14-3, through point C. This line CE then represents the plumb cut.

Tail Of The Rafter

If the rafter tail is to be of the type shown in Figure 14-4, the cut may be made along the measuring line as shown.

The tail of the rafter must be cut differently for the box cornice shown in Figure 14-7. In this case, the level and plumb cuts are laid out as before. Since the sheathing extends into the rafter notch up to the top of the plate, an allowance for the thickness of the sheathing must be made. This can be done by laying out line AB parallel to line CD and the thickness of the sheathing away from CD. The line of the level cut at D is then extended to meet AB. When the rafter is cut on these lines, it will fit over the plate and sheathing.

Assuming that a 12 inch piece is to be used at E, continue the line of the level cut through B to F on the top edge of the rafter. Lay the body of the square along this line with the 12 inch mark of the outside edge directly over point B. Mark a line GH along the outer edge of the tongue across the rafter. The point where this line meets the measuring line at H represents the tip of the rafter. Square the line HJ across the rafter from line GH to locate J.

Rafter Tables

Many steel squares have rafter tables stamped on them. A booklet giving instructions on the use of these tables may be secured from the manufacturer of the square. Several publishers offer rafter tables for almost every span and rise. However, if you have only your steel square to compute the rafter lengths, you should still have no trouble arriving at the correct rafter length.

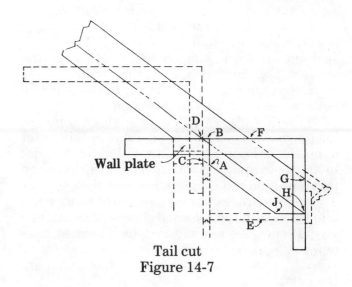

Tail cut
Figure 14-7

After the rise and run of the rafter have been determined, the length of the rafter along the measuring line may be found from the rafter table (Figure 14-8). The inch marks on the outside edge of the square are used to denote rise per foot of run. For example, the figure 8 means 8 inches of rise per foot of run. Directly below each of these figures is the length of main rafters per foot of run, 14.42 in this case. If the run of the rafter is 10 feet, the 14.42 inches should be multiplied by 10. This gives 144.20 inches or 12.01 feet as the length of the rafter.

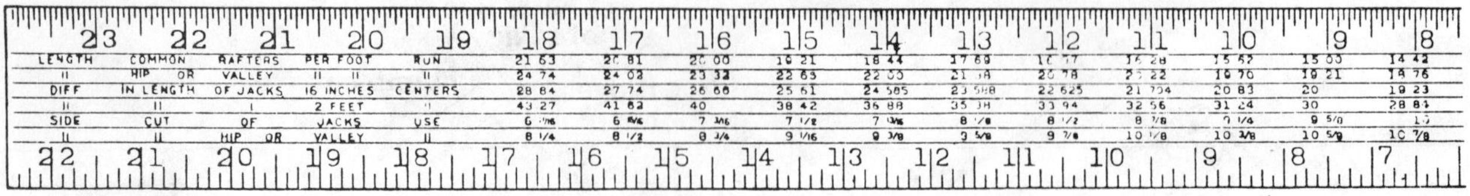

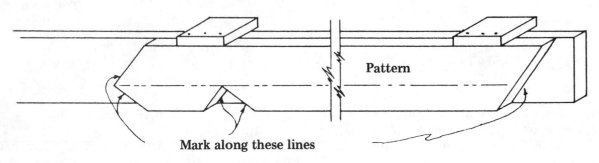

| | | | | | | 23 | 22 | 21 | 20 | 19 | 18 | 17 | 16 | 15 | 14 | 13 | 12 | 11 | 10 | 9 | 8 |

					RUN	21 63	20 81	20 00	19 21	18 44	17 69	16 97	16 28	15 62	15 00	14 42
LENGTH	COMMON	RAFTERS	PER FOOT	RUN												
"	HIP	OR	VALLEY	"	"	24 74	24 02	23 32	22 65	22 00	21 38	20 78	20 22	19 70	19 21	18 76
DIFF	IN LENGTH	OF JACKS	16 INCHES	CENTERS	"	28 84	27 74	26 66	25 61	24 585	23 598	22 625	21 704	20 83	20	19 23
"	"	"	2 FEET	"		43 27	41 62	40	38 42	36 88	35 39	33 94	32 56	31 24	30	28 84
SIDE	CUT	OF	JACKS	USE		6 7/16	6 11/16	7 3/16	7 1/2	8	8 1/2	8 7/8	9 1/4	9 5/8	15
"	"	HIP	OR	VALLEY	"		8 1/4	8 1/2	8 3/4	9 1/16	9 3/8	9 5/8	9 7/8	10 1/8	10 3/8	10 5/8	10 7/8

Rafter table
Figure 14-8

Pattern

Mark along these lines

Rafter pattern
Figure 14-9

The other steps in laying out a rafter are the same as described under the step-off method.

How To Lay And Cut Common Rafters

1. Check the actual width of the building at the attic plate line to find out if the rafters at the seat cut are to butt against the outside of the sheathing or directly against the plate.

2. Lay out and cut one rafter to be used as a pattern, as described in this chapter.

3. Nail a piece of sheathing about 6 inches long near each end on the top edge of this rafter as shown in Figure 14-9. This will make it easier to line up the pattern with the other rafters when they are ready to be marked.

4. Select the required number of rafters and pile them with the crowned edges all facing one way. Do not use any rafters which have serious defects.

5. Lay the pattern on each rafter and mark the cuts.

6. Make the ridge, seat, and tail cuts. These cuts should be square with the side of the rafter. A rip saw or portable electric saw is often used to make these cuts.

How To Space Rafters On The Plates And Ridgeboard

1. Secure a sufficient number of boards to make up the length of the ridgeboard (Figure 14-10). These boards are generally ¾ inch x 6 inches.

2. Lay the boards along the plate on one side of the building.

3. Mark the spacing for the rafters on the plate and on the ridgeboard. Use the same method as in marking the ribbon for the balloon frame. The joints of the ridgeboard should come half way on a rafter. Let the last rafter space come where it will.

4. Cut the last piece of ridgeboard off flush with the outside of the plate at the end of the building.

5. Place the ridgeboard at the side of the plate on the opposite side of the building and mark this plate the same as the ridgeboard (Figure 14-10).

How To Erect And Secure Rafters On A Gable Roof

1. Build a scaffold about 10 feet long for erecting the rafters. The top of the scaffold should be 5 feet from the under side of the ridge and should be so erected that it can be moved along the attic floor (Figure 14-11).

2. Place the ridgeboard on the scaffold with the ends pointing the same way as when they were laid out.

3. Select four of the straightest rafters for gable ends.

Note: When the ridgeboard is made up of several lengths, a group of rafters is nailed on the first length of ridgeboard and to the plate. This section is then braced. The scaffold is then moved along the floor and the next section of rafters is raised.

4. Nail the first length of ridgeboard to the

168

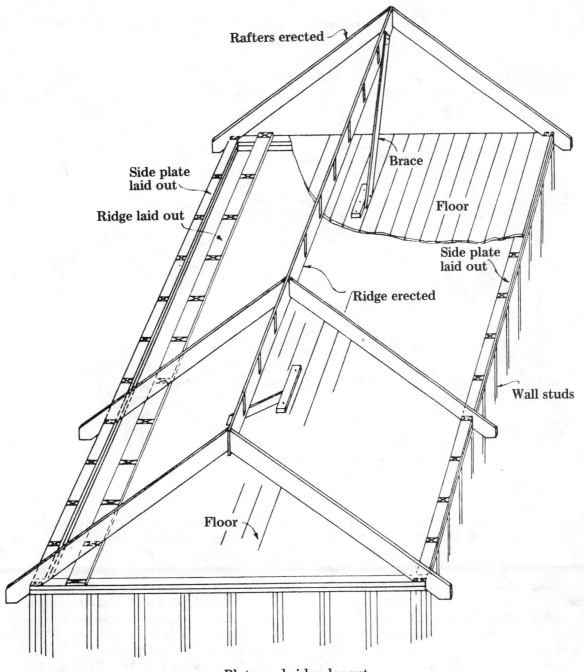

Plate and ridge layout
Figure 14-10

end rafter. The end of the ridge should be flush with the outside of the rafter. The top of the rafter should be on the spacing marks and even with the top of the ridgeboard. Nail through the side of the ridgeboard and into the ridge cut of the rafter with three 8d nails.

5. Nail another rafter on the same side of the ridgeboard but about five rafter spacings away.

6. Raise these rafters until the seat cut fits at the plate and toenail them in place at the bottom marks with 16d nails.

7. Place the opposite rafters on the other side of the ridgeboard and nail them to the plate at the seat cut. Then raise or lower the ridgeboard until it is even with the top of these rafters. Nail the rafters in position through the opposite side of the ridgeboard or toenail them.

8. Plumb and brace this section by nailing one end of a 2 x 4 stay to the ridgeboard at the top of the gable, and the other end to a block spiked to the attic floor. This block should be spiked into at least two joists (Figure 14-10).

9. Nail the rest of the rafters of this section in place. Erect first one rafter and then the opposite one. Nail the bottom end of the rafter first to provide a straighter ridge.

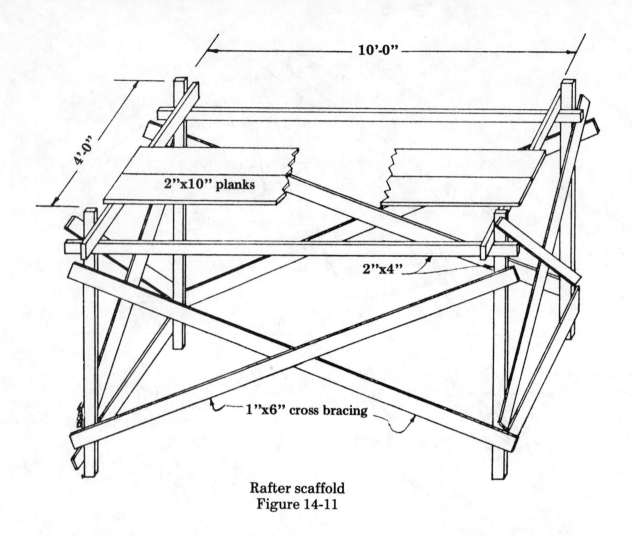

Rafter scaffold
Figure 14-11

Note: One of the rafters at the joint in the ridge should be tacked a little to one side until the rest of the ridge is in place.

10. Move the scaffold along the floor so that the other section of the ridge may be raised in a similar manner. It is assumed that there are only two pieces of ridgeboard.

11. Nail the outer end of the second piece of ridgeboard to the outside or gable rafter, keeping the end flush with the face of the rafter.

12. Nail one rafter near the center of this piece of ridgeboard.

13. Raise these rafters and place the inside end of the ridgeboard tight against the other piece of ridgeboard. Nail the second piece of ridgeboard to the rafter that centers on the joint.

14. Raise the other gable rafter and nail the bottom and then the top.

15. Move the rafter which was tacked to one side of the end of the first piece of ridgeboard and nail it in place over the joint in the ridgeboard.

16. Erect the rest of the rafters as described in step 9.

17. Fasten a brace to the top of the gable as in step 8.

Note: Check the rafters of both gables to see if they are plumb. If one gable is plumb and laid out wrong or the joint was not drawn together, correct this before proceeding.

How To Lay Out, Cut And Install
Gable Studding

1. Square a line across the plate directly below the center of the gable.

2. If a window is to be installed in the gable, measure one half of the opening size on each side of the center line and make a mark for the first stud (Figure 14-12).

3. Starting at this mark, lay out the stud spacing 16 inches on centers to the outside of the building.

4. Stand a 2 x 4 stud upright on the first mark. Place it against the side of the gable rafter and plumb it. See first stud, Figure 14-12 .

5. Mark across the edge of the stud at the underside of the rafter (enlarged view, Figure 14-12).

6. Stand a 2 x 4 on the second stud mark.

7. Plumb and mark it as described for the first stud.

8. Measure the difference in length between

170

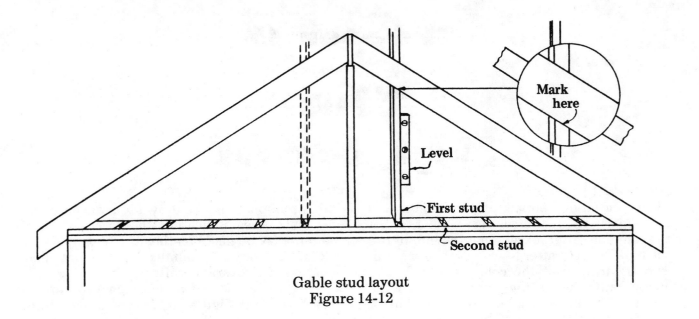

Gable stud layout
Figure 14-12

these two studs. Each remaining stud will be this distance shorter than the preceding one.

9. Mark all the remaining studs.

10. Cut two studs of each length, one for each side of the gable.

11. Nail the studs in position flush with the outside of the rafters and plate.

12. Frame in the window sills and headers as described in Chapter 9.

Chapter 15

The Hip Roof

The method of stepping off the lengths and cuts of hip or valley rafters is very similar to that of the common rafter, except that the unit of run of the hip or valley rafter is greater than the unit of run (12 inches) of the common rafter. The common rafter meets the ridgeboard and the plate at a 90 degree angle, whereas a hip or valley rafter meets these members at an angle other than 90 degrees. This accounts for the greater unit of run of the hip or valley rafter. This also makes it necessary to cut the common rafters where they meet the hip or valley.

A roof plan may be drawn to scale on paper, or chalk lines indicating the location of the rafters may be made on the attic floor. This will show the location, number and approximate lengths of the various rafters.

Figure 15-1 shows a plan view of a hip and valley roof and Figure 15-2 shows a perspective view of the same roof. The six hip rafters are shown at the ends of the main roof and at the end of the projection. The hip jack rafters extend from the hip down to the plate. The valley rafter is shown at the intersection of the main roof and

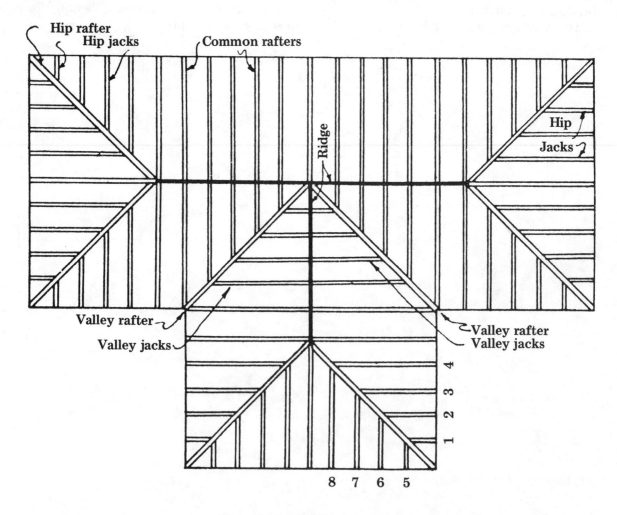

Plan of intersecting hip roofs
Figure 15-1

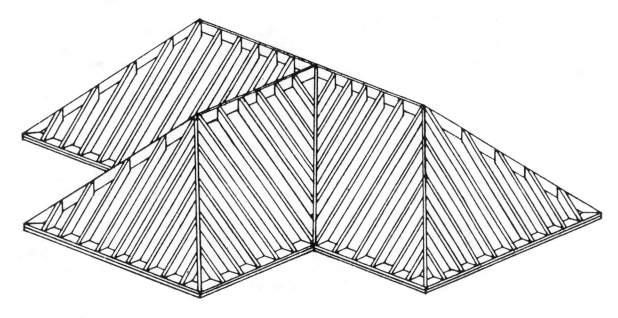

Intersecting hip roofs
Figure 15-2

the side roof. The valley jack rafters extend from this valley rafter up to the ridge.

In Figure 15-1 and 15-2, the pitch of the hip and that of the valley are exactly the same. The pitch of the common rafter of the main roof is the same as that of the common rafter of the projection. This type of roof is called a hip and valley roof of equal pitch.

The common rafter extends from the plate to the ridge with 12 inches used as the unit of run. Figure 15-3 shows this unit of run as two sides of the square ABCD. The hip rafter must also extend from the plate to the ridge, but at an angle of 45 degrees with the plate. See line BD. Therefore, the diagonal of the 12 inch square is used as the unit of run for the hip rafter. This is 16.97 inches or approximately 17 inches. The length of a hip or valley rafter for a roof of equal pitch is found by using 17 inches on the body of the square as the unit of run instead of 12 inches as for a common rafter.

If the rise per foot of run of the common rafter is 8 inches, the length of the hip may be found by following the same step-off method as in Figure 14-4, except that the fence should be set on the body of the square at 17 inches and on the tongue at 8 inches. The number of steps to be taken will be equal to half the span of the building in feet, the same as for a common rafter on the same roof.

Figure 15-3 shows the method of locating the intersection of the hip rafters with the ridge-board at B. Notice that the distance AB is equal to BC, both distances representing the run of the common rafters. The two squares are equal

in size. The distance DB is the diagonal of the square ABCD and represents the run of the hip rafter.

Layout Of A Hip Rafter

Assume that a hip rafter is to be laid out for the roof in Figure 15-3. The rise of the common rafter is 8 inches per foot of run. The rough length of the hip rafter may be found by

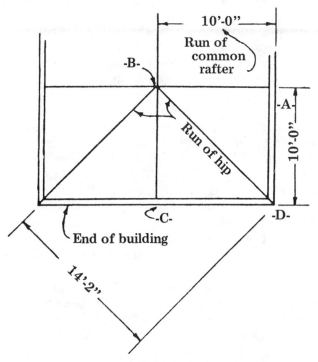

Layout of hip end of roof
Figure 15-3

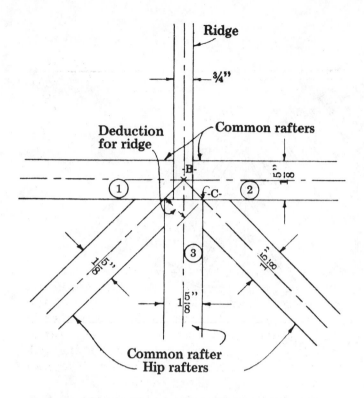

Intersection at ridge of hip and common rafters
Figure 15-4

following the method explained in Chapter 14. The measuring line is gauged the same distance from the top of the hip rafter as from the top of the common rafter.

The length of the hip rafter is found by setting the fence on the square at 17 inches on the body and 8 inches on the tongue and stepping along the rafter for 10 steps. The seat cut is found in the same way as for the common rafter, except that 17 inches is used on the body of the square and 8 inches on the tongue instead of 12 inches and 9 inches as in Chapter 14.

Ridge Cut For Hip Rafters

Figure 15-4 shows an enlarged plan view of the ridge at the intersection of the hip and common rafters. Notice that the common rafters 1 and 2 butt against two sides of the ridgeboard. Rafter 3 butts against the end of the ridge and acts as a 1-5/8 inch ridge between the two hip rafters. The hip rafters fit into the angles made by the three common rafters. The ridge cut at the top of each hip will, therefore, consist of several bevel cuts.

The allowance to be taken off the ridge cut at the top of the hip for the thickness of the ridgeboard is somewhat different from that of the common rafter. To more clearly show this allowance, center lines have been drawn on all the members of the roof at this point (Figure 15-4). The meeting point of these center lines is

located at B. The distance from point B to the intersection of the hip and common rafter, point C, changes as the slope of the roof is changed and as the thickness of the ridge is changed. In this case, where a ¾ inch ridge is used on a ⅓ pitch roof, the distance BC is approximately 1¼ inch. If, however, a ridgeboard 2 inches thick were used for a 1/6 pitch roof, the distance BC would be 1-3/8 inch. If a 2 inch ridge were used for a ½ pitch roof, the distance BC would be 1¾ inch. Therefore, in ordinary pitch roofs 1½ inch may be used as the average distance BC. The distance BC is marked back from the top end of the hip rafter the same way the distance was taken off the common rafter (Figure 14-6) except that the figure on the tongue of the square will be 8 and on the body 17. The mark is then transferred from the side of the hip rafter to the center line on the top edge as at Figure 15-4, C. Another method of construction at the intersection of the hip rafters and ridgeboards is shown in Figure 15-5. This shows a single cheek cut on the hip rafter and a ¾ inch ridgeboard. The deduction for the ridgeboard is also shown.

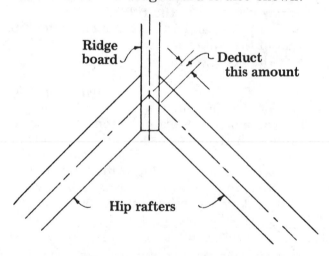

Shortening of hip rafters
Figure 15-5

Before marking out the cheek cuts, it is necessary to find the length of the hip per foot of run. This may be found by taking the rise of the roof per foot of run (8 inches in this case) on the tongue of the square, and the unit of run of the hip (always 17 inches) on the body of the square and then measuring the diagonal (Figure 15-6). In this case it is approximately 18¾ inches.

To obtain the angle of the double cheek cut (Figure 15-4), lay the square on the top edge of the hip rafter (Figure 15-7). The figure on the body representing the run of the hip (17 inches) should lie on the center line, and the figure on the tongue representing the length of the hip per foot of run (18¾ inches in this case) should lie on point C, Figures 15-4 and 15-7. Since the tongue of the square is only 16 inches long, it will be necessary to use figures which are just half of those just given. Therefore, 8½ would be used on the body and 9-3/8 on the tongue. The line CD, Figure 15-7 should then be marked along the outside edge of the tongue of the square. A line should be squared across the edge of the rafter from D to E. A line connecting E and C should then be drawn.

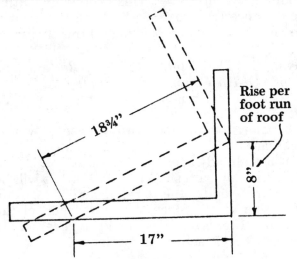

Length of hip per foot of run
Figure 15-6

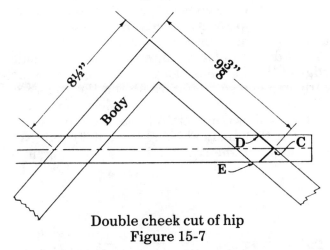

Double cheek cut of hip
Figure 15-7

The plumb cut is marked from D and E down both sides of the rafter. In this case, this cut may be obtained by taking 8 inches on the tongue of the square and 17 inches on the body and marking along the outside of the tongue (Figure 15-8).

The tail cuts for the cornice at the bottom of a hip or valley rafter are found in a similar manner to those of the common rafter. However, the figures 8 and 17 on the square should be used as shown in Figure 15-8.

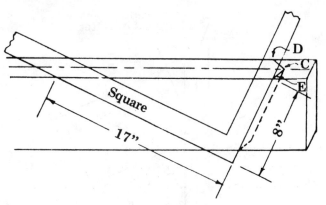

Plumb cut of hip
Figure 15-8

Backing Off The Hip Rafter

Since the top edges of a hip rafter are at the intersection of the side and end slopes of the roof, they project above these slopes as shown in Figure 15-9. The edges must be beveled off to provide an even plane from the top of the jack rafters to the center line of the hip. The roof boards will then lie flat at the hip.

The bevel cuts shown by dotted lines (Figure 15-9) on each side of the top of the hip rafter are rather short and it is difficult to lay them out with the steel square. However, a pattern may be laid out to show the cuts. A center line is gauged along a piece of stock about ¾ inch thick, 1-5/8 inch wide and 30 inches long. The length of the hip per foot of run (18¾ inches in this case) is determined as explained under the bevel or cheek cuts. The square is then placed

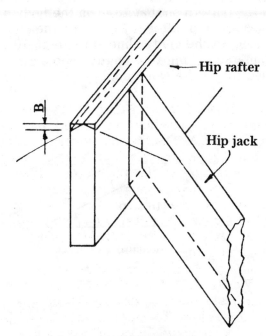

Backing off hip rafter
Figure 15-9

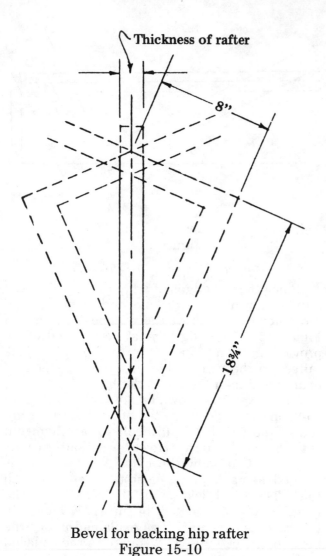

Thickness of rafter

8"

18¾"

Bevel for backing hip rafter
Figure 15-10

side of the center line of the pattern. The pattern is cut along the bevel lines and used as a template to mark the hip section shown in Figure 15-9. A center line is then gauged along the entire length of the top of the hip. Lines are also gauged along the sides of the hip. These marks show the portion of the rafter to be backed off and also mark the location of the tops of the jack rafters.

Some carpenters prefer the method of dropping the hip rafter. This is done by cutting the distance (Figure 15-9, B) off the seat cut of the hip, thus avoiding backing off the hip rafter.

Laying Out The Length Of The Hip Rafter If The Span Of The Common Rafter Is In Odd Feet

Assume that it is necessary to lay out a hip rafter for a building 21 feet wide. The rafter has an 8 inch rise to the foot. In this case, it will be necessary to take 10 and a fraction steps.

It has already been pointed out that the unit of run of a hip rafter is 17. Therefore, it will be necessary to add to the 10 regular steps a certain proportion of 17. Since the extra span of the common rafter is 6 inches, it will be necessary to add 6/12 of 17 inches, or 8½ inches.

This additional length is added to the hip rafter by taking an 11th step in the same manner as the 10th step was taken. A line is marked along the outside edge of the body of the square. The half step length of 8½ is then measured along this line from the point Figure 15-11, A. This is done by placing the square so the outside edge of the body lies along the line made by the 11th step and so the figure 8½ on the square is directly over the point A. This location of the square is shown by dotted lines in Figure 15-11. A line is then marked along the outside edge of the tongue of the square. This gives the plumb

across the pattern so that 18¾ on the body and the rise of the hip rafter (8 inches) on the tongue both come on the center line (Figure 15-10). A line is marked along the tongue. The square is reversed to get the same angle on the opposite

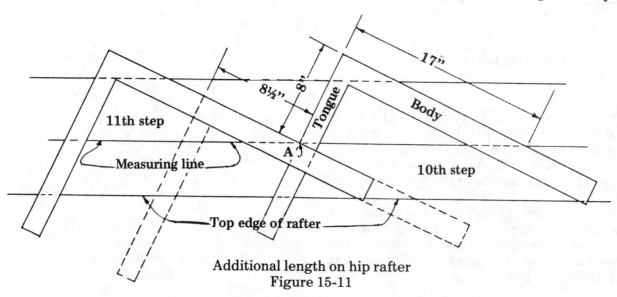

11th step

Measuring line

8½" 8"

Tongue

17"

Body

A

10th step

Top edge of rafter

Additional length on hip rafter
Figure 15-11

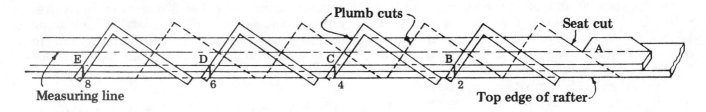

Difference in lengths of jacks
Figure 15-12

or ridge cut of the hip rafter. The deduction for the thickness of the ridge is the same as previously described.

Hip Jack Rafters

Hip jack rafters are those rafters that extend from the plate to the hip and fill in the triangular spaces between the common rafters and the hip rafters. See Figures 15-1 and 15-2. They fit against the plate in the same manner as the common rafters. The plumb cut of a hip jack rafter is the same as the plumb cut of a common rafter except that the cut must be beveled on the hip rafter. This additional angle must be cut because the jack rafter joins the hip rafter at an angle, whereas the common rafter joins the ridge with a square cut. See Figure 15-2. This angle cut on the hip jack is called the cheek or face cut.

The method of finding jack rafter lengths by stepping off a common rafter as shown in Figure 15-12 is similar to that of Figure 14-4. The common rafter should first be laid out, cut and used as a pattern. The length of the first jack rafter is equal to the distance from point A at the seat cut along the measuring line of the common rafter to the second step taken with the square as shown by the mark B, Figure 15-12. The length of the second jack is equal to the distance from point A to the fourth step taken with the square or point C. The length of the third jack is equal to the distance from point A to the sixth step taken with the square or point D. The length of the fourth jack is equal to the distance from point A to the eighth step or point E. The length of any additional jack rafters is found in the same manner. There will be several jack rafters of the same size in any one roof. This method gives the lengths of jacks that are spaced 24 inches on centers which is the most common spacing of rafters for a roof of this type.

The common rafter that is to be used for the pattern should be marked across the top edge at the points where the plumb cuts meet the top edge. See steps 2, 4, 6, and 8, Figure 15-12. The first or shortest rafter is laid out by placing the common rafter pattern on the jack rafter, holding the top edges of both rafters flush. The tail cut and seat cut on the jack rafter should then be marked according to the pattern. The point marked 2 on the pattern should be squared across to the top edge of the jack rafter.

Figure 15-13 shows the top end of a hip jack rafter. The mark 2 corresponds to mark 2 in Figure 15-12. The cheek cut is obtained by taking the unit of run of the common rafter (12 inches) on the body of the square, and the length of the common rafter per foot of run, (14.42 inches or 14-3/8 inches) on the tongue (Figure 15-13). A line is marked on the outside edge of the tongue as shown. This line indicates the cheek cut. It is important to remember that the heel or short point of the bevel is toward the tail of the rafter.

The plumb cut is marked across each face of the rafter from each end of the line just drawn

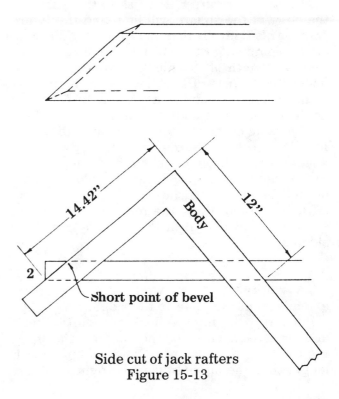

Side cut of jack rafters
Figure 15-13

for the cheek cut. This plumb cut is marked in the same manner as that of the common rafter, using the figures 8 and 12 on the square as in Figure 15-12. This layout is for the shortest jack on the left side of the hip rafter as shown at 5, Figure 15-1. A jack for the right side of the hip rafter (1, Figure 15-1) must also be laid out. This jack is measured and laid out the same as the left jack except that the bevel or cheek cut goes the opposite way, as shown in Figure 15-14. The remaining right side jacks (2, 3 and 4, Figure 15-1) are laid out as described. A corresponding left jack (6, 7 and 8, Figure 15-1) must be provided for each right jack.

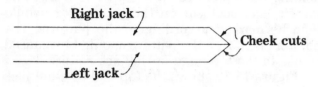

Right and left jack rafters
Figure 15-14

Using The Rafter Table To Check The Lengths And Cuts Of Hip Rafters

Finding the length of a hip rafter by means of the rafter table is similar to finding the length of a common rafter by this method. See Chapter 14. A Stanley Square Number R-100 is used in this case. The row of figures on the square marked *Length of Hip* or *Valley Rafter per Foot of Run* should be used (Figure 14-8).

Under the figure 8 (the rise in this case) on the edge of the square will be found the figure 18.76. This figure is then multiplied by the span of the common rafter, 10 feet for example, to find the length of the hip along the measuring line. 10 x 18.76 = 187.60 inches or 15 feet 7½ inches. The ridge and seat cut may be found by holding the square as in stepping off the rafter.

The side or cheek cuts for the hip rafters are found on the sixth row on the body of the square. Under 8 will be found the figure 10-7/8. This means that 10-7/8 inches should be taken on the body and 12 inches on the tongue. A line is then marked along the tongue in a manner similar to that shown in Figure 15-7 to get the side cut.

Valley Rafters

A valley rafter as well as a hip rafter, extends from the plate to the ridge. The valley rafter forms the intersection between the main and projection parts of the roof as shown in Figure 15-1. The valley rafter is about 1 inch longer than the hip because it is framed against

the ¾ inch ridgeboard whereas the hip is framed against the 1-5/8 inch common rafter. For hip rafters an allowance of 1½ inch is deducted from the stepped off length, and for valleys only ½ inch is deducted for the thickness of the ¾ inch ridgeboard.

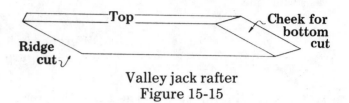

Valley jack rafter
Figure 15-15

Valley Jack Rafters

The valley jacks extend from the ridgeboard down to the valley rafter (Figure 15-1). Their lengths are found in the same manner as those of the hip jacks. The ridge cut of the common rafter is used for the ridge cut of the valley jack and the cheek and plumb cut of the hip jack is used for the bottom cut of the valley jack. See Figure 15-15. These jacks are also laid out in pairs.

How To Lay Out And Cut The Rafters

Note: Assume that a building is to be 20 feet wide and 40 feet long as shown in Figure 15-17 of this chapter.

1. Determine the proper size and number of common, valley and jack rafters.

2. Select straight, sound stock for all the roof members.

3. Lay out the common rafters as described in Chapter 14.

4. Cut the common rafters as explained in Chapter 14.

5. Lay out the hip rafters as explained in this chapter.

6. Cut the hip rafters.

7. Determine the amount to be backed off the hip rafters as described in this chapter.

8. Back off these rafters to the proper shape with a plane.

Note: The hip rafters may be erected before backing them off. The lines on the sides of these rafters may then be used as guides for the location of the tops of the jack rafters. When the jacks are in place, the hip rafters may be backed off with a hand axe.

9. Lay out the jack rafters in pairs as described earlier in this chapter.

10. Cut the jack rafters.

Note: A convenient way to cut the jacks in pairs is to make the cheek cut in the center of the length of the stock and the tail cuts on

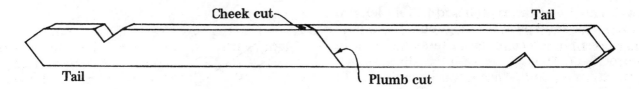

Layout of jack rafters in pairs
Figure 15-16

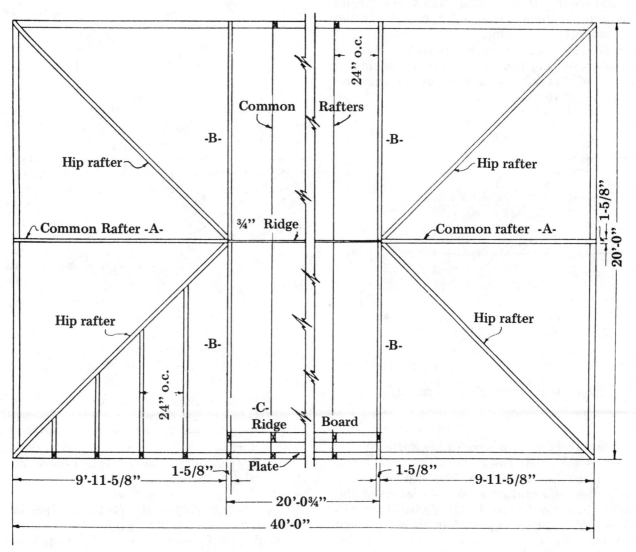

Location of rafters of hip roof
Figure 15-17

the ends (Figure 15-16). If a portable electric saw is to be used, it is advisable to set the saw to the plumb and cheek cut first and to make all of these cuts before the tail cuts.

How To Make The Ridgeboard

1. Measure from the end plate along the side plate toward the center of the building 9 feet 11-5/8 inches on each corner (Figure 15-17).

Note: This distance is equal to the actual run of the common rafter (Figure 15-17, *B*) after one half the thickness of the ridgeboard has been deducted. This point also represents the outside face of the common rafters (B).

2. Mark the side plates at these locations.

3. Along the side plate, measure the distance between the outside faces of the two rafters BB. This distance in Figure 15-17 is 20

179

feet ¾ inch. Cut the ridgeboard to this length.

4. Mark and space the ridgeboard and both side plates for the common rafters as shown in Figure 15-17.

How To Place The Common Rafters

1. Erect and brace the common rafters as explained in Chapter 14.

2. Erect the common rafters (Figure 15-17, A) at the ends of the building. Nail these rafters to the plate at the seat cut and midway between the sides of the building.

3. Nail the tops of these rafters so they butt against the end of the ridge. The center line of the edge of this rafter should be in line with the center line of the ridgeboard.

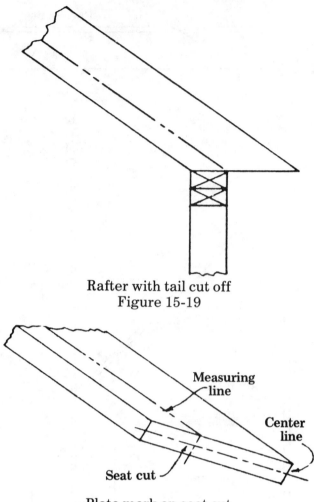

Rafter with tail cut off
Figure 15-19

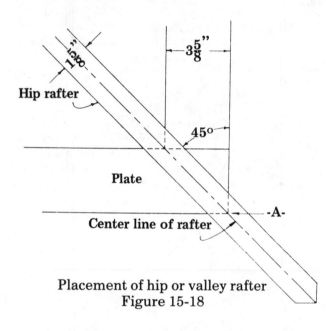

Placement of hip or valley rafter
Figure 15-18

How To Place Hip And Valley Rafters At The Plate

1. Mark the location of the hip or valley rafter at the internal or external corner of the plate as shown in Figure 15-18. These lines will cross the plate at an angle of 45 degrees in all equal pitch roofs.

Note: These lines mark the location of the hip or valley rafters so the center line of the rafter will meet the outside or inside corner of the plate. See Figure 15-18, A.

2. Push the hip rafter against the outside corner of the plate so the heel cut rests tightly against the plate and so the sides of the rafter are in line with the 45 degree lines made across the plate.

3. Toenail the rafter to the plate.

Note: If the seat cut of the hip or valley rafter is cut through to the top edge of the rafter as in Figure 15-19, a center line should be drawn along the seat cut, and the

Plate mark on seat cut
Figure 15-20

measuring line should be squared across this cut as shown in Figure 15-20. The point where the center line and the measuring line meet marks the point that is to be placed at the outside corner of the plate. See Figure 15-18, A.

How To Place The Hip Or Valley Rafters At The Ridge

Nail the hip rafter into the angle formed by the two common rafters (A and B, Figure 15-21), keeping the center line on the top edge of the hip rafter even with the top of the ridgeboard. The other three hip rafters of the roof in Figure 15-17 should be placed in the same manner.

Note: Valley rafters are placed so that the two outside top edges are even with the top of the two ridgeboards.

How To Place The Hip Or Valley Jack Rafters

1. Mark the spacing for the hip jack rafters on the plate at both sides of the four corners of the building. See the lower left corner of Figure 15-17.

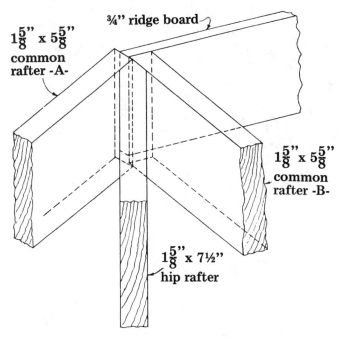

$1\frac{5}{8}$" x $5\frac{5}{8}$" common rafter -A-

¾" ridge board

$1\frac{5}{8}$" x $5\frac{5}{8}$" common rafter -B-

$1\frac{5}{8}$" x $7\frac{1}{2}$" hip rafter

Location of common and hip rafters at ridge
Figure 15-21

2. Check the hip rafter to see if it is straight. If not, fasten one end of a stay to the bottom edge of the hip and the other end to the floor to hold the hip straight.

3. Temporarily nail the hip jack rafters to the plate marks at the seat cut.

Caution: When fastening the jacks in place, do not drive the nails home. Fasten the jacks in pairs, first one on one side of the hip and then one on the other.

4. Temporarily nail the cheek cuts to the hip rafter.

5. After the jacks have been placed and the hip rafter is straight, drive the nails home to fasten the rafters securely to the plate and hip.

How To Erect An Intersecting Gable Roof

Note: Figure 15-22 shows a floor plan of an intersecting roof with the plates marked for the common rafter locations.

1. Find the length of the main ridgeboard by measuring from the outside of the plate at one end of the building to the outside of the plate at the other end.

2. Cut the ridgeboard and space the plate and ridgeboard for the common rafters as explained in Chapter 14.

3. Find the length of the ridgeboard for the projection of the building by measuring from the outside of the plate line of the projection to the center line of the main building as shown at P, Figure 15-22. Deduct one half the thickness of

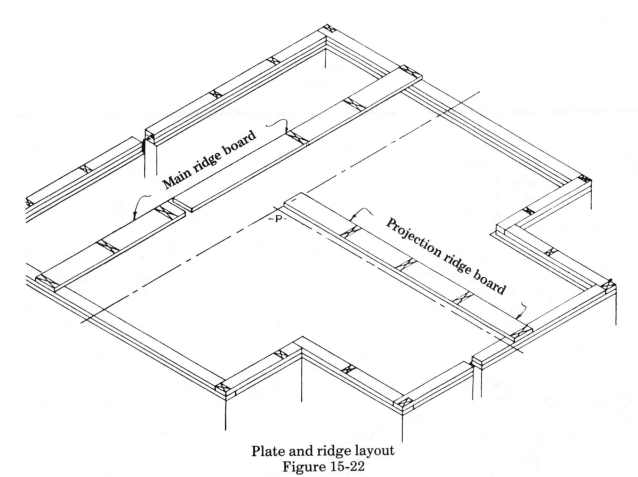

Plate and ridge layout
Figure 15-22

181

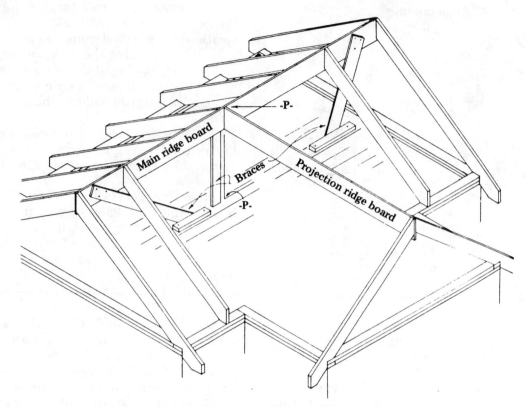

Assembly of common rafters
Figure 15-23

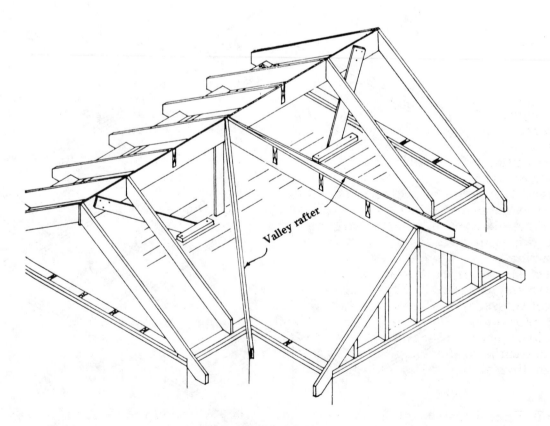

Valley rafters in place
Figure 15-24

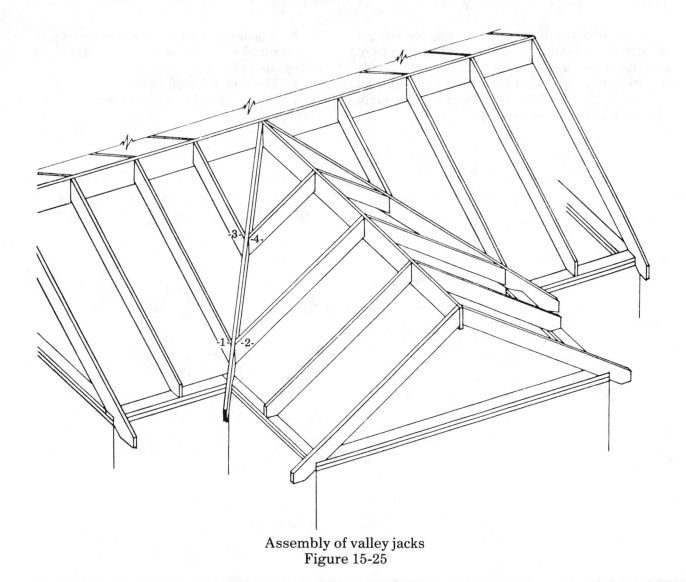

Assembly of valley jacks
Figure 15-25

the main ridgeboard from this length.

4. Cut this ridgeboard to length.

5. Mark the spacing for the common and jack rafters on this ridgeboard, and for the common rafters on the plate.

6. Erect the common rafters and the ridgeboard as explained in Chapter 14 and as shown in Figure 15-24 of this chapter. Be sure to brace the ridges to hold them in a level and plumb position.

7. Erect the valley rafters as explained in this chapter and as shown in Figure 15-24.

8. Erect the valley jack rafters as explained in this chapter and as shown in Figure 15-25. They should be placed in such a position that the top edge of each jack, if continued, would meet the center line of the valley rafter (Figure 15-26).

How To Erect An Intersecting Hip Roof

Note: Figure 15-1, shows a roof plan with two valley rafters at the intersection of the main roof and the projection on the side of the building.

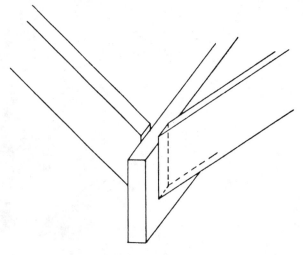

Section of valley rafter and placement of valley jacks on valley
Figure 15-26

1. To find the length of the ridgeboard for the offset of this roof, measure the distance from the outside of the projection to the corner where the projection joins the main roof and subtract half the thickness of the ridgeboard of the main roof (3/8 inch in this case).

2. Space this ridge for rafters 24 inches on centers and erect the common rafters and hip rafters as previously described.

3. Place and fasten the valley rafters.

4. Erect the valley jack rafters as previously described.

Chapter 16

Dormers And Roof Openings

A dormer is a framed structure projecting from a roof surface (Figure 16-1). It might be considered a minor roof in comparison with the major roof span. It may contain a window to admit light and to permit ventilation. In some cases, it may be constructed to improve the exterior appearance of the building and to provide additional space in the interior. It may be built up from the level of the main roof plate or from a point above the plate. The front wall of the dormer may be placed back of the main building line. It may project beyond it or it may be flush with it.

In many cases the dormer roof surfaces are of the same general shape as those of the main roof. The dormer does not necessarily have to carry out the main roof lines unless it is built for appearance only. Dormers constructed for the purpose of providing additional space within the building are generally of the style that will provide the most headroom. In cases where the style of dormer does not provide sufficient headroom, the plate line of the dormer may be raised above that of the main roof so that additional room is provided.

There are many styles and shapes of roof dormers. Space will not permit a description of all the types. However, the basic type, which has the plate line raised above that of the main roof, and whose span is in proportion to the rise and run of the main common rafter will be explained in detail. A study of this type should give you a general understanding of dormer layout.

The wall sections of a dormer are laid out, assembled and erected in the same general way as those of the main building. The rafters of the shed, gable or hip dormer are similar to those of the corresponding type of main roof. The method of framing dormers into the main roof depends on the type and size and is an important consideration in the erection of dormers.

The Shed Dormer

Figure 16-1 shows the outside appearance of a shed dormer and Figure 16-2 shows the framing of such a dormer. If wood shingles are to be used on this type, the rafters must have a rise of at least 6 inches to the foot to permit the roof surface to properly shed rain and snow.

The rafters are laid out and spaced the same as common rafters of the main roof, using the rise from the top of the dormer plate to the top of the header. The run of the rafter is taken from the outside of the dormer plate to the front face of the header.

If the dormer extends up to the ridge, the rise is figured from the top of the dormer plate to the top of the ridge. The run is figured from the outside of the dormer plate to the center of the ridgeboard.

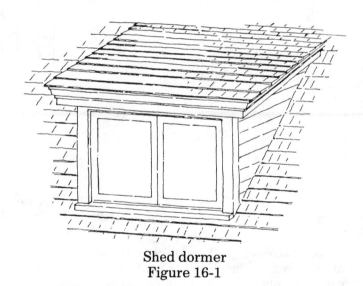

Shed dormer
Figure 16-1

How To Erect A Shed Dormer

Note: It is assumed that the common rafters of the main roof have been placed with an opening left for the dormer.

1. Check the position of the common rafters at the sides of the dormer. See A and B, Figure 16-2.

2. Build the front wall section of the dormer. The studs, plates, and the window opening are framed and erected as in an outside wall of the main building.

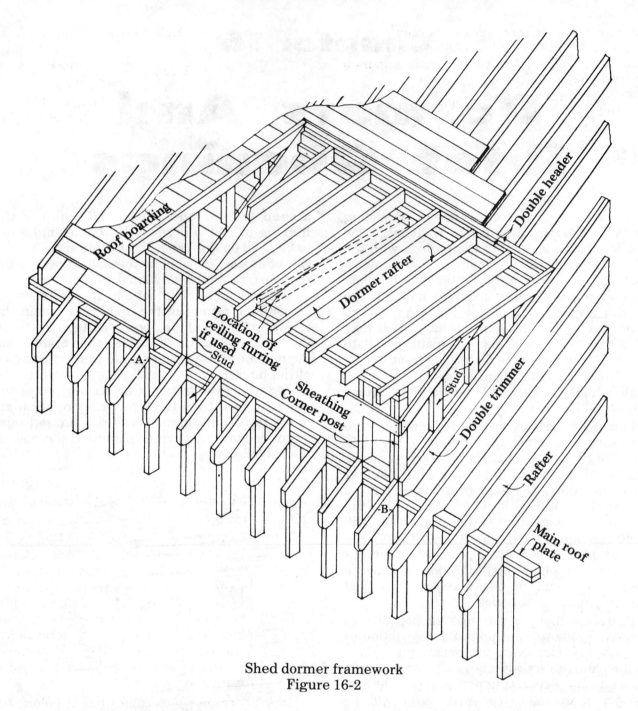

Shed dormer framework
Figure 16-2

3. Raise the wall and toenail the studs to the main plate. Brace and plumb the wall both ways. Apply the sheathing to the front wall.

4. Install the header for the dormer rafters at the point where they will meet the main roof.

Note: Sometimes the header is omitted and the dormer rafters are extended to the ridgeboard of the main roof. In such a case, the ridgeboard should be reinforced to carry the load of the shed roof rafters.

5. Lay out and cut the required number of dormer rafters.

6. Space and spike the rafters in place.

7. Double the header in the same manner as the header is doubled in floor joists.

8. Double the main common rafters on each side of the dormer.

Note: If the side wall studs of the dormer do not rest on the common rafters but are supported from the joists below, it is not necessary to double the rafters.

9. Cut the tail rafters needed to carry the cornice across the lower front of the dormer.

10. Space and spike the rafter tails securely to the front wall of the dormer. Use a straightedge to keep the rafter tails in line with main common rafters.

11. Lay out the spacing for the side studs 16 inches on centers.

12. Hold a piece of 2 x 4 vertically at a

186

spacing mark. Mark the top and bottom cuts on the 2 x 4, using the under side of the dormer rafter and the top side of the main common rafter as guides for the pencil.

13. Repeat this operation for the other side wall studs.

14. Cut the studs and nail them in place.

15. Apply the sheathing to the sides of the dormer.

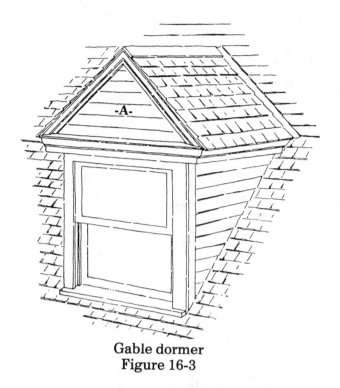

Gable dormer
Figure 16-3

The Gable Dormer

The gable dormer (Figure 16-3) sheds the rain and snow two ways. This type is often used in preference to a shed dormer when it is necessary to provide more pitch for the dormer, or to carry out the general lines of the main roof. A hip end could be used instead of the gable shown at Figure 16-3, *A*. The ridge of this dormer is lower than that of the main roof. In this case a header is installed between the rafters of the main roof at the point where the ridgeboard and the valley rafters of the dormer meet the main roof (Figure 16-5, *C*). The rafters are laid out in the same manner as the common, valley, and valley jack rafters of the main roof except that the rise and span of the dormer is used. To avoid having a different pitch on the dormer roof than on the main roof, the same rise and run per foot should be used on the dormer common rafter as is used on the main common rafter. If the dormer plate is to be raised above the main plate, it should be placed in accordance with the steps marked on the pattern of the main common rafter (Figure 16-4).

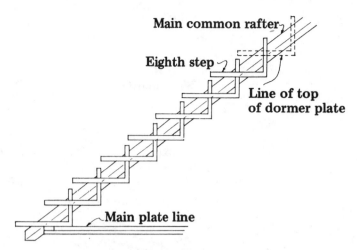

Height of dormer plate
Figure 16-4

Assume that the run of the main roof common rafter is 14 feet and the rise per foot of run is 9 inches. To lay out the common rafter, 14 steps of 9 inches on 12 inches would be taken (Figure 14-4). If the plates of the dormer are to be 6 feet or 72 inches above the main plate, the top of the dormer plate will come at the end of the 8th step taken with the square along the main common rafter (72 inches ÷ 9 inches = 8 steps). The line showing the level cut of the 9th step will show the top of the double plate of the dormer. See Figure 16-4.

If the dormer ridge is level with the main ridge, the length of the common rafter of the dormer can be found by subtracting these 8 steps from the total of 14 steps on the main common rafter. This will leave 6 steps of 9 inches on 12 inches. Thus the run of the dormer common rafters will be 6 feet and the total width of the dormer will be 12 feet. The half thickness of the ridgeboard should be deducted the same as for the main common rafter.

In many cases, the dormer ridge is below the main ridge as shown in Figure 16-5. As in the previous example, 8 steps should be deducted from the length of the main common rafter if the dormer plate is to be 6 feet above the main plate. In addition, a number of steps corresponding to the difference in height between the two ridges should also be deducted. Assuming that this difference is 18 inches, and since the rise per foot of run is 9 inches, this additional deduction would be 2 steps. The run of the dormer common rafter would then be 4 feet and the width of the dormer between the rafters A and B, Figure 16-5 would be 8 feet.

After finding the length of the dormer common rafter, the lengths and cuts of the

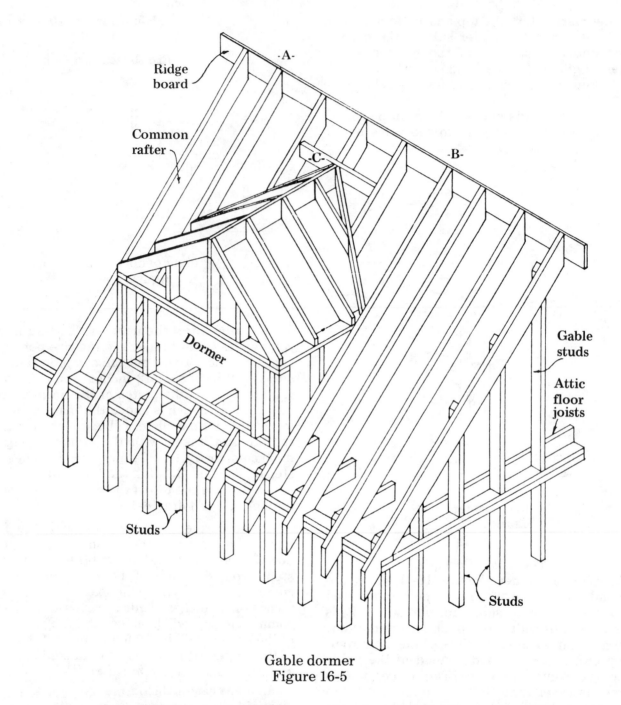

Gable dormer
Figure 16-5

remaining rafters of the dormer may be found in the manner explained for similar rafters of the main roof.

How To Erect A Gable Dormer Using A Header
Note: It is assumed that the common rafters of the main roof have been placed.

1. Check the width of the dormer to see that the common rafters, *A* and *B*, Figure 16-5 are properly placed.

2. Erect the front wall section the same as for the shed dormer.

3. Erect the side wall plates keeping them at the same height as the top of the front wall plate, and allowing them to extend to the sides of the main common rafters. Spike the plates to the sides of these common rafters.

4. Erect the side wall studs.

5. Erect the common rafters of the gable dormer in the same manner as the main common rafters. Allow the ridgeboard to extend beyond the face of the outside plate.

6. Plumb the outside gable rafters with the face of the outside wall section of the gable and brace them in this position.

7. Erect the studs from the top of the outside wall plate to the under side of the gable common rafters.

8. Level the top of the dormer ridgeboard.
Note: If the side wall plates are level and plumb

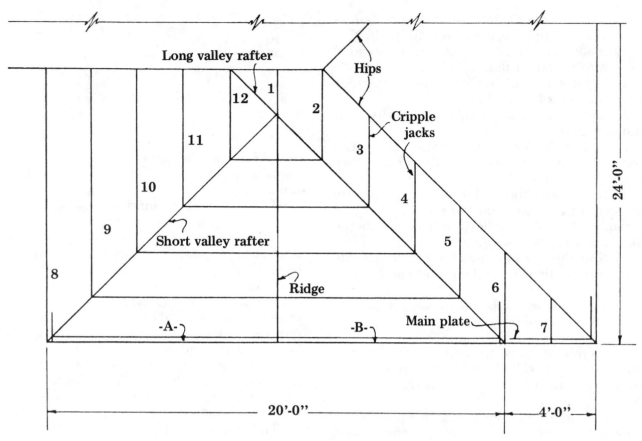

Long and short valley rafter
Figure 16-6

and the common rafters of the dormer have been properly erected, the ridgeboard should be level.

9. Install the valley rafters, following the same general procedure as in erecting a full length valley.

10. Cut the ridgeboard off at the inside end of the valley rafters. This point gives the location of the header (Figure 16-5, *C*).

11. Install the header, keeping the top edge of the header flush with the top of the main common rafters.

12. Install the valley jack rafters of the dormer and the main roof.

13. Install the short common rafter from the header to the main ridge.

14. Sheath the front and side walls of the dormer.

Framing A Gable With A Long And Short Valley Rafter

The upper end of a small gable or dormer may be framed as shown in Figure 16-5 while a large gable or dormer may be framed into the main roof by using a long and a short valley rafter as shown in Figure 16-6. This roof plan shows a main roof with a common rafter having a run of 12 feet and a rise of 9 inches per foot of run. The gable front wall is flush with the main building line. The run of the gable common rafter is 10 feet.

The long valley rafter is laid out in the same general way as a valley rafter of a main roof of 24 foot span (12 steps of 9 inches and 17 inches). A single cheek cut is used at the ridge as shown in Figure 16-7.

The short valley rafter is laid out in a similar manner except that only 10 steps are taken with the square because 10 steps were taken in laying out the gable common rafter.

The top plumb cut of the short valley rafter is found by using the rise and run of the valley rafter (9 on 17). Mark on 9 for the plumb cut. The cheek cut should be square as this rafter butts against the long valley rafter. Half the thickness of the long valley is deducted from the length of the short valley in the same way as half the thickness of the main ridgeboard is deducted from the length of the long valley rafter.

To Erect A Gable Using A Long And Short Valley Rafter

Note: Figure 16-6 shows a plan of a gable with a long and short valley. The long and short valley rafters extend to the main plate of the building so there are no gable side walls. The front wall would be studded and sheathed the same as any gable.

1. Lay out and cut the long and short valley rafters as described earlier.

2. Lay out and cut the valley jack rafters for both the main and gable roofs.

3. Install the long valley rafter first. The procedure is the same as for the main roof valley rafter. See Chapter 15.

4. Install the short valley rafter making sure the seat cut is located on the plate so that the measuring line of the valley rafter intersects the outside edge of the plate.

5. Nail the dormer ridgeboard temporarily at the intersection of the long and short valley rafters.

6. Nail the two common rafters (*A* and *B*, Figure 16-6) to the main plate in the proper location.

7. Plumb up from the outside face of the main plate to the ridgeboard and mark it at this point. This locates the outside face of the common rafter at the ridge.

8. Cut the ridgeboard along this plumb line and nail the common rafters to the ridge.

9. Check the top of the ridgeboard from the outside gable end back to the point where it intersects the long and short valley rafters to see if it is level. When it is level nail the ridgeboard firmly to the valley rafters.

10. Install the remaining gable and main roof valley jack rafters as described in the previous chapters.

Layout Of Ridgeboard

The length of the ridgeboard is found by plumbing down from the intersection of the long and short valley rafters and measuring a level distance to the outside of the main wall plate. The level distance gives the length of the ridgeboard (Figure 16-6)

Lengths And Cuts Of Jack Rafters

Since rafter 1 in Figure 16-6 is parallel to the main roof common rafters, and since the jack rafters are spaced 24 inches on centers, the length of this rafter is equal to 2 steps on the common rafter of this roof. The top or ridge cut is the same as that of the common rafter. The bottom or valley cut is a combination of a plumb cut, which is the same as at the top of the rafter, and a cheek cut. This cheek cut may be found by measuring the diagonal of the rise and run of the common rafter on the square. The rise in this case is 9 inches and the run is 12 inches. The diagonal will be found to be 15 inches. This figure is then taken on the tongue of the square

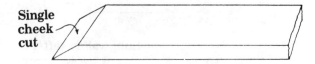

Cheek cut for long valley rafter
Figure 16-7

and 17 inches is taken on the body. A mark made along the outside of the tongue will show the cheek cut.

The cripple jacks marked 2, 3, 4, 5 and 6 in Figure 16-6 are laid out in the same way except that the combination plumb and cheek cut is used at both ends. The faces of these two cuts on each cripple jack must be parallel to each other. The length of these rafters is equal to 4 steps on the main common rafter. This length is used because the corner of the gable is 4 feet from the end of the building and the run of the jacks between the valley and hip is 4 feet.

The hip jack marked 7 is laid out as described in Chapter 15. The cheek cut is the same as for the other jacks and the seat and tail cuts are the same as the common rafter cuts.

The valley jacks (8, 9, 10 and 11, Figure 16-1) are laid out in the same manner as valley jacks of any main roof. The length of rafter Number 8 is 12 steps of the square, rafter Number 9 is 10 steps, rafter Number 10 is 8 steps and rafter Number 11 is 6 steps. Rafter 12 is 4 steps in length. The top and bottom cuts of this rafter consist of a plumb cut and a cheek cut similar to rafter Number 2 except that the cheek cuts converge toward one another.

How To Install Cripple Jack Rafters

1. Lay out and cut the cripple jack rafters as described earlier.

2. Space these rafters by lining them up with the valley jack rafters of the gable on the opposite side of the valley. See Figure 16-6.

Note: If the hip rafter has been dropped, place the tops of the cripple jacks even with the top of the hip rafter. If the hip has been backed, place the cripples against the hip rafter as shown in Figure 15-9, Chapter 15.

Openings In Roof Surfaces

Chimney holes, skylights and scuttles are framed in roof rafters in a similar manner to openings in floor joists. The headers between the rafters are placed so that their faces are plumb (Figure 16-8). The opening in the roof rafters is located by plumbing up from the face of the headers and trimmers of the opening in

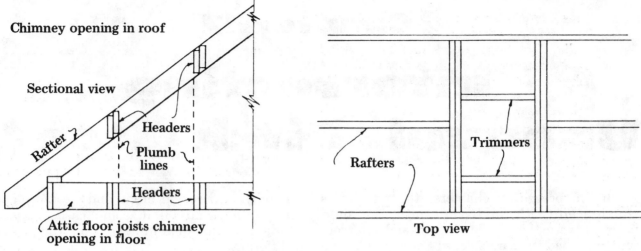

Roof opening location
Figure 16-8

the floor joists and by placing headers between the rafters as shown.

When a chimney hole, a scuttle or a skylight is to be located, a plan of the opening is often drawn full size on the attic floor. The points showing the inside dimensions of the opening are then plumbed up to the rafters or to the boards nailed temporarily to the roof.

How To Provide For A Chimney Opening

Note: Figure 16-8 shows the method of framing a chimney opening in the roof. It is assumed that the opening in the attic floor joists is framed to the size of the chimney.

1. Plumb a line from the corners of the opening in the attic floor joists to the rafters with a spirit level and straightedge or with a plumb bob.

2. Mark a plumb line on the rafters to be cut.

3. Nail a strip across the tops of the rafters to be cut and across at least two rafters on each side. This will support the rafters after they have been cut through.

4. Cut the rafters along the plumb line.

5. Spike the single headers to the rafters.

6. Spike the double headers in place against the single header. Note the arrangement at the top of the rafters in Figure 16-8.

7. Plumb up from the trimmers in the attic floor and mark their position on both the double headers in the roof opening.

8. Lay out and cut the trimmers. The cut on each end of a trimmer is the same as the ridge cut of the common rafter.

9. Spike the trimmers in place, keeping the top of the trimmer on a line with the tops of the common rafters.

Chapter 17

Intersecting Unequal Pitch Roofs

In the preceding chapters, the only roofs considered (with the exception of the shed dormer) were those whose main and intersecting parts had the same slope. For example, the gable dormer had the same rise per foot of run as the main roof.

The width and height of an intersecting roof of the hip or gable type is sometimes restricted. In this case, the rise per foot of run of the main common rafter may be different from that of the gable common rafter. The layout of these roofs presents more difficult problems than the roofs whose parts all have the same slope.

There are three ways in which two intersecting roofs can have unequal pitches: (1) when the ridges of the two parts of the roof are the same height and the spans of the two are different; (2) when the ridges are unequal in height (the rises different) and the two spans are the same; and (3) in some cases, when the ridges are unequal in height and the spans are different.

Layout Of A Roof Of Unequal Pitch

Assume that a building is to be 24 feet wide and 40 feet long with a roof projection on the side as shown in Figure 17-1. The run of the common rafter of the main roof will be 12 feet and that of the projection common rafter will be 6 feet. The rise of each of the common rafters is 8 feet, which brings the ridges of the intersecting roof and of the main roof even. The rafters are to have no overhanging tails for a cornice. It is often advisable to draw a scaled plan of the rafters showing their location in the roof. See Figure 17-1.

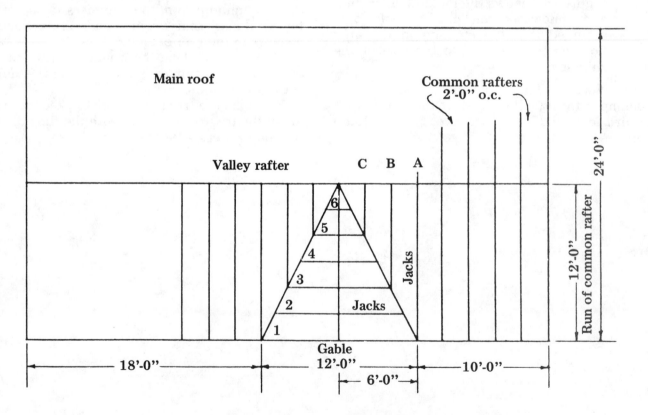

Plan of roofs of unequal pitch
Figure 17-1

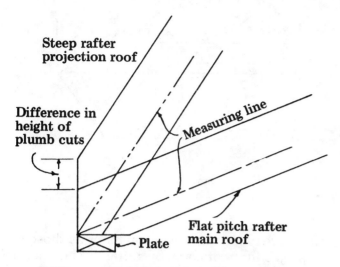

Difference in height of plumb cuts when
using measuring line
Figure 17-2

Layout Of Common Rafters Of
The Projecting Roof

The rise per foot of run of the common rafter is found by dividing the total rise of the rafter by the run of the rafter. (96 inches ÷ 6 feet = 16 inches rise per foot of run). Since 16 inches is too far out on the tongue of the square to permit convenient holding, the 16 inch rise is taken on the body of the square and the 12 inch run is taken on the tongue.

If the measuring line were placed below the top of the rafter as in the equal pitch roof, the tops of the plumb cuts of the main and the projection common rafters at the plate line would not be the same height. The top of the projection rafter would be higher than that of the main roof rafter. See Figure 17-2. Therefore, the measuring line for rafters in unequal pitch

roofs is at the top edge of the rafters if there are to be no rafter tails.

The length of the projection common rafter is found by stepping the square along the top edge of the rafter with 16 on the body and 12 on the tongue (Figure 17-3). Six steps should be taken. If the rise per foot of run comes out in a fraction such as 18⅔ inches, it can be measured accurately by using the back of the body of the square where the inches are divided into twelfths. The ridge or plumb cut is marked along the outside of the body and the seat or level cut along the outside of the tongue. The deduction for half the thickness of the ridge-board is made in the same manner as described in Chapter 14.

When the span of the main or projection roof is in odd feet, the method of finding the common rafter length is the same as described in Chapter 14.

Layout Of Jack Rafters For The
Projection Roof

Six jack rafters will be required on each side of the ridge of the projection as shown in Figure 17-1. Rafter 1 is considered a jack rafter since it has a cheek cut to permit it to fit against the valley rafter.

Figure 17-4 shows the common rafter pattern divided into six equal spaces so it can be used in laying out the jack rafters. For the purpose of finding the length of the rafter, the top edge of the rafter is used.

The longest jack, marked 1 in Figure 17-1, is the same length as the common rafter pattern. The ridge cut and seat cut are the same as those of the common rafter. A cheek cut will have to be made to allow this rafter to fit against the valley rafter (Figure 17-4). The plumb cut is the

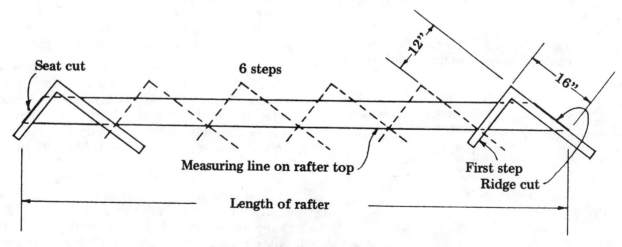

Layout of projection roof rafter
Figure 17-3

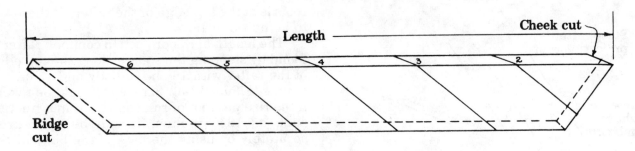

Length

Cheek cut

Ridge cut

Lengths of jack rafters
Figure 17-4

same as the plumb cut at the ridge.

The cheek cut is found by taking one half the span of the projection on the tongue of the square, and one half the span of the main roof on the body of the square. In this case, these figures are 6 for the projection and 12 for the main roof (Figure 17-5). The square is placed on the top edge of the rafter in this position and a line is marked along the outside edge of the tongue for the cheek cut.

The length of the second jack is found by measuring from the top of the ridge cut of the first jack along the edge of the rafter to the line 2 shown in Figure 17-4.

The length of the third jack is measured from the ridge cut to the line marked 3. The remaining jacks are measured in a like manner.

The deduction for half the thickness of the valley rafter at the valley end of the jack is found by gauging a center line along the top edge of the jack and squaring a line across this center line at the point where it meets the cheek cut. The distance from this line to the short or long tip of the cheek cut (see A or B, Figure 17-6) is the deduction to be made. This deduction is measured along the center line and back from the face or cheek cut. See the dotted line in Figure 17-6.

Layout Of Rafters Of Main Roof
The length and cuts of the main roof common

rafters are found in the same manner as those of the projection roof common rafters. The length of the main roof jacks is found in the same way as those of the projection roof. In Figure 17-1, the valley jack rafters needed for the main roof at one side of the projection are marked A, B, and C. The jack rafter A is the full length of the common rafter of the main roof. The lower end fits against the valley rafter in the same way as the corresponding jack of the projection.

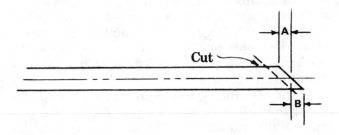

Cut

Deduction for thickness of valley rafter
Figure 17-6

The lengths of the other two main roof jacks are found by dividing the length of the main common rafter into three equal parts. This is because there are only three valley jacks required for this section of the roof instead of six as in the case of the projection roof jacks.

The same general procedure is followed in finding the cuts of the main roof valley jacks as for the projection roof valley jacks. The only difference is that the cheek cut for a main roof jack is marked along the body of the square instead of along the tongue (Figure 17-5).

Layout Of Valley Rafters
The length of a valley rafter of an unequal pitch roof is found in the same way as that of any valley rafter except that the unit of run for a valley rafter of an unequal pitch roof is figured in terms of the two unequal pitches of the roof. The unit of run for a valley rafter in a roof of

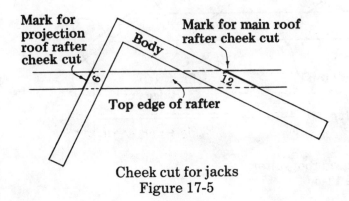

Mark for projection roof rafter cheek cut

Body

Mark for main roof rafter cheek cut

6

12

Top edge of rafter

Cheek cut for jacks
Figure 17-5

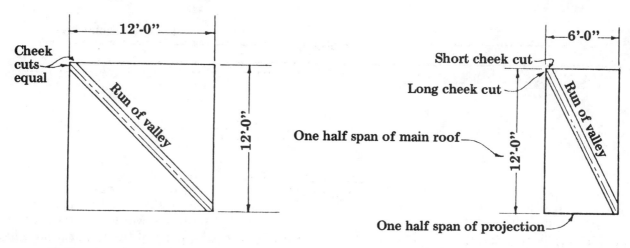

Valley for roofs of equal pitch
Figure 17-7

Valley for roofs of unequal pitch
Figure 17-8

equal pitch is 17 inches. The unit of run of the valley rafter for an unequal pitch roof is the diagonal of one half the span of the main roof and one half the span of the projection roof (Figure 17-8). In this case, one half the projection roof span is 6 feet and one half the main roof span is 12 feet. Therefore, the diagonal is measured from 6 on the tongue of the square to 12 on the body and will be found to be 13-7/16 inches. The square is stepped along the top edge of the rafter 12 times using a rise of 8 inches on the tongue of the square and 13-7/16, the unit of run, on the body. The ridge plumb cut of the valley is marked along the outside edge of the tongue and the seat cut is marked along the outside edge of the body.

The method of finding the cheek cuts at the ridge for the equal pitch valley rafter is described in Chapter 15. The cheek cuts on this valley rafter are of equal length because the run of the valley is the diagonal of the main roof and projection roof spans which are equal (Figure 17-7).

The two parts of a roof of unequal pitch have unequal spans. Therefore, the run of the valley rafter is the diagonal of a rectangle instead of a square. The cheek cut at the ridge on one side of the rafter will be different in length from the cheek cut on the other side (Figure 17-8).

One of the cheek cuts at the ridge end of the valley rafter is the same as the cheek cut of the main roof valley jack. The other cheek cut is the same as the cheek cut of the projection roof valley jack. See Figure 17-5. In the roof illustrated in Figure 17-1, the long cheek cut will come against the projection roof ridge and the short cheek cut against the main roof ridge (Figure 17-8). The same cuts are used at the lower end of the valley rafter at the plate. However, the position of the two cuts at the plate end are reversed from the position of the two cuts at the ridge end (Figure 17-8).

These cuts at the two ends of the valley rafter can be more easily laid out by using the jack rafters as patterns than by using a square. The reason for this is that the intersection of the two cheek cuts on the valley rafter is not on the center line as one cut is longer than the other. The cheek cut of the main roof jack rafter is used as a pattern for marking the long cheek cut on the valley rafter. After the length of the valley rafter has been stepped off, this long cheek cut should be marked as shown in Figure 17-9. The cheek cut of the projection roof jack rafter is used for marking the short cheek cut on the valley rafter. Figure 17-10 shows this short cheek cut marked on both ends of the valley rafter.

The deduction from the length of this valley rafter for the thickness of the ridgeboard is

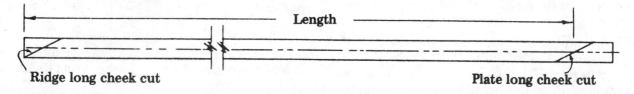

Long cheek cut of valley rafter
Figure 17-9

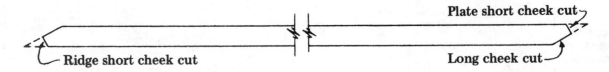

Plate short cheek cut

Ridge short cheek cut

Long cheek cut

Short cheek cut of valley rafter
Figure 17-10

shown in Figure 17-11. An amount equal to one half the thickness of each ridgeboard is taken off the corresponding cheek cut.

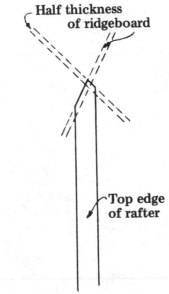

Half thickness
of ridgeboard

Top edge
of rafter

Shortening the valley for ridgeboard
Figure 17-11

Layout Of Rafters With Overhang For Cornice

When it is necessary to have a continuous cornice on both sections of the unequal pitch roof, the problem of laying out rafters with tails for the cornice is somewhat different from the layout of rafters without tails. This is because the rafters with no tails are laid out according to the span measured from the outside edge of the plates (A, Figure 17-12) whereas the rafters with tails are laid out according to the span measured from the outside vertical end of the rafter tail (A, Figure 17-13). The rafters with tails are laid out from this point to provide a level point common to both pitches of the roof.

For example, assume that the main roof common rafter has a pitch of 8 inches on 12 inches or ⅓ pitch. The projection common rafter has a pitch of 16 inches on 12 inches or ⅔ pitch. If the rafters have no tails, they could be laid out from the outer edge of the plate as shown in Figure 17-12. If they are to have tails, it would

be impossible to run a continuous cornice from the ⅓ pitch rafter to the ⅔ pitch rafter because the tops of the rafters would not be level with one another at the end of the rafter tails (Figure 17-12).

Added Run And Rise Of The Common Rafters

Figure 17-13 shows a section of the same roof shown in Figure 17-12, but provision is made for an overhanging cornice. If the cornice is to have a 12 inch run as shown in Figure 17-13, the rafter tail will project one step of the square beyond and below the plate line. It will be necessary to add one step of the square to the common rafter when it is stepped off. This is sometimes referred to as the added run of the rafter to provide for the run of the cornice.

The added rise of the main rafters will be 8 inches because both ridges are even. The added step to the main rafter will bring the rafter tail 8 inches below the main plate or seat cut (Figure 17-13).

The tail of the projection roof rafter, which has a rise of 16 inches per foot of run, will also be one step below the seat cut. The plate for this

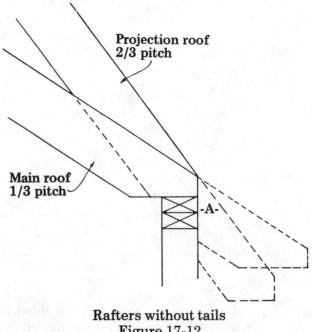

Projection roof
2/3 pitch

Main roof
1/3 pitch

-A-

Rafters without tails
Figure 17-12

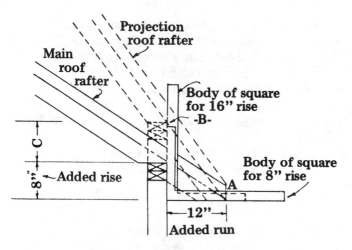

Added step of square on rafters
Figure 17-13

rafter should be raised 8 inches above the main roof plate so both rafter tails will be on the same level.

Layout Of The Main Roof Rafter

The measuring line of this rafter may be located in the same way as described for the common rafter of the equal pitch roof. The figures on the square will be 8 inches rise and 12 inches run. The method of finding the length, cuts and allowance for the ridgeboard is the same as described in Chapter 14.

Layout Of Projection Roof Rafter

The measuring line of the projection roof rafter is found by placing the square on the rafter, using a rise of 16 inches on the body and a run of 12 inches on the tongue, and marking a plumb cut at the end of the rafter stock (A, Figure 17-13). The distance from the top of the rafter down along this plumb line must be the same as the corresponding distance on the main roof rafter. This point establishes the measuring line of the projection roof rafter.

The remaining layout of this rafter is similar to the layout of the main roof rafter. The seat cut is marked out on the projection roof rafter at the first step of the square where the outside edge of the body of the square crosses the measuring line of the rafter (B, Figure 17-13).

The distance the plate of the projection is raised over that of the main roof is shown at C. This distance is the difference between the rise per foot of run of the two rafters, 8 inches in this case.

Layout Of Valley Rafter

The run of the valley rafter is also increased the length of the overhanging cornice, but this added distance is measured along the diagonal of the two unequal spans of the main and projection roofs.

A simple method of showing the location and diagonal length of this rafter is to draw a plan view of the roof, showing the plate, ridge and cornice lines at the valley intersections. The drawing may be made to a convenient scale, or chalk lines may be snapped on the attic subfloor of the building showing a full size plan of the rafters.

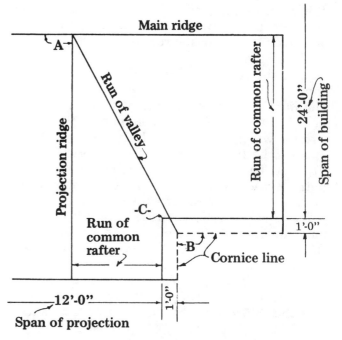

Run of valley rafter
Figure 17-14

Figure 17-14 shows this type of plan. Notice that the run of the valley rafter is measured along the diagonal line that extends from the intersection of the two ridgeboards (A) to the intersection of the two cornice lines (B). The valley rafter does not pass over the intersections of the two plate lines (C), because the run of the two common rafters is unequal but the run of the cornice is equal on both the projection and the main roofs.

The method of locating the measuring line on the valley rafter is the same as that described for the projection common rafter in this chapter, except that the rise and run of the valley rafter is used instead of the rise and run of the projection roof rafter. If a measuring line is used below the top edge of the projection and main common rafters, the measuring line on the valley rafter should be located in the same way.

The length of the valley rafter is found in the same manner as the length of the unequal valley

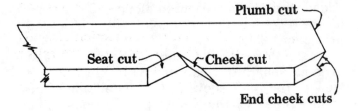

Valley rafter tail and seat cuts
Figure 17-15

rafter with no tail except that the added run on the two common rafters would form a rectangle from which the increased run of the valley rafter would be found.

Since the run of the projection roof rafter in Figure 17-1 is 6 feet, the run of this rafter including the overhanging tail of 1 foot will be 7 feet. The run of the main common rafter with overhang will be 13 feet. The length of the valley rafter will be the diagonal of a rectangle 7 feet by 13 feet. The remaining problem of finding the length of the rafter is similar to finding the length of the valley rafter with no tail.

The cheek cuts at the ridge are found in a similar manner to those of the rafter without a tail. The seat cut is found as described for the main common rafter with a tail. The plumb cut, which fits against the plate at the seat of the rafter, also has a cheek cut (Figure 17-15). When the valley rafter lies on the main roof side

of the corner of the plate line as it does in Figure 17-14, this cheek cut is parallel to the ridge cheek cut as the main ridgeboard. When the rafter lies on the projection roof side, the cheek cut is parallel to the projection ridge cheek cut. The cheek cuts at the end of the rafter tail are found in the same way as those at the ridge (Figure 17-15). However, the left cheek cut at the top of the valley rafter is used for the right cheek cut at the lower end and vice versa.

How To Frame Intersecting Roofs Of Unequal Pitch

The erection of a roof of unequal pitch differs slightly from that of the equal pitch roof because the rafters are laid out somewhat differently. The ridgeboards, plates, common rafters and valley jacks are laid out and spaced in the same general way as for any equal pitch roof. The cuts of the various rafters have been described in previous units pertaining to each particular type of rafter.

1. Lay out the ridgeboards and plates of the main and projection roofs in the same general way as described in Chapter 15.

2. Lay out and cut the valley jacks in the same manner as the common rafters of the gable roof.

3. Lay out and cut the valley jacks in the same manner as described in Chapter 15.

Note: Some mechanics prefer to use the sliding T bevel to transfer the bevel cuts from

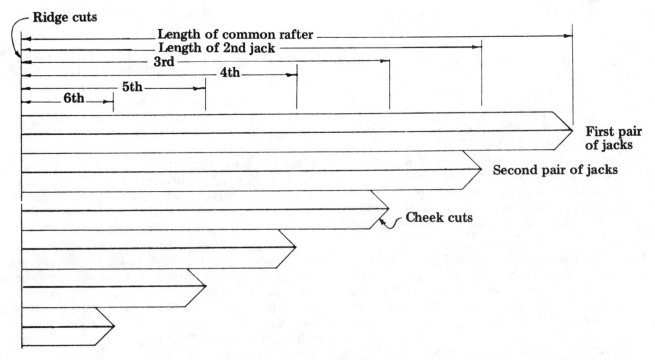

Layout of jack rafters
Figure 17-16

one rafter to another. This method makes for more accurate work.

Figure 17-16 shows a convenient way of arranging the rafters on edge in order to lay out the lengths, ridge cuts and valley cuts and to mark them in pairs for both sides of the projection roof shown in this chapter.

4. Lay out and cut the valley rafters as described in this chapter.

Note: The above procedure applies to the rafters with or without rafter tails.

5. Erect and brace the ridgeboards and common rafters of the main section of the roof as shown in Chapter 15.

Note: Figure 17-1 shows a roof plan where there is no need to raise the plate of the projection roof section, because the projection does not extend beyond the side wall of the building and there are no rafter tails. When the projection extends beyond the side wall and there are rafter tails, that section of the plate must be raised to form a seat support for the common rafters.

6. If necessary, raise the plate for the steeper pitched rafters by doubling up the plate. If the rise is over 6 inches, build up the plate. See Figure 17-13.

7. Erect and brace the ridgeboard and common rafters of the projection part of the roof.

8. Erect the valley rafters as described in Chapter 15.

9. Install the valley jack rafters as shown in Chapter 15.

Chapter 18
Special Roof Problems

There are some special types of roof construction which have not been described in the previous units on roof framing. These deviations from the regular type of roof are often used by the designer to carry out certain lines of the roof and to modify the exterior appearance of the house. You should have a general knowledge of their layout and construction. The procedures are similar to those described for the main roofs in previous units. Only the processes which differ from those of the main roof construction will be explained.

Odd Pitch Rafter

An odd pitch rafter is one where the total rise is not an even fractional part of the total span of the building. For example, a rafter may have a total rise of 12 feet 8 inches and the span of the building may be 26 feet. The cuts for the common rafter would be difficult to lay out by using the usual 12 inch unit of run as described in the previous units on roof framing. This is because the rise per foot of run will not come out in even inches and fractions of an inch.

The outside edge on the back of the square is divided into twelfths of an inch for the purpose of laying out rafter cuts of fractional rise rafters. When 1 inch on the square represents 1 foot of rise or run on the rafter, 1/12 inch will represent 1 inch of rise or run.

The method employed to lay out the rafter for the building in the above example is to use the total rise of 12 feet 8 inches or 12 and 8/12 of an inch on the tongue of the square and 13 inches on the body. The 13 inches represents the total run of the rafter. The square is stepped along the rafter 12 times using 12 8/12 on 13 in the same manner as for the even pitch common rafter.

When the figures of the total rise and run are greater than the length of the square, both figures may be divided by two. The diagonal must then be doubled to find the true length per foot of run of the rafter.

Assume that the total rise of the rafter is to be 20 feet 4 inches and the total run 14 feet. Dividing these figures by two and changing the

result to inches would give 10 2/12 inches on the tongue of the square and 7 inches on the body. The diagonal of these measurements (12 3/8) would be doubled to find the true line length of the rafter.

The square could also be reversed and 20 4/12 inches used on the body of the square and 14 inches on the tongue, the square being stepped along the rafter twelve times.

There should be no difference in spacing the ridgeboard and plates and in cutting and erecting the rafters from the similar operations described for an even pitch roof.

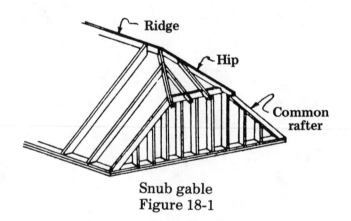

Snub gable
Figure 18-1

The Snub Gable

A snub gable is made up of two shortened common rafters running into a level plate. This plate is located at some point between the main plate and the ridge (Figures 18-1 and 18-2). The size of the snub gable is determined by the height of the gable plate above the main plate (Figure 18-3).

The height of the gable plate above the common rafter plate is found by using the step-off method on the common rafter. The number of steps is determined by the rise in feet of the common rafter.

Assume that a main roof is to have a span of 24 feet and a rise of 8 feet. The top of the gable plate is to be 6 feet above the common rafter plate. A common rafter would be laid out with 12 steps of 8 inches on 12 inches.

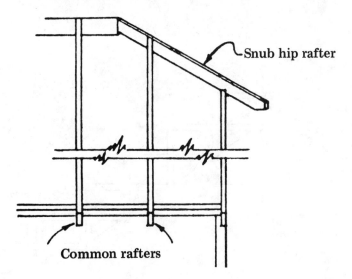

Snub hip rafter

Common rafters

Side elevation

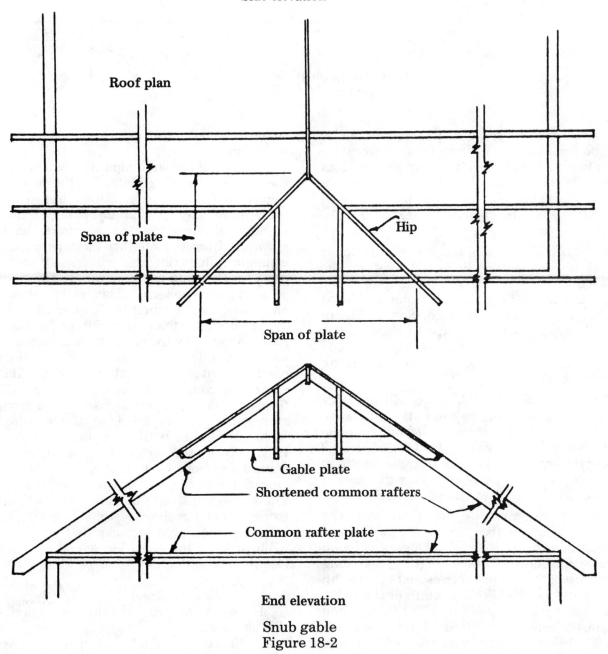

Roof plan

Span of plate →

Hip

Span of plate

Gable plate

Shortened common rafters

Common rafter plate

End elevation

Snub gable
Figure 18-2

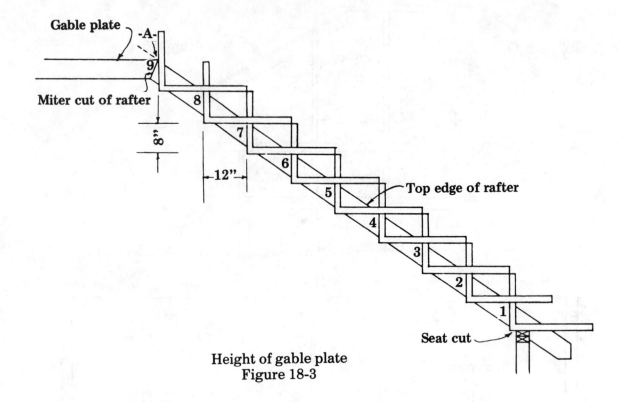

Height of gable plate
Figure 18-3

The number of steps of the square to be taken on the shortened common rafter from the main plate to the gable plate may be found as follows: 72 inches (total rise to top of gable plate) ÷ 8 inches (rise per foot of run) = 9. Therefore, 9 steps should be taken (Figure 18-3).

When the top edge of the common rafter is used as the measuring line of the rafter, the intersection of the 9th step on the tongue of the square and the top edge of the rafter will be at the top edge of the gable plate. See A, Figure 18-3. When the measuring line of the rafter is below the top edge, an allowance must be made for the vertical distance to the top of the common rafter.

The length of the gable plate will be twice the difference between the run of the shortened common rafter and the run of the full common rafter. Since in this case this difference is 3 steps of the square or 3 feet of run, the length of the gable plate will be 6 feet. This distance is shown in Figure 18-2 as span of plate.

A miter cut must be made on both the shortened common rafter and on the gable plate at the point at which they join. This cut is found by measuring down from A, Figure 18-4 along the outside of the tongue of the square at the 9th step, a distance equal to the width of the gable plate. This will locate point B. A line should then be squared from line AB through B to the lower edge of the rafter to locate point C. A line

connecting points A and C will then show the miter cut.

The run of the snub hip rafter will be one half the span of the gable plate. See the roof plan in Figure 18-2. The length of this rafter in steps of the square may be found by subtracting the number of steps on the shortened common rafter (9) from the number of steps on the full common rafter (12). This will leave 3 steps of 8 on 17 for the snub common rafter. The ridge cut is a single cheek cut. This rafter should be shortened as described in Chapter 15.

The length of the jack rafters for the snub gable is found by using the main common rafter as a pattern. In the plan in Figure 18-2, the main common rafter has 12 steps of 8 on 12. The shortened common rafter has 9 steps. Therefore, if a common rafter were used from the snub gable plate to the ridge, the length would be the difference between 9 and 12, or 3 steps. In this case, the jacks of the snub gable are spaced one foot each side of the ridge. This would be one step down from the ridge. Therefore, the length of these jacks would be 3 steps minus 1 step or 2 steps of 8 on 12. The length of the opposite jacks that run from the main plate to the snub hip would be one step less than the common rafter or 11 steps.

How To Erect A Snub Gable

1. Lay out and cut the rafters as described in this chapter.

2. Space the ridgeboard as usual, marking

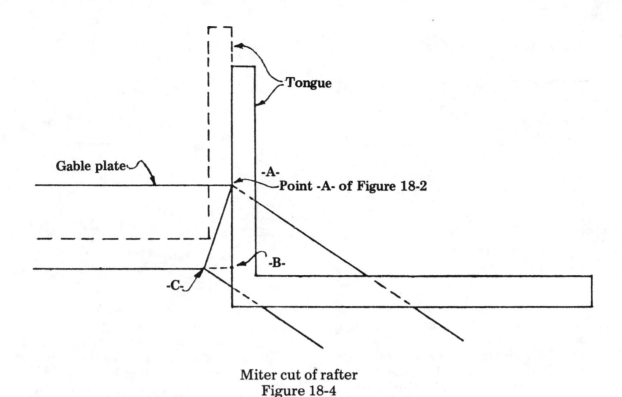

Miter cut of rafter
Figure 18-4

the location of the snub hip rafters the same distance from the end of the building as there are feet or inches of run in the snub hip rafter. See plan in Figure 18-2.

3. Lay out and cut the snub gable plate and shortened main common rafters (Figures 18-2, 18-3 and 18-4).

4. Nail the snub gable plate and the shortened common rafters together at the mitered joints.

Note: This process is best accomplished by nailing the plate and both rafters together on the subfloor, and raising the plate and rafters as a unit.

5. Nail the common rafters to the plates at the seat cuts. Plumb and brace the rafters to the floor as in Chapter 14.

6. Erect the snub gable hip rafters and jacks as described in Chapter 15.

7. Fill in the main roof jacks which run from the snub hip rafters down to the main roof plate.

Dormers Built On The Roof

Small dormers without windows and used as decorative features only, are often built on the surface of the roof after it has been sheathed as shown in Figure 18-5. Larger dormers with windows may also be built on the roof using a header as for the shed roof. However, the cuts on the rafters that fit on the main roof surface differ from the seat cuts of other rafters.

When the dormer and the main roof are of equal pitch the bottom cut of a dormer rafter which rests against the main roof sheathing may be described as a level cheek cut (Figure 18-5). This cut is obtained by laying out the level or seat cut of the main common rafter on a piece of dormer rafter stock (Figure 18-6). A distance equal to the thickness of the rafter stock (1-5/8 inch) is measured from the top edge of the rafter along this level line. A line is then squared from the top edge of the rafter down to this point (A, Figure 18-6). From the top of this line, a line is then squared across the top edge of the rafter to point B. The cheek cut line is then drawn from point B to point C. The lengths of the jack rafters are found from the common rafter in the same way as the jacks of the dormer described in Chapter 15.

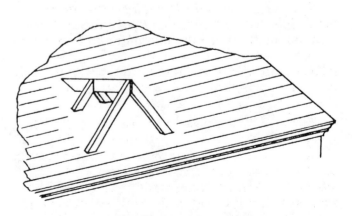

Dormer rafter cuts
Figure 18-5

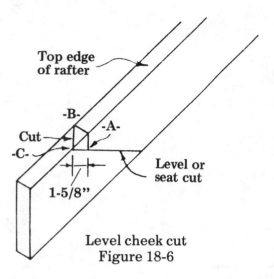

Level cheek cut
Figure 18-6

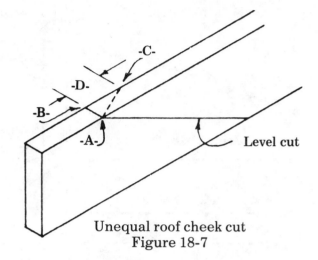

Unequal roof cheek cut
Figure 18-7

When the dormer and the main roof are of unequal pitch, the general shape of the cheek cut for the dormer rafter is similar to that shown in Figure 18-6 but the layout is slightly different. To lay out this cheek cut, it is necessary to determine how much the main roof rises in a level distance equal to the thickness of the dormer rafter. Assume that the rise of the main roof is 9 inches and the run is 12 inches. On a horizontal distance of 2 inches (the approximate thickness of a rafter), the rise would be 1/6 of 9 inches or 1½ inch. This distance is referred to as D, Figure 18-7. The level cut line of the dormer rafter is laid out using the rise and run of the dormer rafter. See Figure 18-7. From the point where the level line meets the top edge of the rafter (A, Figure 18-7), a line is squared across the top of the rafter to point B. The distance D, which was found to be 1½ inch in this case, is then measured from point B to obtain point C. From point C a line is drawn to point A to give the cheek cut for the bottom end of the rafter. The lengths and ridge cuts of the dormer rafters are found as described in Chapters 15 and 16.

How To Build A Small Dormer On The Main Roof

Note: In erecting small dormers on the sheath-ing of the main roof, it may be more convenient to place the longest dormer rafters on the main roof first by tacking them temporarily to the roof and to the dormer ridgeboard.

1. Cut the ridgeboard so that it fits against the main roof in a level position. This cut is the same as the bottom cheek cut of the dormer rafters. Make the ridgeboard long enough so that it will extend beyond the face of the longest rafters of the dormer.

2. Lay out and cut the jack rafters.

3. Tack the long jacks to both sides of the ridgeboard, keeping the tops of the rafters even with the top of the ridgeboard.

4. Level the ridgeboard from the top of the rafters back to the main roof. Tack it to the roof at this point.

5. Plumb the long jacks from the main roof up to the ridgeboard.

6. Recheck the ridgeboard and rafters for plumbness and nail the rafters.

7. Fill in the remaining dormer jacks that extend from the ridge to the main roof.

Note: Be sure the ridgeboard is not forced out of line when nailing the jacks to the ridge and main roof.

Chapter 19
Roof Sheathing

Roof sheathing is the covering over the rafters or trusses and usually consists of nominal 1 inch lumber or plywood. In some types of flat or low pitched roofs with post and beam construction, wood roof planking or fiberboard roof decking might be used. Diagonal wood sheathing on flat or low pitched roofs provides racking resistance where areas with high winds demand added rigidity. Plywood sheathing provides the same desired rigidity and bracing effect. Sheathing should be thick enough to span between supports and provide a solid base for fastening the roofing material.

Roof sheathing boards are generally the third grades of species such as the pines, redwood, the hemlocks, western larch, the firs, and the spruces. It is important that thoroughly seasoned material be used with asphalt shingles. Unseasoned wood will dry out and shrink in width causing buckling or lifting of the shingles along the length of the board. Twelve percent is a desirable maximum moisture content for wood sheathing in most parts of the country. Plywood for roofs is commonly standard sheathing grade.

Installation of board roof sheathing, showing both closed and spaced types
Figure 19-1

Lumber Sheathing

Board sheathing to be used under such roofing as asphalt shingles, metal-sheet roofing, or other materials that require continuous support should be laid closed (without spacing). See Figure 19-1. Wood shingles can also be used over such sheathing. Boards should be matched, shiplapped, or square-edged with joints made over the center of rafters. Not more than two adjacent boards should have joints over the same support. It is preferable to use boards no wider than 6 or 8 inches to minimize problems which can be caused by shrinkage. Boards should have a minimum thickness of ¾ inch for rafter spacing of 16 to 24 inches, and be nailed with two eightpenny common or sevenpenny threaded nails for each board at each bearing. End-matched tongued-and-grooved boards can also be used and joints of adjoining boards be made over the same rafter space. Each board should be supported by at least two rafters.

Use of long sheathing boards at roof ends is desirable to obtain good framing anchorage, especially in gable roofs where there is a substantial rake overhang.

The placing of roof boards should begin at the eaves with a wide board of good quality. Where composition or asbestos roofing is used, the surface should be carefully matched or shiplapped from the eaves to a line well inside the stud spaces as there is a tendency for moisture to condense between such roofing and the roof boards and then run down between the stud spaces.

Spaced Sheathing

When wood shingles or shakes are used in damp climates, it is common to have spaced roof boards (Figure 19-1). Wood nailing strips in nominal 1 by 3 or 1 by 4 inch size are spaced the same distance on centers as the shingles are to be laid to the weather. For example, if shingles are laid 5 inches to the weather and nominal (surfaced) 1 by 4 inch strips are used, there would be spaces of 1-3/8 to 1½ inches between each board to provide the needed ventilation spaces.

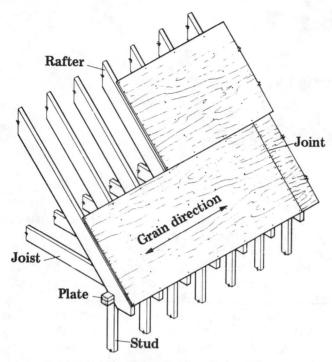

Rafter

Joint

Joist

Grain direction

Plate

Stud

Application of plywood roof sheathing
Figure 19-2

Plywood Thickness (inch)	Plywood Panel Identification Index	Maximum (4) Span (3) (inches) (2)
5/16	12/0	12
5/16	16/0	16
5/16	20/0	20
3/8	24/0	24
5/8	30/12	30
1/2	32/16	32
3/4	36/16	36
5/8	42/20	42
3/4	48/24	48

Notes:

(1) Applies to Standard, Structural I and II and C-C grades only conforming to U.S. Commerce Dept. PS 1-66.

(2) Use 6d common smooth, ring-shank or spiral-thread nails for ½-inch thick or less, and 8d common or 8d ring-shank or spiral-thread for plywood 1-inch thick or less.

(3) These spans shall not be exceeded for any load conditions.

(4) Provide adequate blocking, tongue and groove edges, or other suitable edge support such as metal fasteners. Use two metal fasteners for 48 inches or greater spans.

(5) Plywood roof sheathing continuous over two or more spans; grain of face plys perpendicular to supports (1), (2).

Table 19-3

Plywood Roof Sheathing

When plywood roof sheathing is used, it should be laid with the face grain perpendicular to the rafters (Figure 19-2). Standard sheathing grade plywood is commonly specified but, where damp conditions occur, it is desirable to use a standard sheathing grade with exterior glueline. End joints are made over the center of the rafters and are usually staggered by at least one rafter.

Plywood should be nailed at each bearing, 6 inches on center along all edges and 12 inches on center along intermediate members. Unless plywood has an exterior glueline, raw edges should not be exposed to the weather at the gable end or at the cornice, but should be protected by the trim. Allow a 1/8 inch edge spacing and 1/16 inch end spacing between sheets when installing.

Select the minimum permissible thickness of plywood roof sheathing that will effectively span roof framing members. Unsupported plywood edges that are perpendicular to the roof framing should be connected with metal fasteners designed for this purpose. When metal fasteners are used, the minimum thicknesses of plywood permissible for roofs covered with asphalt shingles, wood shingles, shakes, and built-up roofing are listed in Table 19-3. For slate and similar heavy roofing materials, ½ inch plywood is considered minimum for 16 inch rafter spacing.

Plywood roof sheathing edges that run perpendicular to the roof framing must be supported by wood blocking or fastened together with metal fasteners. Figure 19-4 illustrates the use of metal fasteners that are acceptable substitutes for edge blocking. Replace wood blocking used for edge supports with metal fasteners, or use plywood roof sheathing with tongue-and-groove edges. No edge support is necessary when plywood is 1/8 inch thicker than the minimum thickness required for the given span and plywood grade.

Studies conducted by the National Association Of Home Builders Research Foundation, Inc., indicate that scrap and waste can be reduced by about two-thirds if a plywood sheathing layout is planned and a cutting schedule is developed. In addition to scrap savings, preplanning the schedule will minimize

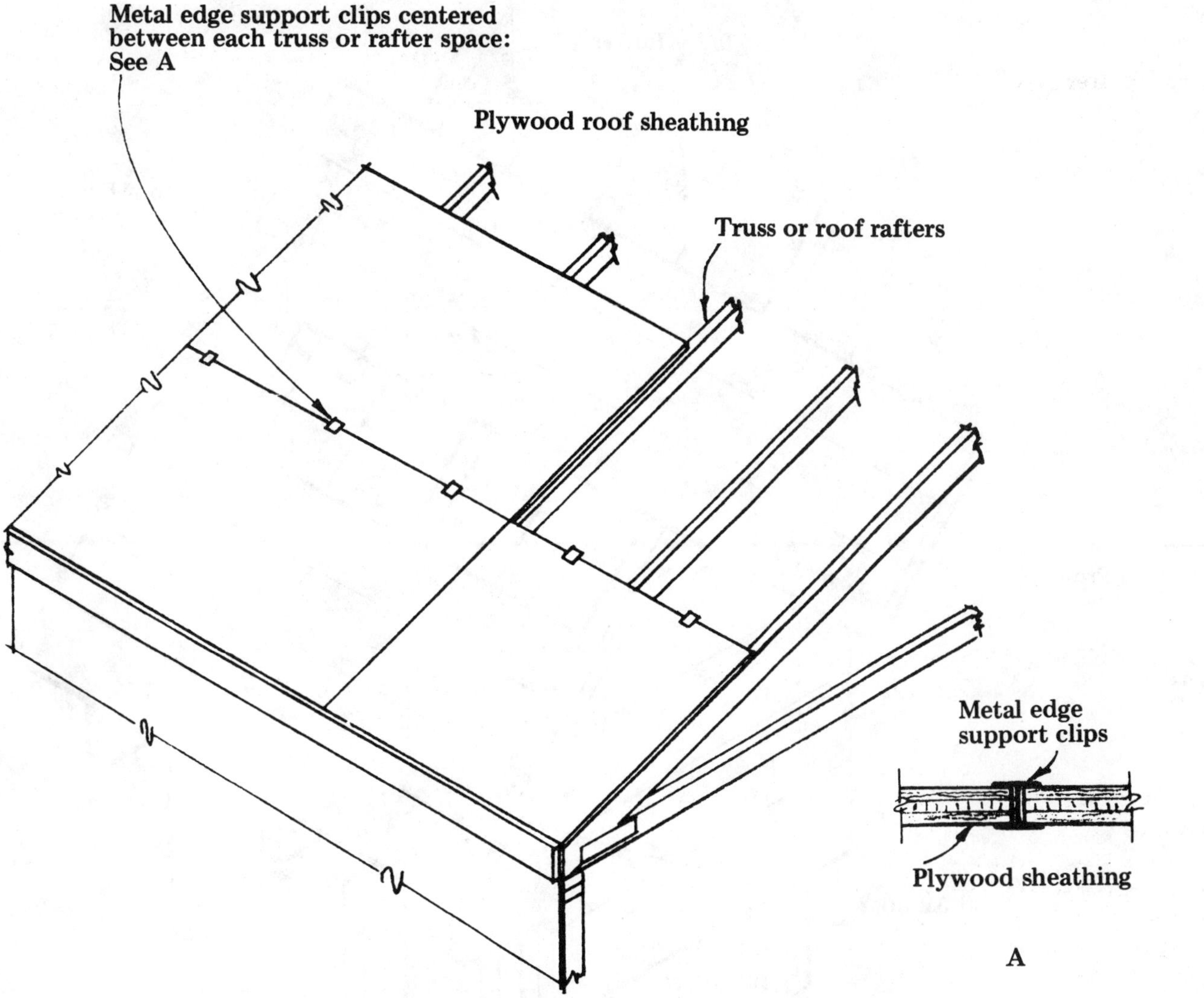

Metal edge support clips centered between each truss or rafter space: See A

Plywood roof sheathing

Truss or roof rafters

Metal edge support clips

Plywood sheathing

A

Method of eliminating plywood edge blocking
Figure 19-4

cutting errors and plywood misuse and reduce the necessity of delivering surplus plywood to the site.

Also remember that when plywood is fastened to framing in a deflected condition, it tends to retain such deformation after nailing. Therefore, it is important that workmen do not stack materials or kneel on the plywood between supports while it is being fastened to the roof framing.

Plank Roof Decking

Plank roof decking, consisting of 2 inch and thicker tongued-and-grooved wood planking is commonly used in flat or low pitched roofs in post and beam construction. Common sizes are nominal 2 by 6, 3 by 6, and 4 by 6 inch V-grooved members, the thicker planking being suitable for spans up to 10 or 12 feet. The maximum span for 2 inch planking is 8 feet

when continuous over two supports, and 6 feet over single spans in grades and species commonly used for this purpose. Special load requirements may reduce these allowable spans. Roof decking can serve both as an interior ceiling finish and as a base for roofing. Heat loss is greatly reduced by adding fiberboard or other rigid insulation over the wood decking.

The decking is blind nailed through the tongue and also face nailed at each support. In 4 by 6 inch size, it is predrilled for edge nailing. For thinner decking, a vapor barrier is ordinarily installed between the top of the plank and the roof insulation when planking does not provide sufficient insulation.

Fiberboard Roof Decking

Fiberboard roof decking is used the same way as wood decking, except that supports are

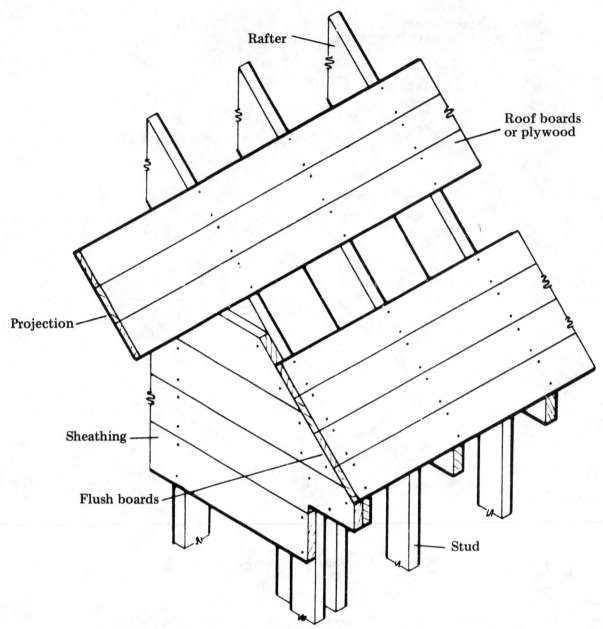

Board roof sheathing at ends of gable
Figure 19-5

spaced much closer together. Planking is usually supplied in 2 by 8 foot sheets and with tongued-and-grooved edges. Thicknesses of the plank and spacing of supports ordinarily comply with the following tabulation:

Minimum thickness (Inches)	Maximum joist spacing (Inches)
1½	24
2	32
3	48

Manufactuers of some types of roof decking recommend the use of 1-7/8 inch thickness for 48 inch spacing of supports.

Nails used to fasten the fiberboard to the wood members are corrosion-resistant and spaced not more than 5 inches on center. They should be long enough to penetrate the joist or beam at least 1½ inch. A built-up roof is normally used for flat and low pitched roofs having wood or fiberboard decking.

Extension Of Roof Sheathing At Gable Ends

The method of installing board or plywood roof sheathing at the gable ends of the roof is shown in Figure 19-5. Where the gable ends of the house have little or no extension (rake projection), roof sheathing is placed flush with the outside of the wall sheathing.

208

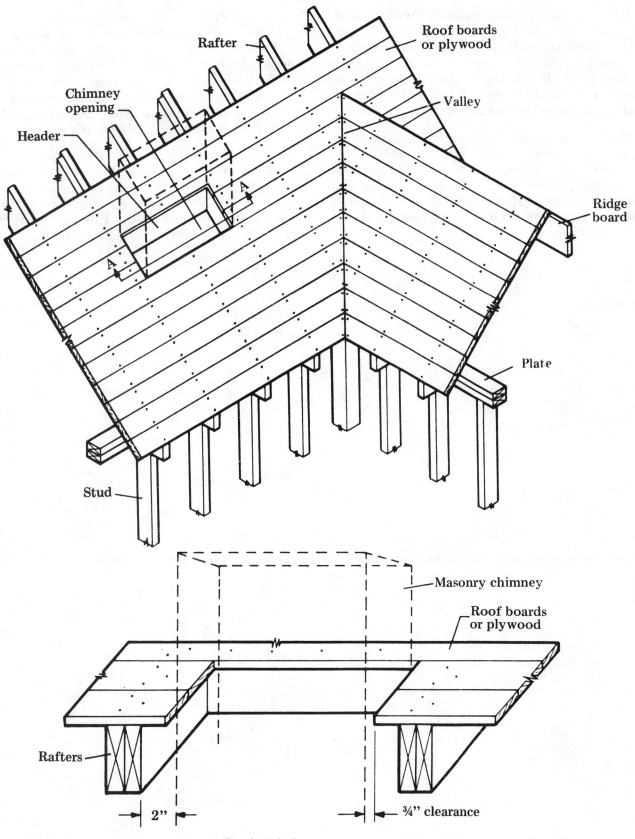

Roof boards or plywood

Rafter

Chimney opening

Valley

Header

Ridge board

A

A

Plate

Stud

Masonry chimney

Roof boards or plywood

Rafters

2"

¾" clearance

Section A-A

Board roof sheathing detail at valley and
chimney openings. Section A-A shows
clearance from masonry
Figure 19-6

Roof sheathing that extends beyond end walls for a projected roof at the gables should span not less than three rafter spaces to insure anchorage to the rafters and to prevent sagging (Figure 19-5). When the projection is greater than 16 to 20 inches, special ladder framing is used to support the sheathing.

Plywood extension beyond the end wall is usually governed by the rafter spacing to minimize waste. Thus, a 16 inch rake projection is commonly used when rafters are spaced 16 inches on center. Butt joints of the plywood sheets should be alternated so they do not occur on the same rafter.

Sheathing At Chimney Openings

Where chimney openings occur within the roof area, the roof sheathing and subfloor should have a clearance of ¾ inch from the finished masonry on all sides (Figure 19-6, Section A-A). Rafters and headers around the opening should have a clearance of 2 inches from the masonry for fire protection.

Sheathing At Valleys And Hips

Wood or plywood sheathing at the valleys and hips should be installed to provide a tight joint and should be securely nailed to hip and valley rafters (Figure 19-6). This will provide a solid and smooth base for metal flashing.

Chapter 20
Framing Bay Windows

A bay window is a projection built on the outside wall of a building. It is designed to add more floor space to a room, to admit more light, or to improve the appearance of the building.

The many styles and shapes of bay windows may be classified as square, three sided or circular. The circular bay window is rarely used today. It has been replaced by various shapes composed of straight lines in keeping with the interior and exterior style of the building. A description of one type of straight line bay window will illustrate the basic principles of this type. The framing of the projection into the main floor joists of the building is more or less common to all types. The side wall and roof construction will be described in terms of the more commonly used types.

The procedure used in the erection of a bay window is very much like that used in the erection of similar parts of the main building. The layout, assembly and bracing are the same in the rectangular bay as in the main building. The three sided bay differs slightly in the plate and stud assembly and for this reason it will be used as an example in this unit.

Floor Framing

When the floor joists of the bay window are to run parallel to the joists of the main building, the framing is rather simple. See Figure 20-1. The joists are allowed to project beyond the outer face of the foundation the same distance the end wall of the bay extends beyond the main wall of the building. The floor plan of the bay is laid out on the tops of the floor joists. The joists are cut along these marks, and a sole plate is nailed to them to form a support for the wall studs of the bay.

When the joists of the bay run at right angles to the main joists, it is necessary to use headers and trimmers in the main joists as shown in Figure 20-2. This type of floor construction is necessary for the proper support of the outside walls of the bay.

Figures 20-1 and 20-2 show the three sided bay. However, the rectangular bay is often used and, as far as the framing is concerned, it is similar to the corner construction of the main building.

Framing Bay Window Floor Joists

1. Determine the location and size of the bay from the blue print.

2. Place the floor joists in the usual manner, but allow them to project beyond the foundation line as described in this chapter.

3. Mark the shape and angle of the bay on the tops of the joists.

4. Square the lines down along the sides of the joists and cut them on these lines.

Note: When a box type sill is used on the main building, the vertical member of the sill should be run around the bay.

Wall Framing

The sole plate and the double plate at the second floor line are laid out and spaced for studs in the same way as the main walls of the building. The double plate is lapped at the angles of the bay and is attached to the main wall as shown at E, Figure 20-1.

The same length studs are used for the bay window as for the walls of the main building. The studs at the angles of the bay are turned so the edges are parallel to the outside edges of the upper and lower plates of the bay.

The plates and studs are nailed together and raised in the same manner as the other bearing partitions of the building.

Framing Bay Window Studding, Plates And Openings

1. Cut pieces of 2 x 4 for the sole plate to follow the outline of the ends of the joists.

2. Cut the studs of the bay the same length as those of the rest of the building.

3. Lay out and cut the top plates so that they will lap over each other at the corners.

4. Frame in the window openings in the same manner as for the other window openings of the main building. Apply the sheathing to the side walls of the bay.

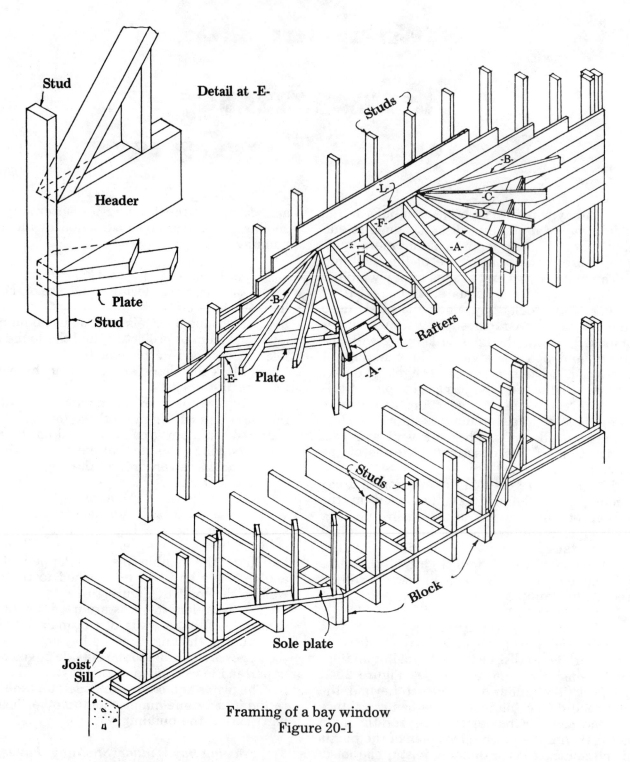

Detail at -E-

Stud

Header

Plate

Stud

Studs

-L-

-B-

-C-

-D-

-F-

-A-

-B-

Rafters

-E- Plate

-A-

Studs

Block

Sole plate

Joist
Sill

Framing of a bay window
Figure 20-1

Roof Framing

The roof shown in Figures 20-1 and 20-3 is typical of roofs on bay windows. Each of the pairs of rafters at the ends must be laid out as individual hip rafters because they have different runs.

In Figure 20-3, the distance from the face of the 2 x 4 (F, Figure 20-1) on the main building to the sheathing line at the outside of the bay window is shown at D. This distance is the run of the rafter. It may be assumed to be 20 inches in

this case. The vertical distance from the top of the double plate to the top of the 2 x 4 F is the rise of the rafter. In Figure 20-1, this is 11 inches. In small roofs it is often more convenient to change the rise and run to inches.

The common rafter is laid out by locating a measuring line on the rafter as described in Chapter 14. A 2 x 4 is sufficiently strong for these rafters and the measuring line should be at least 2½ inches below the top edge.

This rafter is laid out with one step of the

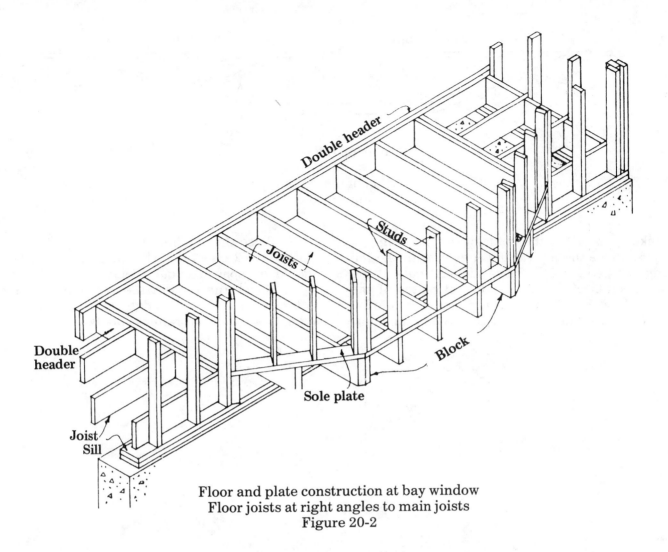

Floor and plate construction at bay window
Floor joists at right angles to main joists
Figure 20-2

square. The figures 20 on the body and 11 on the tongue should be placed on the measuring line. A mark along the outside of the tongue will give the plumb cut and one along the body will give the seat cut.

For the hip rafter at A, Figure 20-3, the rise will also be 11 inches. To find the run of this rafter, a full size plan of the roof should be made on the subfloor (Figure 20-3). The run can be measured on this plan.

The run of this rafter most likely will be greater than 24 inches and therefore greater than the length of the square. Assuming that the run of rafter A is 26 inches and that the rise is 11 inches, it will be necessary to divide these figures by 2. Two steps of 13 inches and 5½ inches should be taken to find the length and cuts of the hip rafter. The cheek cuts are found from the full size roof plan (Figure 20-3). The bevels are transferred from the drawing to the rafter by using a sliding T bevel.

The lengths and cuts of the other hip rafters (B, C, and D, Figure 20-3) are found in a similar manner, the run being measured from the

spaced position of the rafter on the plate. The backing of the hip rafters, in this case, is negligible and may be omitted.

The rafters for a square or rectangular bay are laid out in about the same manner as those of a main roof.

Bay Window Roof Framing

1. Lay out and cut the required numbers of rafters as described in this chapter. The hip rafters should be cut in pairs.

2. Mark a level line along the face of the main building showing the height of the common rafters. See L, Figure 20-1.

3. Place the hip rafters B against the building so that the tops of the rafters are even with the level line, and so the seat cuts fit over the plate. Tack these rafters to the sheathing of the main building.

4. Place the 2 x 4 F against the sheathing of the building with the top edge in line with the level line. Its length should be the distance between the plumb cuts of the two hip rafters B.

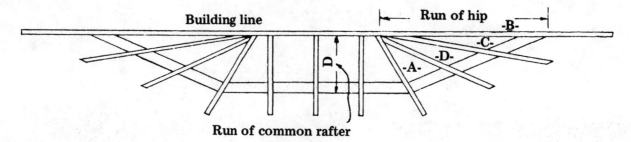

Run of hip rafters
Figure 20-3

Spike the 2 x 4 to the studs of the building.

5. Place the hip rafters A even with the top edge of the 2 x 4 F. Place the seat cuts at the intersection of the side and front plates of the bay. Tack these rafters in position temporarily.

6. Place rafters C and D in a like manner.

7. Place the common rafters from the front plate to the 2 x 4 F on the building. These rafters should be spaced about 16 inches on centers.

8. Check the tops of the hip and common rafters with a straight edge to see if they are even. If so, nail the rafters solidly.

Chapter 21
Framing Interior Partitions

The interior walls in a house with conventional joist and rafter roof construction are normally located to serve as bearing walls for the ceiling joists as well as room dividers. Walls located parallel to the direction of the joists are commonly nonload-bearing. Studs are nominal 2 by 4 inches in size for load-bearing walls but can be 2 by 3 inches in size for nonload-bearing walls. However, most contractors use 2 by 4's throughout. Spacing of the studs is usually controlled by the thickness of the covering material. For example, 24 inch stud spacing will require ½ inch gypsum board for dry wall interior covering.

The interior walls are assembled and erected in the same manner as exterior walls, with a single bottom (sole) plate and double top plates. See Figure 21-1. The upper top plate is used to

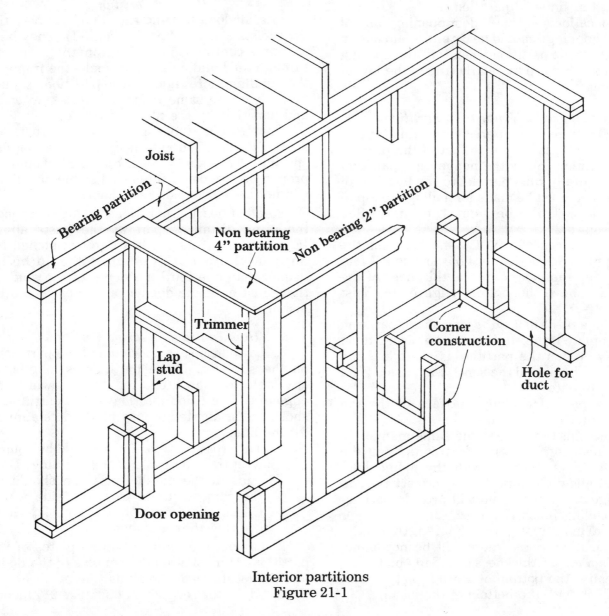

Interior partitions
Figure 21-1

tie intersecting and crossing walls to each other. A single framing stud can be used at each side of a door opening in nonload-bearing partitions. They must be doubled for load-bearing walls, however. When trussed rafters (roof trusses) are used, no load-bearing interior partitions are required. Thus, location of the walls and size and spacing of the studs are determined by the room size desired and type of interior covering selected. The bottom chords of the trusses are used to fasten and anchor crossing partitions. When partition walls are parallel to and located between trusses, they are fastened to 2 by 4 inch blocks which are nailed between the lower chords.

It is important that measurements be taken carefully from the plans. These plans should show the dimensions of rooms from the center of one wall partition to the center of the opposite partition. They should also show the thickness of the rough or finished partition.

The location and size of door openings and of all other openings should be carefully laid out on the plates of the partition when the stud spacing is marked. This is particularly true in the layout of bearing partitions.

How To Frame Nonbearing Partitions

Note: The framing and erection of this type of partition is similar to that of the bearing partition. Since the nonbearing partition generally runs parallel to the joists, provision must be made for nailing the top of the partition between the joists. See Figure 21-2.

1. Cut pieces of 2 x 4 to fit between the ceiling joists. Space the 2 x 4 blocks about 4 feet apart. Keep the blocks up, the thickness of the top plate, above the bottom edge of the joist (Figure 21-2).

2. Cut a piece of 2 x 6 as long as the partition for a top plate if the studs are to be placed the 4 inch way. When the partition is short and the studs are to be placed the 2 inch way, use a 2 x 4 for the top plate.

3. Gauge a line along the face of the top plate about 1 inch from the edge.

4. Use this top plate in the same manner as the top plate of a bearing partition. Keep the edges of the studs even with the gauge line. This will allow the top plate to project beyond each edge of the studs and will provide backing for the ceiling finish (Figure 21-2).

Note: In measuring for the length of the partition studs, there will be no deduction for the thickness of the top plate, because the bottom of the top plate comes flush with the bottom of the joists.

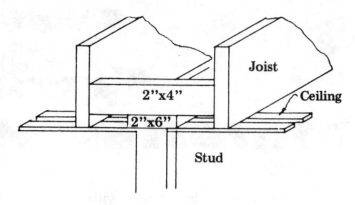

Securing the top of a partition parallel to joists
Figure 21-2

5. Assemble, raise, plumb and fasten the partitions in the same manner as explained for the bearing partition.

Bearing Partitions

Interior load-bearing partitions support the floor joists above them (Figure 21-1). They help to form a continuous bearing from the girder up through the building and also make the frame of the building more rigid. These partitions should be built in the same manner as the side walls and partitions in the platform frame.

Openings in these partitions for doors or supply ducts often interfere with the regular spacing of the studs. Such openings should be properly trussed and supported so the partitions will not be weakened. See Chapter 8.

Studs of bearing partitions should be spaced the same distance apart as the joists above (Figure 21-1). If the studs are to be notched for horizontal pipes, the notches should be no deeper than one third the depth of the stud. If they must be made deeper, a header should be installed.

How To Frame Bearing Partitions

1. Lay out the exact location of the partitions on the subfloor. If the partition is long, it is advisable to snap a chalked line showing one edge of the partition location. If the partition is short, the straightedge may be used as a guide for the line.

Note: The cross partition lines should be square with the side walls of the building. Both ends of the parallel partition should be the same distance from the side walls. Check all corners of the room layout to see that they are square.

2. Lay out the top and bottom plates of the partition. Mark the rough openings of the doors and the locations of the studs (Figure 21-3).

Note: Make the rough door openings 2½ inches

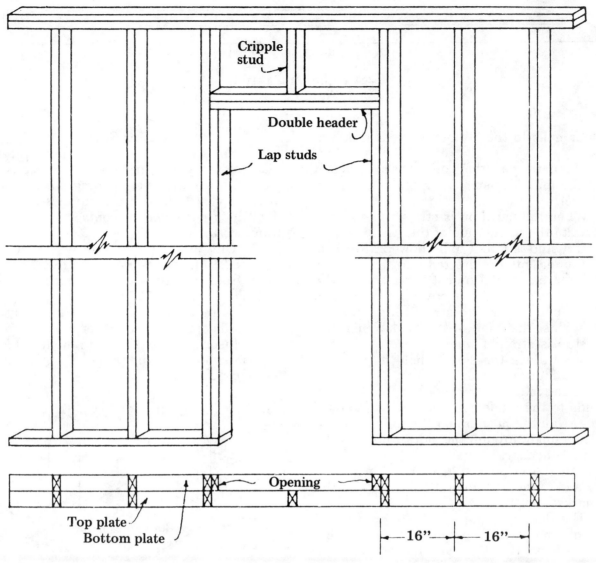

Partition layout
Figure 21-3

wider than the finish mill size of the door. This allowance is for the thickness of the two side jambs and allowance for plumbing the jambs. The allowance for the jamb is 1½ inch and for the finish floor, 1 inch.

The window rod used to mark the height of window headers in laying out window openings in side walls may be used to mark the door headers in partitions. Refer to Chapter 8.

3. Cut the top plate to the proper length.

4. Cut the bottom plate to the correct length. It may be cut in two pieces, one for each side of the door. See Figure 21-3.

Note: Some mechanics prefer to leave the top and bottom plates the full length of the partition. The plates are then spiked to the top and bottom of the studs, the partition is raised and the bottom plate at the door opening is later cut out.

5. Nail the bottom plates to the subfloor, keeping them in line with the chalk line snapped on the floor and spacing them correctly at the door opening marks. Spike the plates to the joists with 16d spikes.

6. Lay the top plate on the bottom plate and with an extension measuring stick, measure the distance from the top of the two plates to the bottom of the joists above. This distance is the length of the studs.

Note: If a double plate is to be used at the top of the partition, lay an extra 2 x 4 block on the plates when taking the measurement for the studs.

7. Cut the required number of studs and spike them to the upper plate at the stud marks.

8. Raise the partition and toenail each stud

Straightening a stud
Figure 21-4

with four nails to the bottom plate at the stud marks.

9. Plumb the partition both ways at each end and in the middle, using a straightedge and a spirit level.

10. When the partition is straight, spike the top plate to the under side of the joists above.

Note: If a stud is found to be bowed after it is framed into the erected partition, it may be straightened by cutting a kerf in the stud and forcing it straight. It may be checked with the strightedge placed against the edge of the stud. When the stud is straight, nail a piece of ¾ inch stock along the side to hold it in place. See Figure 21-4.

11. Frame in the door headers by nailing the lap studs to the studs at each side of the door opening. Nail the headers on the top of the lap studs and spike through the side studs into the ends of the header.

12. Nail the cripple stud to the top of the header and to the bottom of the top plate at the stud marks.

Note: The door opening is sometimes framed into the partition before it is raised. This method saves time and makes the work easier.

Openings In Partitions

Openings for doors in partitions are framed with trimmer and lap studs (Figure 21-3). The lap studs run from the shoe to the under side of the double header. These studs support the header and also stiffen the stud at the side. They also provide a better nailing surface for the interior trim.

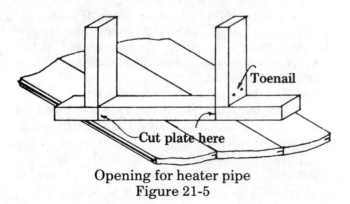

Opening for heater pipe
Figure 21-5

Openings For Heating Pipes

1. Mark the size of the heating duct on the plate of the partition where the duct is to be placed. Make the duct opening large enough to allow at least ½ inch on all four sides so that the duct does not come in contact with the wood framework.

2. Cut the plate out between the studs (Figure 21-5). If possible, make the cuts ½ inch away from the studs to avoid hitting the nails with the saw.

3. Bore two holes through the subfloor diagonally opposite each other. See Figure 21-6.

4. Insert a compass saw in the holes and cut out the subfloor to the proper size to admit the heating duct.

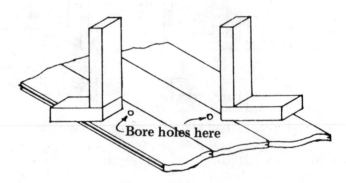

Opening for heater pipe
Figure 21-6

Saving Time And Material On Interior Framing

Overall dimensions of partitions and ceilings usually are not multiples of 24 inches or 48 inches. Thicknesses of exterior and interior wall facings and framing vary. If total thicknesses of all walls were modular, interior partition dimensions could be made modular more easily.

Except for load-bearing partitions, little structural engineering is involved in interior partitions and ceilings. To minimize partition and ceiling lumber or plywood, concentrate on eliminating wood or substituting other material for wood.

Except for interior decorative wall panelling, little plywood is used in interior partitions and ceilings. Interior wall framing members can usually be nominal 2 x 2's or 2 x 3's, and

218

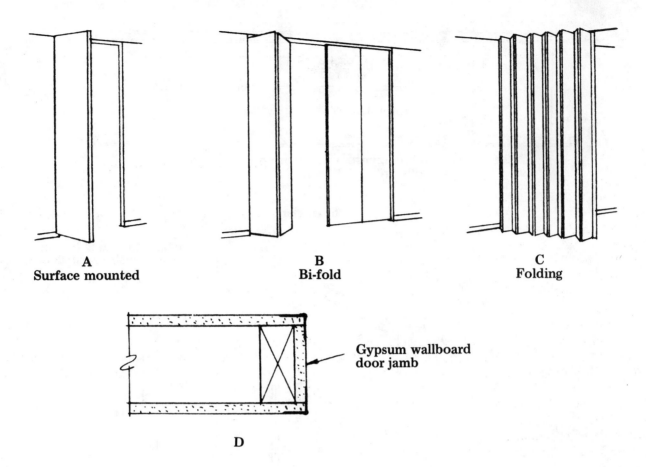

A
Surface mounted

B
Bi-fold

C
Folding

Gypsum wallboard
door jamb

D

Full interior wall height closet doors
Figure 21-7

nonload-bearing partitions can use 1 inch thick top and bottom plates. Backup blocking for wall-facing material should be eliminated or reduced in size.

Use a drawing to locate most of the interior partition studs on modular spacings whenever possible. If possible, match interior door openings to the stud spacings.

Save On Closet Door Framing

Full wall height closet doors, illustrated in Figure 21-7, may be used to reduce interior wall framing. The door head, cripples, and trim used above conventional 6 foot 8 inch high doors can be eliminated. A surface mounted, full wall height door shown in *A* is an example of this. Many styles of folding and bi-fold closet doors are available that can be mounted on the inside of a wall opening with gypsum wallboard jambs (see *B* and *C*). Gypsum wallboard with metal corner reinforcement may be used for door jambs of the opening (see cross section, *D*). This eliminates wood trim and jamb on each side of the door opening.

When 4 x 8 foot sheet material is fitted around door openings, it is necessary to cut pieces out of one or more sheets the size of the door. These cut-outs are often thrown away. Use of the full wall height door eliminates this costly practice.

Eliminate Mid Height Interior Wall Blocking

Top and bottom plates of framed walls act as fire stops. As a result, mid height wall blocking usually is not required between studs and should be eliminated.

Adjacent pieces of gypsum wallboard applied horizontally do not require blocking between stud spaces (3/8 inch gypsum-board on 16 inch stud spacing or ½ inch gypsum-board on 24 inch stud spacing). The taped joint is strong enough to resist usual occupancy loads placed on the gypsum-board without wood back-up blocking.

Space 2 x 3 Interior Wall Studs 24 Inches on Center

Use 2 x 3 rather than 2 x 4 stud framing for nonbearing interior partitions. The framing amount can be reduced more if the maximum stud spacing permissible is used. Space 2 x 3

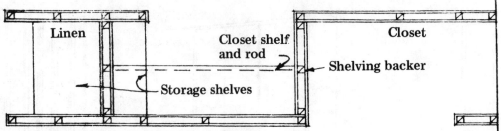

Linen   Closet shelf   Closet
                     and rod
                                    Shelving backer

        Storage shelves

**2x2 framing with 3/8" gypsum wallboard on 16" centers
or ½" gypsum wallboard on 24" centers**

Closet Plan

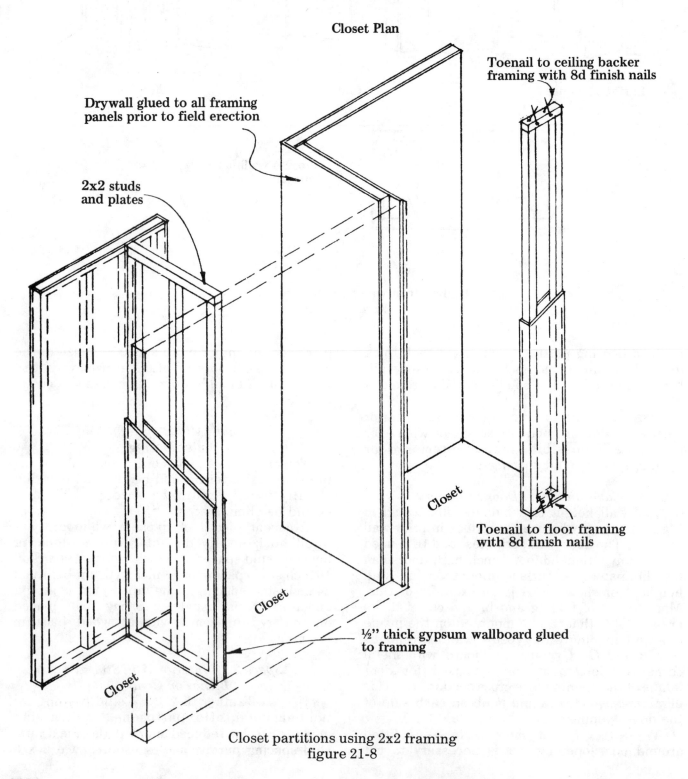

Toenail to ceiling backer
framing with 8d finish nails

Drywall glued to all framing
panels prior to field erection

2x2 studs
and plates

Closet

Toenail to floor framing
with 8d finish nails

Closet

Closet

½" thick gypsum wallboard glued
to framing

Closet

Closet partitions using 2x2 framing
figure 21-8

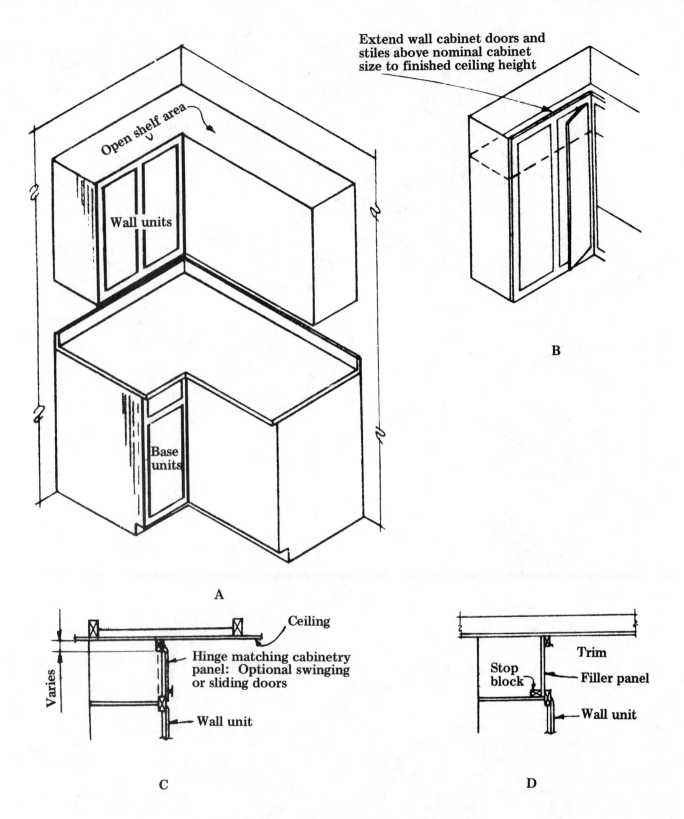

Extend wall cabinet doors and stiles above nominal cabinet size to finished ceiling height

Open shelf area

Wall units

Base units

A

B

Ceiling

Hinge matching cabinetry panel: Optional swinging or sliding doors

Varies

Wall unit

C

Trim

Stop block

Filler panel

Wall unit

D

Details requiring minimum framing
over kitchen wall cabinets
Figure 21-9

studs 24 inches on center when using ½ inch thick gypsum wallboard facings, or if 3/8 inch thick gypsum wallboard facing is used, the 2 x 3 studs must be spaced 16 inches on center.

Use Nominal 2 x 2's For Nonload-Bearing Closet Wall Framing

Rather than using 2 x 4 studs spaced 16 inches on center for closet partition wall

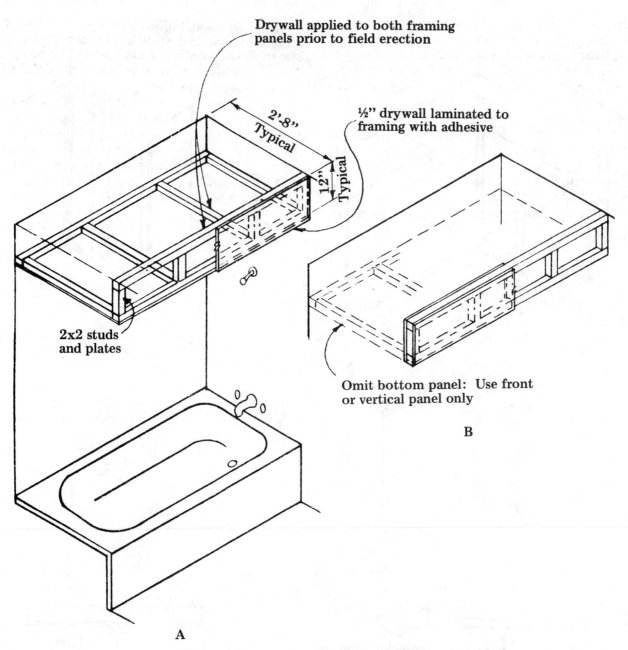

Drywall applied to both framing panels prior to field erection

2'-8" Typical

½" drywall laminated to framing with adhesive

12" Typical

2x2 studs and plates

Omit bottom panel: Use front or vertical panel only

B

A

Bulkhead using 2x2 framing over bathtub
Figure 21-10

framing, use 2 x 3 studs on 24 inch centers or preassembled panels with 2 x 2 framing on 16 inch or 24 inch centers depending on facing thickness (see Figure 21-8). Since 2 x 2 framing is somewhat resilient when wallboard is being nailed to it, walls with 2 x 2 framing may be prefabricated with glued facings. The glued facings provide extra stiffness to the wall as an added benefit.

To install panels, toenail 8d finishing nails into the wall facing, bottom or top panel plates, and floor or ceiling framing. Preboring during panel fabrication can be used to facilitate the toenailing operation. If gypsum wallboard

facings are used, mechanical fasteners are not required to anchor intersections between panels and adjacent walls. Then, floating, taped, and spackled joints can be used.

Full wall height bi-fold or folding closet doors are suggested for this type of system. Hardware for such doors mounts directly to the floor and ceiling framing.

Eliminate Bulkheads

Eliminate 2 x 3 and 2 x 4 bulkhead framing over kitchen cabinets and bath tubs. Sometimes 2 x 3 or 2 x 4 bulkhead framing is used over kitchen cabinets or bath tubs to provide a mounting surface for wall facing. Wall facing material usually has sufficient strength to act as a filler over kitchen cabinets without requiring 2 x 3 or 2 x 4 wood framing behind it. The concept that uses the least amount of material is shown in Figure 21-9, *A* where the space over the kitchen cabinets is open and can be used as a shelf. Figure 21-9, *B* shows the cabinet extending to the ceiling and uses no wood framing. *C* illustrates a method of using cabinetry paneling to close off the space and maintain its usability. *D* shows a method of using a filler panel over the cabinets. This filler panel can be the same material as the wall facings or any other cost effective product. If this filler material is very thin or of low strength, glue 1 x 2 stiffener strips to the back of the filler before it is installed.

Bulkheads over bath tubs are sometimes required to cover heating ducts, vent ducts, or plumbing lines. Bulkheads also serve as a mounting surface for sliding door tracks. Figure 21-10, *A* shows use of 2 x 2 bulkhead framing over a bath tub, and *B*, a let down panel for mounting a door track that is framed out of 2 x 2's.

Chapter 22
Furring, Grounds And Backing

You should have a general understanding of the methods used by the other building tradesmen to fasten their various fixtures to the frame of the building. You should know how to read from the building drawings the locations of plumbing, electrical, and heating fixtures which are to be installed. This information is necessary so that you may provide proper backing to which these fixtures may be attached after the wall finish has been applied to the walls.

Furring strips, grounds and backing must be installed before the interior wall finish is applied. They should be of regular framing material and should be free of knots and cracks. They must be firmly secured so that they will not become loosened when the nails or screws holding the fixtures or trim enter them.

Furring Strips

Furring strips are fastened to masonry walls or chimney surfaces which are to be plastered or they may be spaced away from such a wall. Their function is to provide a backing upon which the wall cover is to be nailed and also to provide an air space between the face of the masonry and the back of the finish wall cover. This space should be at least ¾ inch so that the key of the plaster will not touch the masonry. This is to prevent the moisture in the masonry from staining the wall surface. Furring strips also help to prevent cracks due to the difference in the rate of settling between the masonry and the framework of the building.

How To Apply Furring

Note: Figure 22-1 shows the proper method of furring around the surfaces of a chimney.

1. Cut the required number of studs to reach from the floor up to the trimmer and header joists of the chimney opening.

2. Erect the studs at the corners of the brickwork as shown in Figure 22-1. Keep the studs 2 inches from the face of the brickwork.

3. Square the corners horizontally and plumb the studs vertically.

4. Nail the bottom of the studs firmly to the subfloor and the tops to the headers.

Minimize Size And Amount Of Furring Strips On Concrete Block Walls

Furring strips on concrete block walls should be placed at the maximum spacing allowed for the facing material they support. For example, spacings of 16 inches on center for 3/8 inch thick gypsum wallboard and 24 inches on center for ½ inch thick gypsum wallboard are used.

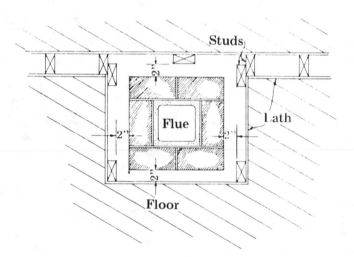

Furring at chimney
Figure 22-1

Furring strip lumber should be nominal 1 x 2 rather than 1 x 3 or 1 x 4, and 1 inch thick furring strips are adequate even when batt type insulation is used (see Figure 22-2). Insulation can be compressed from a thickness of 1½ inch to ¾ inch and still provide significant thermal insulation for walls in most areas of the country. Compressed insulation lends some support to wall facing, and for this reason, it is permissible to use 3/8 inch thick gypsum board and ¼ inch thick wood paneling on furring strips spaced 24 inches on center.

Horizontal wood furring strips at the base, mid height, and top of the wall are not required and should be eliminated.

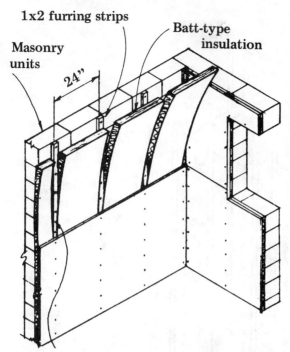

1x2 furring strips

Masonry units

Batt-type insulation

24"

1x2 furring 24" on center for most facing types when backed up with ½" thick batt type insulation compressed to ¾". Otherwise use support spacing specified for facing material. Attach furring strips with 6d masonry nails 16" on center.

Details showing use of 1x2 furring strips on concrete block walls
Figure 22-2

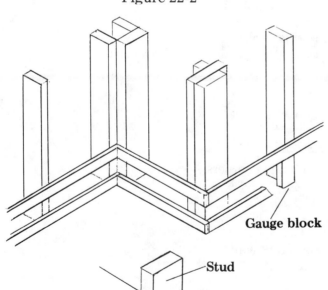

Gauge block

Stud

Ground

Finish base

Floor

Detail -A-

Base grounds
Figure 22-3

Grounds

Grounds are rough dimensioned stock, generally the thickness of the combined lath and plaster coats. Their function is to form a straight rigid surface to which the plasterer may bring the finish coat of plaster and to which the carpenter may secure the interior trim.

How To Install Base Grounds

Note: Figure 22-3 shows how base grounds are nailed to the studs.

1. Nail the top ground strip to each stud 1 inch below the point where the top of the finish baseboard will be. See *A*, Figure 22-3. Use 6d nails.

Note: Use a gauge block placed between the bottom of the ground strip and the floor.

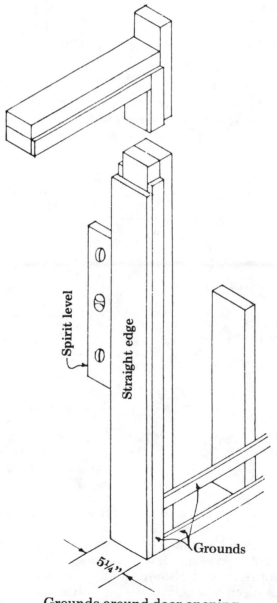

Spirit level

Straight edge

Grounds

5¼"

Grounds around door opening
Figure 22-4

225

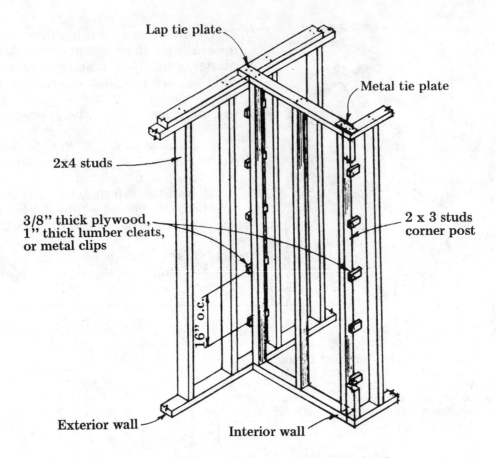

Lap tie plate

Metal tie plate

2x4 studs

3/8" thick plywood,
1" thick lumber cleats,
or metal clips

2 x 3 studs
corner post

16" o.c.

Exterior wall

Interior wall

For 3/8" - thick plywood use two 4d annular threaded nails in each cleat.

For 1" - thick lumber use two 5d annular threaded nails in each cleat.

For metal clips use nails as specified by manufacturers.

A

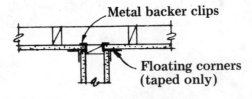

Metal backer clips

Floating corners
(taped only)

B

Details showing elimination of wall
facing back-up blocking
Figure 22-5

This will avoid the necessity of repeated measurements to get the ground straight.

2. Check the face of the ground strip with a straightedge to see that it is straight throughout its entire length. If there are any irregularities, use a piece of wood to wedge between the stud and the back of the ground. Drive the wedge down until the face of the strip is straight.

3. Nail the bottom ground strip to the studs, keeping the strip tight against the top of the subfloor.

Grounds Around Door Openings

1. Nail the ground strips around the edges of the door to the vertical studs and to the header (Figure 22-4).

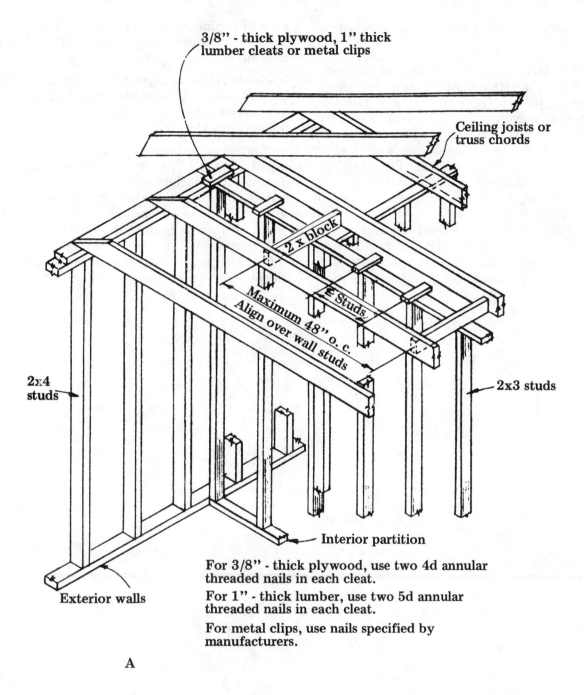

3/8" - thick plywood, 1" thick lumber cleats or metal clips

Ceiling joists or truss chords

2 x block

Studs

Maximum 48" o. c.
Align over wall studs

2x4 studs

2x3 studs

Interior partition

Exterior walls

For 3/8" - thick plywood, use two 4d annular threaded nails in each cleat.

For 1" - thick lumber, use two 5d annular threaded nails in each cleat.

For metal clips, use nails specified by manufacturers.

A

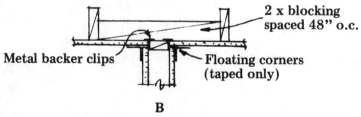

2 x blocking spaced 48" o.c.

Metal backer clips

Floating corners (taped only)

B

Details showing elimination of ceiling
facing back-up blocking
Figure 22-6

Note: Use a straightedge if necessary to straighten the faces of the strips and to maintain a uniform width between these faces on each side of the opening. A piece of uniform width similar to that shown in Figure 22-4 may be used in plumbing and keeping the strip straight and uniform.

Backing

Where interior partitions intersect exterior walls or other interior partitions, it is necessary to provide supports for wall facing materials. Traditionally, this has been accomplished by using additional studs or pieces of dimension lumber. Methods for eliminating wall facing backup studs are illustrated in Figure 22-5. Figure 22-5, *A* illustrates the use of wood cleats nailed to the end stud of an interior partition to support wall facing materials.

Cleats can be 3/8 inch thick plywood or 1 inch thick lumber, but plywood cleats are preferable because they tend to split less. If 1 inch thick lumber is used, select soft woods such as white pine or spruce for the cleats. These woods are nailed easily and have little tendency to split.

Metal clips designed for gypsum wallboard backup support or other wall facing material also may be substituted for 2 x 4 studs used for backup blocking (see Figure 22-5, *B*).

Gypsum wallboard is not fastened to the wood cleats. The wallboard sheet resting against the backup cleats should be installed before the wallboard sheet on the adjacent wall. In this way, the second sheet can lock the first sheet into place against the backup cleats.

Figure 22-6 illustrates similar details for ceiling backers using wood or metal backup cleats.

Chapter 23

Stairs

The construction of stairs is usually left to the most experienced carpenter on the job. Yet most journeymen carpenters would agree that more material and time have been wasted tearing down poorly designed or poorly built stairways than in any other framing job. No matter how experienced the craftsman, each stairway presents its own design and construction problems. Even in highly repetitive jobs where many similar stairways are constructed on one site, each stairway to be built may be somewhat unique. The floor to floor rise of each stairway built must be measured with accuracy and the craftsman must select the right tread and riser combination so that every rise and run is within ¼ inch of every other rise and run. Many times the plans are inadequate or the actual floor to floor dimension does not correspond with the floor to floor dimension on the plans. Consequently, the craftsman who actually builds the stairway is required to design the stairway before he builds it. More and more stairways are being designed since the increasing value of land recommends multiple story structures.

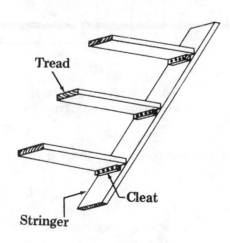

Cleat stairway
Figure 23-1

There are many different kinds of stairs but all have two main parts: treads people walk on and the stringers, carriage or horse which supports the treads. A very simple type of stairway consisting only of stringers and treads is shown in Figure 23-1. Treads in the type shown here are called plank treads, and this simple type of stairway is called a "cleat" stairway because of the cleat attached to the stringers to support the treads. A more finished type of stairway has treads mounted on two or more sawtooth-edged stringers and includes risers as shown in Figure 23-2. The stringers shown here are cut out of solid pieces of lumber (usually 2" x 12") and are therefore called "cutout" or "sawed" stringers.

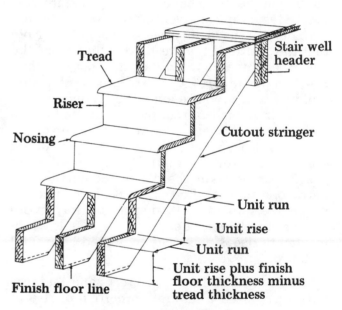

Cutout stringer stairway
Figure 23-2

Stairways may be straight, curved, "L" shaped, "U" shaped or a combination of several shapes. Straight stairs, that is stairs that rise from one floor level to the next without changing direction, are most common. Straight stairs, however, may not be practical where horizontal space is limited. The chance of a harmful fall is greatest on long straight stairways. Any long straight run should include a landing to break falls and give the climber a place to rest while ascending.

Figure 23-3 shows three views of the common "L" shaped stairway with one landing. Notice the 90 degree change of direction in the stairway. Where space is limited "winders" or

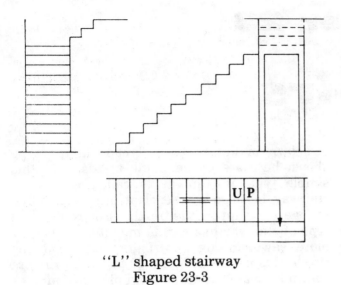

"L" shaped stairway
Figure 23-3

"pie shaped" treads may be substituted for the landing in the "L" shaped stairway. Figure 23-4 illustrates a typical layout of the tread where a "winder" is used in place of a landing. Note that the width of the tread 18 inches from the narrow end of each tread should be not less than the tread width of the straight run. Many building codes require a full 9 or 10 inch run to each stair 12 inches from the narrow end of the stair. Some codes also require that the run of the stair be at least 6 inches at the narrow end of the stair. "Winders" should be avoided when possible because of the danger of falls on the narrow portions of the treads. Space limitations may make spiral or "U" shaped stairways more advisable. These designs are used when maximum rise is necessary in limited horizontal distance.

Since ancient times, architects and builders have recognized a natural proportion of rise to run of each stair. The rise of a stair is the vertical distance from the top of one horizontal level or tread to the top of the next tread. For most purposes, a rise of about 7 inches per stair is best. A rise much more than 7 inches per stair seems to tire the climber unnecessarily. A rise of much less than 7 inches per stair makes the stairway longer than necessary. Just as important as the rise of each stair is the width of each tread. Treads too wide or too narrow don't have the right "feel" to the climber and seem awkward. There is a length of rise and run which seems most comfortable to the largest number of adults. Usually this length is thought of as some riser dimension added to some tread dimension which equals 17½ inches. For example, where the riser is 7 inches the tread would have to be 10½ inches. By this "Rule of 17½" most any combination of rise and run will

have a comfortable "feel" so long as the rise is not less than 6 inches or more than 8 inches per stair. A 6 inch rise (with an 11½ inch tread width) will result in an angle of climb of about 27 degrees. A rise of this type would be best suited for persons of restricted physical ability or for stairways of monumental character such as an entrance to a public building. A rise of 8 inches (with a tread width of 9½ inches) produces an incline of about 40 degrees. Many basement stairs or stairs built in restricted areas may have up to a 40 degree angle of incline. Most interior stairs are designed with an incline of between 33 and 37 degrees as this produces a safe, economical stairway with the natural "feel" most adults find comfortable. Where enough horizontal space and head room are available, the designer should select a rise and run combination which yields an incline of between 33 and 37 degrees. Most building codes require a rise of no more than 7½ inches and a run of no less than 10 inches in other than private residential construction. In private homes most codes permit a rise of as much as 8 inches and a run of as little as 9 inches.

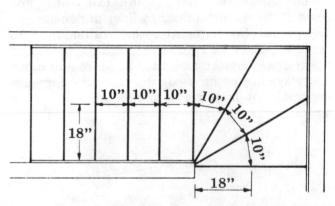

Note: Some building codes require a minimum run of 6 inches at the narrow end of each stair.

Winders used in an "L" shaped stairway
Figure 23-4

Several additional fundamentals of stair construction are widely recognized. First, when building a staircase where a door opens at the top of a stairway, such as in a basement, a landing is always provided at the top of the stairs. This landing should be long enough so that the door, when fully open, does not extend over the first step. The door should be hung so that it does not reduce the width of the landing by more than 3½ inches. All landings should be at least as long as the stairway is wide but should not exceed 4 feet in length if there is no

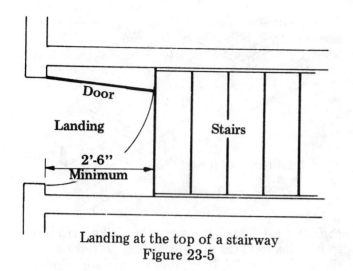

Landing at the top of a stairway
Figure 23-5

more than four risers. Most building codes require at least two handrails on stairways in public buildings regardless of how few risers there are in the stairway. In public structures, stairways more than 88 inches wide should have an intermediate handrail approximately half way between the two side handrails. When stairs are open on two sides, a protective railing should be provided on each side even in private residential construction. The top of the handrail should be not less than 30 inches and not more than 34 inches above the nosing of each stair. The handrail must extend the full length of the stairs and, in other than private residential construction, one of the handrails should extend 6 inches beyond the top and bottom riser. The

change in direction of the stairway. See Figure 23-5. Also, landings should be used to break any stairway which rises 12 feet or more. Landings are placed half way between the top and bottom of the staircase when possible.

Adequate head room must be maintained while ascending the stairs and many building codes prescribe minimum head room requirements. Generally, head room of 6 feet 8 inches will be enough for main stairs and 6 feet 6 inches will be enough for basement stairs. See Figure 23-6. The angle of incline of the stairway, which is determined by the rise and run of each stair, will dictate the well opening required to maintain adequate head room.

Most stairs are built with "nosing" or a protruding edge on the front of each tread. The projection of one tread over the tread below is usually about 1¼ inches and is designed to give the climber a wider base of support on each stair. See Figure 23-7. The nosing is not considered when calculating the run (horizontal distance) of each tread but must be considered when ordering materials.

The width of main stairs should be not less than 2 feet 8 inches clear of the handrail. Many stairs are designed with a distance of 3 feet 6 inches between the enclosing side walls. This will result in a stairway with a width of about three feet. Split level entrance stairs are even wider. For basement stairs, a clear width of 2 feet 6 inches may be adequate. Most codes provide widths of at least 36 inches for other than private residential construction and at least 30 inches for private residential stairways. The handrails can project up to 3½ inches into the prescribed width.

A continuous handrail should be used for at least one side of the stairway when there are

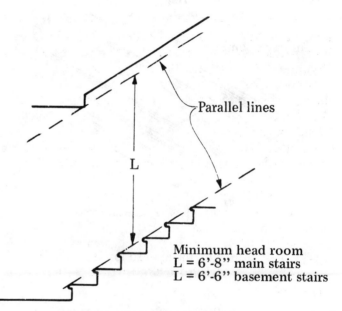

Minimum head room requirement
Figure 23-6

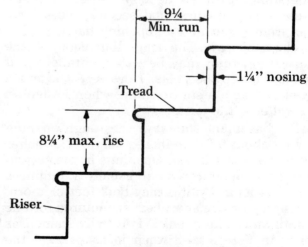

Nosing, rise and tread dimensions
Figure 23-7

end of each handrail should terminate in a newel post or safety terminal. On landings or horizontal areas, the height of the handrail should be 2 feet 10 inches. Handrails which project from the wall should allow at least 1½ inch clearance between the wall and the handrail.

Finally, every stairway should be designed so that each riser is equal in height and each tread is equal in width. Most building codes require that the rise and run of each stair be within ¼ inch of the rise and run of every other stair in that flight. Often stairways in public use are required to meet tolerances of less than ¼ inch.

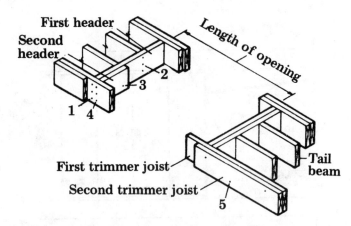

Framing for floor openings: (1) Nailing trimmer to first header; (2) nailing header to tail beams; (3) nailing header together; (4) nailing trimmer to second header; (5) nailing trimmers together

Figure 23-9

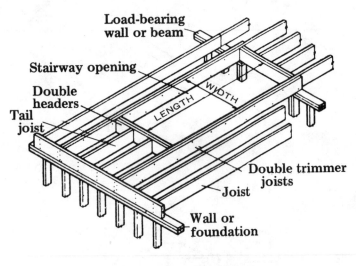

Stairway parallel to joists
Figure 23-8

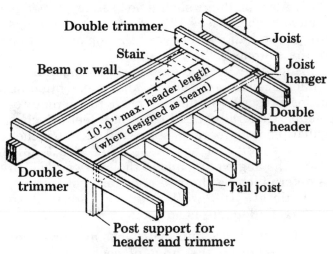

Stairway perpendicular to joists
Figure 23-10

Building Stairs

Stair construction is not complex and can be mastered by anyone with patience and an understanding of elementary carpentry. Let's assume that the first and second floors have been framed out and an opening has been left for the stairway. The long dimension of the stairway opening may be either parallel or at right angles to the joists. However, it is much easier to frame a stair opening when its length is parallel to the joists.

For basement stairways the rough opening may be about 9 feet 6 inches long by 32 inches wide (two joist spaces). Openings in the second floor for main stairways are usually a minimum of 10 feet long. Widths may be 3 feet or more. Depending on the short header required for one or both ends, the opening is usually framed as shown in Figure 23-8 when joists parallel the length of the opening. Nailing should conform to that shown in Figure 23-9.

When the length of the stair opening is perpendicular to the lengths of the joists, a long doubled header is required as in Figure 23-10. A header under these conditions without a supporting wall beneath is usually limited to a 10 foot length. A load-bearing wall under all or part of this opening simplifies the framing immensely, as the joists will then bear on the top plate of the wall rather than be supported at the header by joist hangers or other means.

The framing for an "L"-shaped stairway is usually supported in the basement by a post at the corner of the opening or by a load-bearing wall beneath. When a similar stair leads from the first to the second floor, the landing can be framed out. See Figure 23-11. The platform frame is nailed into the enclosing stud walls and provides a nailing area for the subfloor as well as a support for the stair carriages.

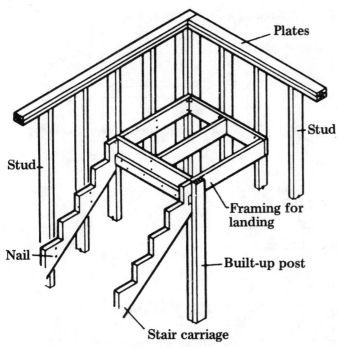

Framing for stair landing
Figure 23-11

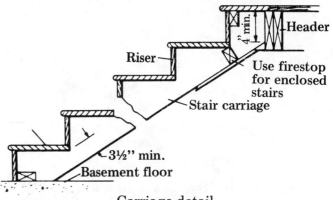

Carriage detail
Figure 23-12

Once the rough opening is framed, measure the exact distance from the first floor to the second floor. If an intermediate landing is planned, measure the vertical distance between the first floor and the landing. Estimate how much the finish flooring material on both the upper and lower floors will increase or reduce this distance and adjust the distance you measured accordingly. Select the riser and tread combination that best suits the well opening, total run available and materials on hand.

The carriage supports the load on the stairs and can be made in either of several ways. No matter whether you are building a "cleat", "sawed" or "housed" stairway, you must cut the carriage to fit and locate the position of

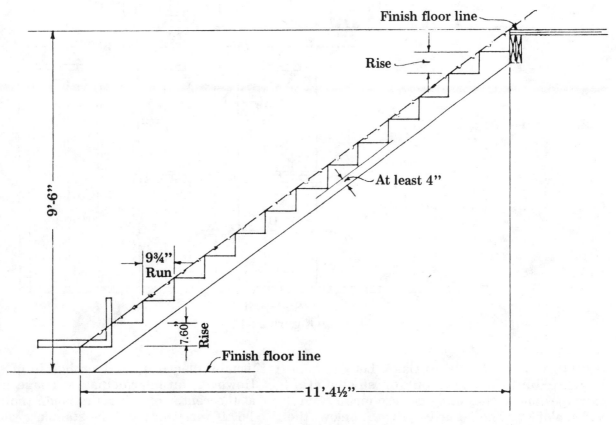

Layout of stair template or carriage
Figure 23-13

233

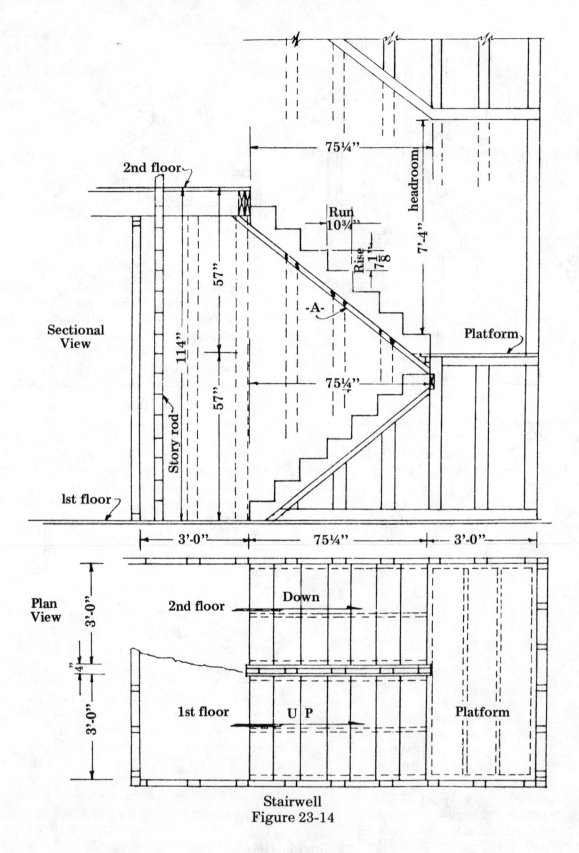

Sectional View

Plan View

Stairwell
Figure 23-14

treads and risers on the carriage. Let's assume you are building a cutout or sawed stair carriage. These stair carriages are made from 2 x 12 planks. The effective depth below the treads and riser notches must be at least 3½ inches (see Figure 23-12). Such carriages are usually placed only at each side of the stairs. However, an intermediate carriage is required at the center of the stairs when the treads are 1-1/16 inch thick and the stairs are wider than 2 feet 6 inches. Three carriages are also required when treads are 1-5/8 inch thick and stairs are

234

Angle
Of Incline And Cosine Of The Angle

Riser Height	Tread Width	Angle Of Incline	Cosine Of Angle	Riser Height	Tread Width	Angle Of Incline	Cosine Of Angle	Riser Height	Tread Width	Angle Of Incline	Cosine Of Angle
6"	11½"	27°33'	.88658	6 21/32"	10 27/32"	31°33'	.85225	7 5/16"	10 3/16"	35°40'	.81239
6 1/32"	11 15/32"	27°44'	.88507	6 11/16"	10 13/16"	31°44'	.85048	7 11/32"	10 5/32"	35°52'	.81035
6 1/16"	11 7/16"	27°56'	.88356	6 23/32"	10 25/32"	31°56'	.84869	7 3/8"	10 1/8"	36°4'	.80830
6 3/32"	11 13/32"	28°7'	.88202	6 3/4"	10 3/4"	32°7'	.84689	7 13/32"	10 3/32"	36°16'	.80625
6 1/8"	11 3/8"	28°18'	.88047	6 25/32"	10 23/32"	32°19'	.84508	7 7/16"	10 1/16"	36°28'	.80417
6 5/32"	11 11/32"	28°29'	.87891	6 13/16"	10 11/16"	32°31'	.84326	7 15/32"	10 1/32"	36°40'	.80209
6 3/16"	11 5/16"	28°41'	.87733	6 27/32"	10 21/32"	32°43'	.84142	7 1/2"	10"	36°52'	.80000
6 7/32"	11 9/32"	28°52'	.87575	6 7/8"	10 5/8"	32°54'	.83956	7 17/32"	9 31/32"	37°4'	.79789
6 1/4"	11 1/4"	29°3'	.87415	6 29/32"	10 19/32"	33°6'	.83771	7 9/16"	9 15/16"	37°16'	.79579
6 9/32"	11 7/32"	29°15'	.87255	6 15/16"	10 9/16"	33°18'	.83583	7 5/8"	9 7/8"	37°40'	.79152
6 5/16"	11 3/16"	29°26'	.87093	6 31/32"	10 17/32"	33°30'	.83395	7 11/16"	9 13/16"	38°5'	.78717
6 11/32"	11 5/32"	29°37'	.86929	7"	10 1/2"	33°41'	.83206	7 3/4"	9 3/4"	38°29'	.78281
6 3/8"	11 1/8"	29°49'	.86764	7 1/32"	10 15/32"	33°53'	.83014	7 13/16"	9 11/16"	38°53'	.77842
6 13/32"	11 3/32"	30°0'	.86598	7 1/16"	10 7/16"	34°5'	.82822	7 7/8"	9 5/8"	39°17'	.77396
6 7/16"	11 1/16"	30°12'	.86430	7 3/32"	10 13/32"	34°17'	.82628	7 15/16"	9 9/16"	39°42'	.76946
6 15/32"	11 1/32"	30°23'	.86262	7 1/8"	10 3/8"	34°29'	.82434	8"	9 1/2"	40°6'	.76492
6 1/2"	11"	30°35'	.86093	7 5/32"	10 11/32"	34°41'	.82237	8 1/16"	9 7/16"	40°30'	.76031
6 17/32"	10 31/32"	30°46'	.85922	7 3/16"	10 5/16"	34°53'	.82040	8 1/8"	9 3/8"	40°55'	.75569
6 9/16"	10 15/16"	30°58'	.85749	7 7/32"	10 9/32"	35°4'	.81841	8 3/16"	9 5/16"	41°19'	.75100
6 19/32"	10 29/32"	31°9'	.85576	7 1/4"	10 1/4"	35°16'	.81642	8 1/4"	9 1/4"	41°44'	.74628
6 5/8"	10 7/8"	31°21'	.85400	7 9/32"	10 7/32"	35°28'	.81441				

NOTE: All tread and riser combinations above follow the "rule of 17½"-riser height plus tread width equals 17½ inches.

The total run of the stairway divided by the cosine of the angle of incline equals the carriage length.

Table 23-15

wider than 3 feet. The carriages are fastened to the joist header at the top of the stairway or rest on a supporting ledger nailed to the header.

When you lay out the plank stringer for the stairs, the steel square should be used as shown in Figure 23-13. If the stairs are to be made in a mill, it will only be necessary to lay out a template on a piece of 1 x 8 so that the head room, line of pitch, and lines necessary for the erection of the bridgework can be determined.

How To Make And Install Carriages

1. Find the exact distance from the top of one floor level to the top of the floor above.

2. Lay off this distance on a story rod. This rod may be made of any smooth faced board. A piece of finish flooring makes a good rod. See sectional view, Figure 23-14.

3. Space the required number of risers with dividers equally over the distance marked on the rod. The number of risers is found by dividing the total height of the story from the top of one floor to the top of the next floor by the height of the stair riser. The higher each riser is, the fewer risers and treads will be needed, and consequently the less space the stairs will take up. This is the reason why the width of the risers is often increased and the width of the treads decreased. This practice is permissible in some instances but the rise should not exceed 8 inches

and the tread should not be less than 9½ inches. Assume that the total height from the top of one floor to the top of the next floor is 9 feet 6 inches, or 114 inches, and that the riser is to be approximately 7½ inches. 114 + 7½ would give 15-1/5 risers. However, the number of risers must be a whole number. Since the nearest whole number to 15-1/5 is 15, it may be assumed that there are to be 15 risers. 114 inches ÷ 15 = 7.6 or approximately 7-5/8 for the height of each riser. Now use Table 23-15 to find the tread width. There is always one less tread than riser, so in this case there will be 14 risers, and 14 times 9-7/8 inches is 11 feet 6¼ inches, the total run or horizontal distance. Many professional stair builders use a book of stair design tables to find the correct tread and riser combination for the floor to floor rise they are figuring.

4. Select a sound 2 by 12, 18 to 20 inches longer than the carriage length. First you must cut the floor line which will rest squarely on the lower floor. Lay the body or long end of the framing square across one corner of the plank. Move the square so that the tread distance intersects the top of the plank on the body and the rise distance intersects the tongue on the top of the plank. See Figure 23-16, *A*.

5. Sound craftsmanship requires that these dimensions be marked and cut accurately. A set

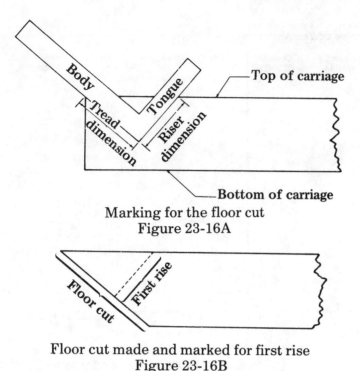

Marking for the floor cut
Figure 23-16A

Floor cut made and marked for first rise
Figure 23-16B

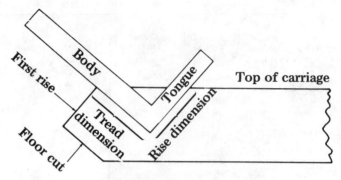

Floor and first rise cuts made and measuring
for first tread and second rise
Figure 23-16C

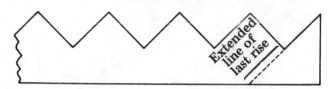

Cutting top end of carriage
Figure 23-16D

of stair gauges or some device to hold the correct dimensions will be useful here. Mark the outline of both the rise and tread dimensions on the plank. Extend the tread dimension in a straight line to the bottom edge of the material and cut along this line. This is the floor cut. If this is to be a cutout carriage, cut along the line marked for the first rise (See Figure 23-16, B).

6. Next, place the tread dimension on the square body so that it meets the top edge of the first rise and the rise dimension on the tongue so that it meets the top edge of the carriage. See Figure 23-16, C.

7. Repeat this process marking and cutting tread and rise dimensions for as many steps as there are in the stairway. When the last rise is reached, extend the line of the last rise to the bottom edge of the carriage. See Figure 23-16, D. Cut along this extended line if the carriage is to meet the end of the well hole as illustrated in Figure 23-17, C. If the carriage is to meet the well hole as in 23-17, A or 23-17, B, allow the appropriate additional material beyond the last rise and cut parallel to the last rise. For additional strength, the carriage may be supported below the header or extend beyond the end of the well opening below the upper floor.

8. After the carriages have been cut, they should be "dropped" or lowered to allow for the thickness of the tread and the finish flooring. You will recall that every rise dimension must be the same within a ¼ inch tolerance. The carriage has been designed to reach between the finished first floor and the finished second floor or landing. If the carriage is installed on the subfloor, it will be lower than the height you planned for by the thickness of the finished flooring material. However, the tread material will raise the level of the tread surface above the top of each cutout on the carriage. Usually the tread thickness will be greater than the depth of the finish flooring material. When this is the case, this difference must be cut from the bottom end of each carriage parallel to the floor cut. In Figure 23-18, 5/16 inch has been marked to be cut off the lower end of the carriage. The height of the finish flooring material, ¾ inch, has been subtracted from the thickness of the tread material, 1-1/16 inch. When the stairway is completed, the rise of each step, including the

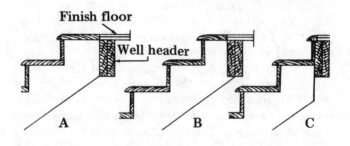

Three methods for anchoring the upper end of a stairway
Figure 23-17A Full thread width extension
Figure 23-17B Partial tread width extension
Figure 23-17C Top riser flush against header

236

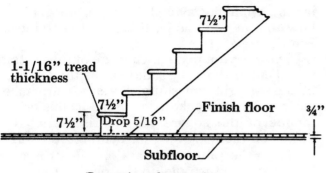

Dropping the carriage
Figure 23-18

first step and last step, will be 7½ inches. When the carriage rests on the finished floor level, the carriage should be "dropped" the full thickness of the treads. Once the carriage is cut and "dropped" it should be checked for fit. Each tread cutout should be level when the carriage is in place.

Note: If a single run is to be used with no platform between the floors, a template may be laid out on a sheathing board. This is to be used as a guide for the erection of the bridgework. See Figure 23-19.

9. Lay out and cut at least two planks for the carriages. The blocks cut out of one carriage may be nailed on the top of a 2 x 4 to make a third carriage.

10. Space the three carriages as shown by dotted lines in plan view, Figure 23-14.

11. Install the three carriages against the header of the floor above as shown at A, Figure 23-19.

Note: Keep the side carriages about ½ inch from the face of the partition studs if the stairs are to be enclosed. This is to allow a space between the outside face of the carriage and the inside face of the stair partition. See plan view, Figure 23-14. If the stairs are to be enclosed on both sides, it is best to erect the partitions before placing the stair carriages so that the carriages may be spiked to the studs for support.

12. Nail boards on the carriages at the tread cuts to provide temporary treads.

Layout Of A Return Or Platform Type Of Stairway

Assume that the staircase shown in Figure 23-20 is to be laid out. The total rise between the top of the first floor and the top of the second floor is 9 feet 6 inches. See Figure 23-14. The width of the platform is 3 feet and the run of each flight of stairs is 75¼ inches.

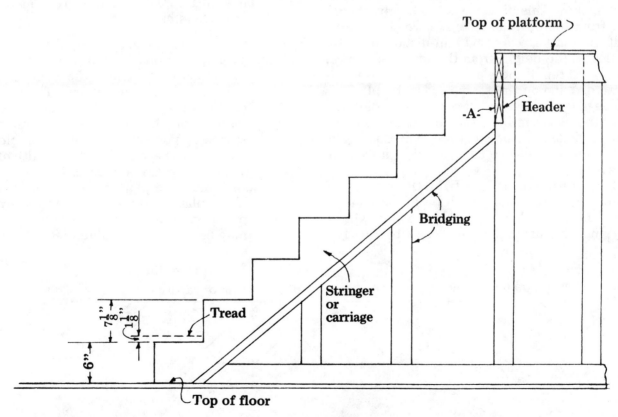

Placement of stringer
Figure 23-19

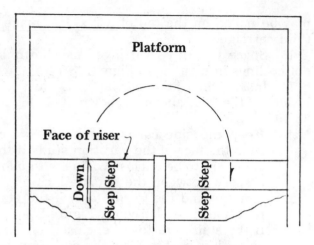

Half turn or U shape stair arrangement
Figure 23-20

A story rod is first laid out by marking the exact distance between the top of the first floor and the top of the second floor on a rod about 11 feet long. This distance is 114 inches. A pair of dividers is set at a little over 7 inches and the rod is stepped off. If there is not an even number of steps, the setting of the dividers should be changed and the process repeated until the last division falls on the 114 inch mark. If the spacing is correctly performed, there should be 16 divisions of 7.13 inches or approximately 7-1/8 inches. This distance is the height of each riser from top of tread to top of tread.

If the platform is to be mid-distant between the floors, the height from the top of the lower floor to the top of the platform is taken from the eighth division on the story rod. If the platform is to be placed above or below the mid-point, the top of the platform floor should line up with one of the divisions on the story rod. The flooring of the platform should be considered as a tread of the stairs.

The approach at the first floor and the platform landing should each be at least 3 feet wide. The entire length of the stairway will be 75¼ inches plus 3 feet for each of the two landings making a total of 12 feet 3¼ inches between the two end partitions of the staircase. See Figure 23-14.

The stair template or carriage of these stairs is laid out in about the same way as the template of the straight run stairs. Six steps are taken along a 2 x 10 plank, using 7-1/8 inches on the tongue of the square for the rise, and 10¾ inches on the body for run. See Figure 23-21.

From the top tread mark on the stringer, a line is squared parallel to the riser marks. This line is shown as the top header cut at *A*, Figure 23-19 and is the portion of the stringer which is fastened to the platform header.

The bottom stringer cut, which fits on top of the floor, is found by squaring from the last riser mark. This line is shown as the floor cut and is the dotted line at *B*, Figure 23-21. The solid line parallel to this dotted line shows the deduction to be made when the treads are to be nailed above the tread marks where the stringer is cut out. See the dotted lines showing the tread in position in Figure 23-19.

When the finish stairs are to be built at a mill, temporary stairs are built of two or three cutout stringers which are nailed to the header of the platform. Temporary treads are nailed in place to provide a stairway for the mechanics working in the building. The stringers may also be used as a template for the erection of the bridging of the stairs.

How To Build A Stair Platform

Note: In placing a platform in a stairway, the partitions that enclose the stairs at this point should be erected first. The location of the top of the platform may then be marked on the studs and the platform spiked in position when it is completed.

1. Determine the length and the width of the platform by taking the distances from between the studs where the platform is to be located.

Note: If the platform is to be supported by the partition studs on all four sides, 2 x 4's may be used for building the platform. If

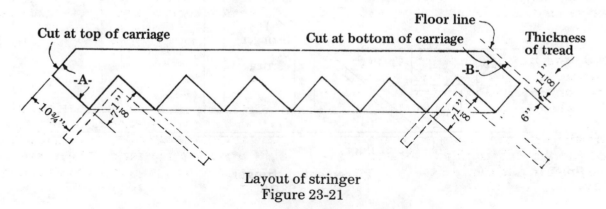

Layout of stringer
Figure 23-21

the platform cannot be well supported by the studs or if there is a span of over 6 feet in the platform, 2 x 6's should be used in making the platform.

The platform is framed as shown by the dotted lines in the plan view, Figure 23-14.

2. Build the framework of the platform to the proper size with 2 x 4's or 2 x 6's. Use 20d nails and space the middle members 16 inches on center.

3. Mark the height of the platform on the studs from the story rod.

4. Temporarily spike the platform in position at these marks.

Caution: Remember that the marks on the rod represent the top of the treads and also represent the top of the finish floor of the platform.

5. Spike the platform to all the supporting partition studs when it is leveled and in its proper position.

6. Lay the flooring on the platform.

Note: The carriages for the platform type of stairs should be laid out, cut and erected in the same manner as those of the straight stairs and as shown in Figures 23-19 and 23-21.

Stair Bridging

Stair bridging is the 2 x 4 framing that supports the staircase. The method of building the bridging depends upon the type of staircase. In Figure 23-20, the 2 x 4 framing at each side of the staircase may completely enclose the stairs. The studs would extend from the floor to the ceiling.

In Figure 23-19, the studs are shown extending from the floor to the bottom of the stairs only. The staircase is open from the top of the treads to the ceiling. This bridging is framed the same as a partition except that the top plate is run parallel to the slope of the stairs. The 2 x 4 furring at the underside of the stairs is shown in the sectional view of Figure 23-14. This furring provides a base to which the lath and plaster may be secured, thus forming an enclosure for the underside of the stairs.

The partitions or bridging are often laid out, erected, and finished before the stairs are constructed. The finish stairs are then built as a separate unit and fitted to the partitions or bridging when the rest of the interior finish is applied.

Sometimes the partitions and bridgework are erected and the stair carriages are installed so that the finish treads may be fitted to them after the walls have been plastered.

When the finish stairs are installed at the time the partitions are set, they should be thoroughly covered and protected until after the interior trim is finished.

For any of these methods, stair templates should be laid out and the bridging constructed in accordance with them.

How To Erect The Bridging Of Platform Stairs

Note: The bridging of the platform type of stairs may be more difficult to install than that of the straight type because very often one of the enclosing partitions runs only from the first floor to the bottom of the stair stringer of the upper flight as shown by dotted lines in sectional view, Figure 23-14. The bridging of the lower flight runs from the bottom of the lower stair carriage to the floor line. See the solid lines.

1. Lay out and erect the partition that runs from the bottom of the stair carriage of the lower flight to the floor.

Note: If the rough stair carriage is to be permanent, the top plate of the partition may be built to support it. If the finish stairs are to be made in the mill and an open stringer is to be built at this time, the bridging should be 2 inches below the bottom of the carriage or template. This allowance is to provide clearance between the bottom back of the stair risers and the top of the stair bridging.

2. Set the stair carriages and partitions so that they are the same width at the platform level and at the floor line. This is important if an open stringer is to be used in the finish stairs. See the dotted lines showing carriage spacing in the plan view of Figure 23-14.

3. Place 2 x 4's that extend from the upper bridging to the floor as shown at A in the sectional view of Figure 23-14. These 2 x 4's should be nailed to the bottom edges of the bridging and to the flooring of the first floor. Find the cuts of these 2 x 4's by placing them in position and marking along the bottom edges of the bridging for the top cut.

4. Spike the 2 x 4's in position to the floor and to the bridging.

Layout Of Winders

Generally the L and U types of stairs are built with a straight platform but where the stair space is further restricted, risers and treads are built upon the platform. These are called winders. Winders are not recommended and should be avoided if possible. However if

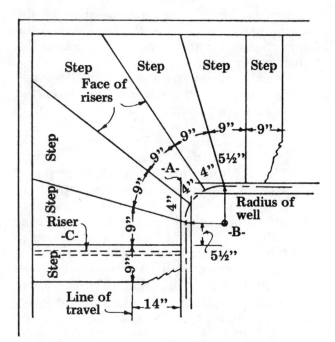

Quarter turn winder arrangement
Figure 23-22

the straight flight. See the 9 inch spacings in Figure 23-22. The spacing is started from the face of the riser C of the straight flight. These points on the circumference are then connected by straight lines to the 4 inch spacings on the small circumference. These straight lines give the locations and show the arrangement of the winders.

Figure 23-23 shows the arrangement of the winders in a U-shaped stairway. The diameter of the cylinder is shown but the construction at this point could also be square as shown by dotted lines. The circular lines in this case are merely to show the method of spacing the risers at the inside of the turn. In this type of stairs, only a certain number of risers may be included in the half turn. The horizontal distance between the two inside stringers would necessarily be 12 inches in order to provide 10½ inch treads at the line of travel. The method of spacing the risers is similar to that of the L type of stairs.

winders are necessary they can be built so they are reasonably safe.

The arrangement of the winders at the platform of the stairs is dependent upon the number of risers that are included around the turn. However, the fundamental principle is that the winders should be so placed that the width of the tread at the line of travel should be the same as the width of the treads throughout the entire staircase. The line of travel is taken to be 14 inches from the inside edge of the internal stringer or hand rail (Figure 23-22). Since each of the straight treads of these stairs is 9 inches wide, each of the winders should be 9 inches wide at this 14 inch point.

The lines of the risers should not converge to the point A, Figure 23-22 but rather to a point B which lies beyond the intersection of the inside stringers of the stairs. The distance from B to the stringer is the radius of the stair cylinder. This does not necessarily mean that the intersection of the stair stringers at this point must be curved. It merely locates the center of the quarter circle and shows the quarter circumference which is divided by the number of risers in the platform. These divisions give the width of the treads at the inside stringer. See the 4 inch spacings in Figure 23-22.

With a radius equal to the distance from point B to a point 14 inches from the inside stringer, another arc is drawn using B as a center. The circumference of this arc is spaced in divisions equal to the width of the treads of

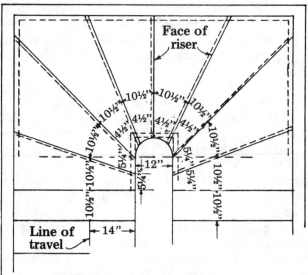

Half turn winder arrangement
Figure 23-23

How To Build Winders On The Platform

Note: In building winders, a level platform is first built as in an L or U shaped stairway. The winder layout is made on the level platform and the winder framework is then secured to the top of this platform.

1. Lay out the locations of the winders as described in this chapter.

2. Build a framework of 2 x 4's to represent the outline of the winders. Keep back from the winder layout line the thickness of the finish riser that is to be nailed on the edge of the framework. See the dotted lines showing the outlines of the winder platforms in Figure 23-23.

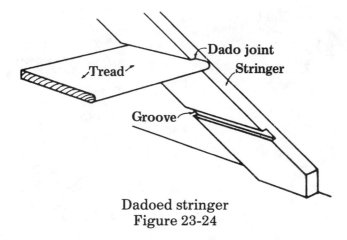

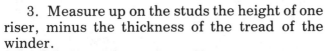

Dadoed stringer
Figure 23-24

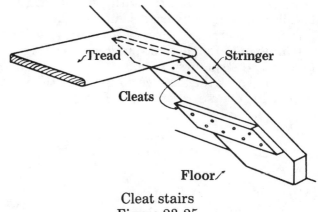

Cleat stairs
Figure 23-25

3. Measure up on the studs the height of one riser, minus the thickness of the tread of the winder.

4. Temporarily nail the framework to the stud at the mark and level the framework.

5. When it is level, spike it securely to all the supporting studs.

6. If there are more winders in the platform, follow the same procedure, being sure to follow the winder layout marked on the platform.

7. Brace the winder platforms by running cross bracing 2 x 4's where the spaces are more than 16 inches in the framework of the riser platforms.

Cleat And Dado Stairways

Cleat stairways (consisting only of a carriage and tread) should have 2 by 10 inch stock for the carriage. This design is not as desirable as the cutout carriage style where a wall is adjacent to one side of the stairway or where appearance of the finished stairway is important. The stringers should be at least 1-5/8 inch thick and wide enough to give a full width bearing for the tread. If cleats are used, they should be at least 25/32 inch thick, 3 inches wide and as long as the width of the tread. The treads can be only 1-1/16 inch thick unless the stairs are more than 3 feet wide.

A similar open stairway uses dado joints instead of cleats to support the treads. See Figure 23-24. This type of construction is quite common in steep stairs or ladders for attic or scuttle openings. If dado joints are used, they should be only one third as deep as the stringer is thick.

How To Build Cleated Or Dadoed Stairs

1. Select the required stock. Use clear dressed stock free from defects.

2. Lay out the stringers in the same way as in laying out carriages.

Note: If the treads are to butt against the stringers and are to be supported by cleats, the tread marks on the stringers represent the tops of the finish treads and the top of the cleat should be the thickness of the tread below this line. See Figure 23-25. If the treads are to be dadoed into the stringers, assume that the tread marks on the stringer represent the tops of the dado cuts.

3. Determine the length of the treads and cut the required number to this length.

4. Cut the required number of cleats and chamfer the edges that will show.

5. Nail the cleats to the proper marks below the tread marks on the stringers. Use nails long enough to reach within ½ inch of the combined thickness of the stringers and cleats.

6. Assemble the treads in place on the cleats or in the dadoes and nail through the stringers into the ends of the treads. Use 16d common nails if the stringers are 1-5/8 inch thick and cleats are used, or 10d casing nails if dado joints are used.

7. Square the assembled stairs and fasten them in place in about the same way as stair carriages are fastened.

Note: If the lower floor and side wall is of masonry, some means should be used to fasten the stringers firmly to these surfaces. Expansive shields and lag screws or wood blocks inserted into the masonry may be used for this purpose.

Full Stringer Stairways

A somewhat more finished staircase for a fully enclosed stairway combines the rough notched carriage with a finish stringer along each side (Figure 23-26). The finish stringer is fastened to the wall before carriages are fastened. Treads and risers are cut to fit snugly between the stringers and are fastened to the rough carriage with finishing nails (Figure 23-26). This may be varied somewhat by nailing the rough carriage directly to the wall and notching the finished stringer to fit (Figure 23-27). The stringers are laid out with the same rise and run as the stair carriages, but they are cut out in reverse as shown in Figure 23-27.

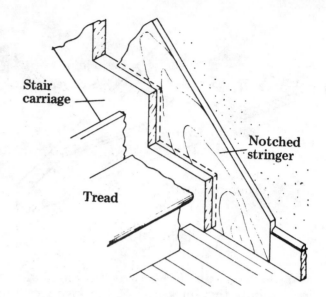

Cut out or notched stringer
Figure 23-27

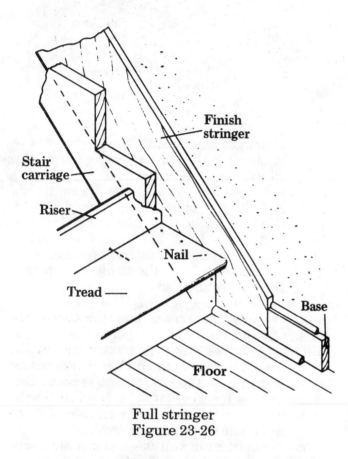

Full stringer
Figure 23-26

The risers are butted and nailed to the riser cuts of the wall stringers and the assembled stringers and risers are laid over the carriage. The assembly is then adjusted and nailed so that the tread cuts of the stringers fit against the tread cuts of the carriages. The treads are then nailed on the tread cuts of tne carriage and butted to the stringers the same as in Figure 23-26. The finish stringer may be 25/32 inch or 1-1/16 inch thick and wide enough to cover the intersection of the tread and riser cut in the carriage and to reach about 2 inches beyond the tread nosing. This width is usually 12 inches to 14 inches. If the stringer is not cut out at the riser and tread marks, it is laid out in about the same way as a rough carriage. Only the level cut for the floor, the plumb cut for the header, and the plumb cuts at the top and bottom of the stringer for the baseboards are made. This is perhaps the best type of construction to use when the treads and risers are to be nailed to the

carriages. This method saves time and labor in installing stairs but it is difficult to prevent squeaks when the stairs are stepped on. Another disadvantage is that any shrinkage of the frame of the building causes the joints to open up and to present an unsightly appearance. The notched stringer method has the advantage of having the two stringers tied together by the nailing of the risers to them. This prevents the two side stringers from spreading and showing open joints at the ends of the treads. Since the risers are nailed to the stringers, the face nailing of the risers to the carriage as in Figure 23-26 is eliminated. This type of stringer is used where the carriages were fitted permanently to the bridgework at the time the building was framed.

Sometimes the treads are allowed to run underneath the tread cut of the stringer. This makes it necessary to notch the tread at the nosing to fit around the stringer.

How To Build Stairs With A Full Stringer
Note: It is assumed that the temporary carriages have been removed and that the finish lumber to be used in the stairs is clear stock and sanded.

1. Lay out and cut the stair carriages as outlined above for cut out stringer carriages. Use 1-5/8 inch stock.

2. Select the stair stringers of the proper thickness, width and length and lay out the exact length and the bottom and top cuts. Make the line of the bottom cut level to meet the floor line and the line of the baseboard cuts plumb to

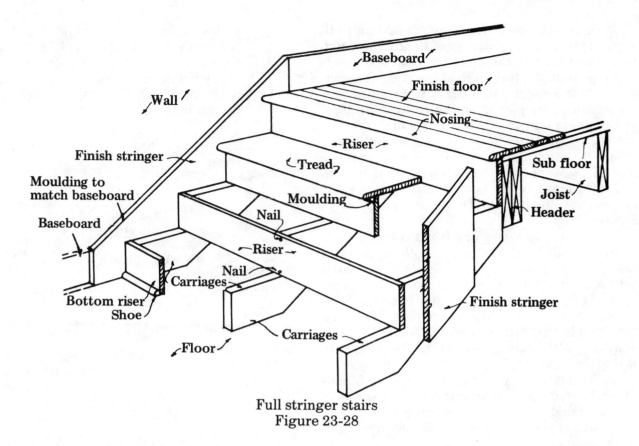

Full stringer stairs
Figure 23-28

meet the baseboard at the top and bottom of the stair. See Figure 23-28.

3. Lay out a right and a left hand stringer and cut along the bottom and top marks on the stringer.

4. Nail the right hand stringer to the stair carriage, keeping the bottom level cut of both pieces even and the top edge of the stringers at least 4 inches above the cutouts of the carriage. See Figure 23-28.

Note: Be sure the bottom edge of the carriage is parallel with that of the stringer.

5. Nail the left hand stringer to a carriage in the same manner.

6. Place the built-up stringers in their proper places against the header and side walls of the stair opening. Nail them to the headers and side walls.

7. Select the required number of riser boards. These are generally 25/32 inch thick, 7½ inches wide and as long as the distance between the inside faces of the two stringers.

8. Rip a riser board for the first riser to a width equal to the height of the first tread cut of the carriage above the floor. See the bottom riser of Figure 23-28. Square the ends of this riser to length so that it fits tightly against the finish stringer on each side of the stairs.

9. Face nail the riser board to the riser cuts of the three carriages with two 8d finishing nails in each carriage. Keep the nails about 1 inch

from the top and bottom edges of the board. See Figure 23-28.

Note: The nails at the top will be covered by the moulding that is to be placed underneath the tread nosing and the nails at the bottom of the riser will be covered by the floor shoe, or on the other risers by the thickness of the tread.

10. Cut, fit and nail the remaining riser boards in a similar manner.

11. Cut the treads to the same length as the riser boards and fit them in place. Face nail them with three 8d finishing nails at each tread cut of each carriage.

Note: After each tread is face nailed, drive 8d common nails through the back of each riser board into the back edge of each tread. Space the nails every 8 inches between the carriages.

12. Cut a piece of rabbeted nosing stock to the same length as a tread and face nail it to the top edge of the top riser and to the subfloor. Set all nails that will show and sand the surfaces where necessary.

13. Fit and nail cove moulding under the nose of each tread.

How To Build Stairs With Cut Out Stringers [Figure 23-27]

Note: When cut-out stringers are used, the same general procedure is followed except that the finish stringers are cut to fit against the tread cuts of the carriage and the face of the riser boards. The risers extend the full width of the stair opening. It is assumed that the carriages are permanently placed so the finish stringers, risers and treads will fit them.

1. Lay out the finish right and left hand stringers using the same figures as on the carriages. Mark the cuts at the top and bottom of the stringers for the floor and baseboard cuts. These cuts should be the same as shown in Figure 23-28.

2. Lay out the riser and tread cuts.

3. Cut along these marks with a crosscut saw. Be careful not to break the wood where the riser and tread cuts meet.

4. Temporarily nail a stringer to the wall on the left hand side of the stair opening. Keep the riser cuts of the stringer approximately 1¼ inch from cuts of the stringer on top of the tread cuts of the carriage.

5. Measure the distance between the finished walls at the top and bottom of the stairs to find the lengths of the riser boards.

6. Cut riser boards ½ inch shorter than these lengths.

7. Place the top riser board between the riser cut of the left hand stringer and the riser cut of the carriage.

8. Place the right hand stringer in the proper position on the right hand wall of the opening.

9. Mark the face of the riser board along the inside surface of both stringers. Be sure that there is a space of ¼ inch on each side between the outside of the stringers and the wall.

10. Follow the same procedure for the bottom and intermediate riser boards.

11. Remove the stringers and risers. Face nail the risers to the riser cuts of the stringer, keeping the tops of the risers tight against the tread cuts of the stringers and the face of the stringer in line with the marks on the faces of the risers. Nail the riser boards to both stringers in the same way.

12. Replace the assembled stringers on the carriages and adjust them so that the riser boards are tight against the riser cuts of the carriage and the tread cuts of the stringers are tight against the tread cuts of the carriages.

13. Nail the stringers to the walls in this location.

14. Cut, fit and nail the treads in the same way as for the built-up stringer.

Note: Some carpenters prefer to allow the treads to run under the stringers the same as the risers. See Figure 23-29. The

stairs in this case are built in practically the same way except that in laying out the cut-out stringer no deduction is made at the bottom riser mark for the thickness of the tread. The treads are notched to fit underneath the stringers at the nosings.

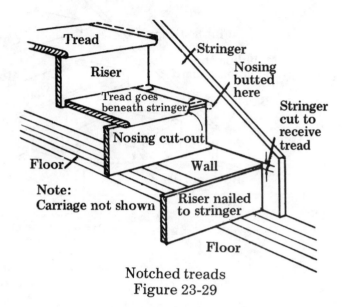

Notched treads
Figure 23-29

The Housed Stringer Stairway

Housed stair stringers are frequently considered a mill job but these stringers may be housed by the carpenter even if modern power woodworking tools are not available. The methods used in laying out the stringers, cutting the risers and treads and assembling the stairs are quite similar to these processes in other stairwork.

A closed type of housed stringer staircase is enclosed by walls on both sides of the staircase. The stringers are housed out to receive the ends of the treads and risers. This type is similar to the cut-out stringer stairway in which the treads and risers extend through the thickness of the stringer. In this case they only extend approximately 3/8 inch into the stringers. The treads and riser boards are then wedged into the dado joints of the stringers.

Figure 23-30 shows a housed stringer in various stages of construction. The layout of the treads and risers is similar to that of the cut-out type of stair stringer except that before the stringer is laid out, a mark is gauged about 1½ inch from the bottom edge on the face of each stringer. See line X-Y, Figure 23-30. This line acts as a measuring line the same as in laying out roof rafters. The purpose of using this line instead of the edge of the stringer is to provide room for the riser and tread boards to be supported by the wedges. The steel square at A.

244

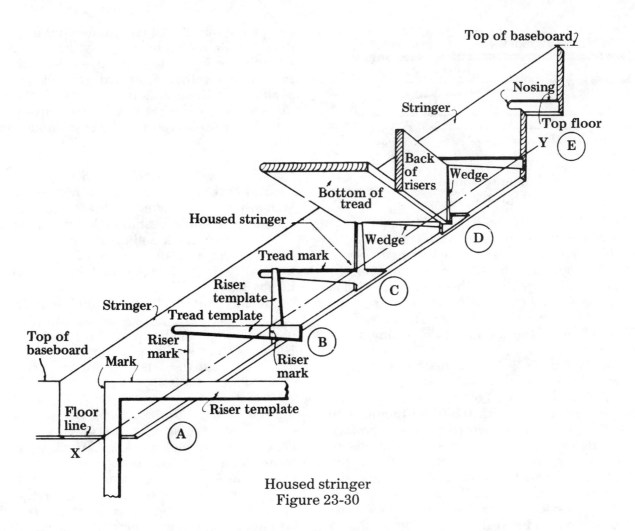

Housed stringer
Figure 23-30

Figure 23-30, shows the position in which the square is used on the gauge line in laying out the stringer.

The marks of the individual risers and treads represent the outside faces of the tread and riser boards when they are placed in the stringers. In order to form the proper dado outlines for the risers, treads and wedges, riser and tread templates are made and placed at the riser and tread marks on the stringer as shown in B, Figure 23-30.

The tread template is laid out by drawing a straight line on the face of a board about ¼ inch by 2 inches by 14 inches. On this line is drawn the exact end section of the tread stock to be used and also the outline of the wedges to be used to tighten the tread boards in the dado joints of the stringer. See Figure 23-31. The exact width of the tread stock including the nosing is measured and marked on the template as shown by the riser mark in Figure 23-31. This mark is also placed on the other face of the template so that the template may be used for both the right and left hand stringers. The center of the nosing profile is also located as shown in Figure 23-31. This mark is transferred

to the other side by drilling a small hole through the template.

The riser template is laid out in the same way by using the end section of a riser board and the wedge outline on a thin piece of board for the tread template. See Figure 23-32. This template is placed with its straight edge along the riser marks on the stringer. The outline of the template is then marked along the tapered side of the template on the stringer face. See B, Figure 23-30.

The tread template is placed in a similar manner with its straight edge on the tread mark

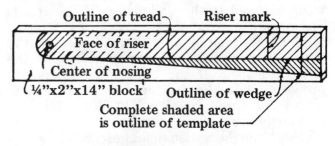

Tread template
Figure 23-31

245

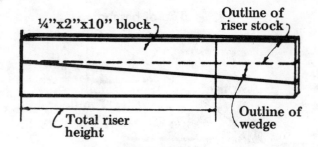

¼"x2"x10" block

Outline of
riser stock

Total riser
height

Outline of
wedge

Riser template
Figure 23-32

of the stringer and the riser mark (Figure 23-31) in line with the riser mark on the stringer. See B Figure 23-30. The location of the hole in the template is marked on the stringer with a scratch awl. This point shows the center of the hole that is to be bored in the stringer to form the round end of the dado to fit the nosing of the tread.

The layout at the top end of the stringer where the baseboard and the stringer meet is shown in E, Figure 23-30. The nosing at the floor line is laid out with the tread template to show the same nosing projection from the face of the riser as on the other treads. The nosing should be housed out in the same manner as the other treads. The riser cut directly below the nosing is cut completely through the stringer and the top riser is nailed to this surface. The depth of the tread and riser dado cuts is shown at C, Figure 23-30.

The length of the treads is determined and they are inserted into the tread dadoes. They are then wedged, glued, and nailed from the outside of the stringers and into the ends of the treads. All risers, except the top one, are cut to the same length as the treads. The top riser is about 1¼ inch longer as it does not fit into the dado cuts but extends to the outside of the stringers. The other riser boards are inserted, wedged, glued and nailed into the riser dadoes the same as the treads. Moulding is sometimes fitted between the stringers under the nosing of the treads. Figure 23-30 at D shows one riser and tread in place in the housed dadoes and the wedges glued to the undersides of the treads and risers.

Some stairs are built with a rabbeted joint at the back of the tread and also at the top of the riser. However, if the treads and risers are properly jointed, driven up tight, wedged and nailed in this position, the butt joint is satisfactory and saves much labor.

Stock for the various parts of stairs is generally obtained from a mill in partially finished form. Treads and risers may be obtained

completely machined and sanded but somewhat oversize. Rabbeted nosing stock for the edge of a landing or for the top step is usually obtainable in rough lengths. Standard wedges are also available. If the mill is furnished with the exact dimensions of the stair well, the parts can be completely machined and then assembled on the job.

How To Build Housed Stringers

The first step in building a housed stringer stairway is the making of the tread and riser template. It will be assumed that the tread stock is 1-1/16 inch thick and is nosed.

1. Select a straight piece of stock approximately ¼ inch thick, 2 inches wide and 14 inches long.

2. Lay out and plane one edge of the template straight and square and taper the opposite edge as shown in Figure 23-31.

Note: Select a similar piece of wood and lay out and cut the template for the risers. See Figure 23-32.

How To Lay Out And House The Stringers

1. Select the stringer stock. This should be at least 1-1/16 inch thick and from 10½ to 14 inches wide. The length depends on the length of the stairs. Allow about one foot at each end for the top and bottom cuts of the stringer.

2. Sometimes regular tread stock is used for stringers. If so, plane the nosed surface flat so that moulding may be fitted to this surface after the stairs are in place.

3. Plane the bottom edge of the stringer straight.

4. Set a marking gauge to about 1½ inch and gauge a line from this edge. See line X-Y, Figure 23-30.

5. Lay out the tread and riser marks with a steel square. Use a scratch awl or a fine hard lead pencil in marking the stringers.

Note: It is well to make a pitch board or to use a "fence" on the steel square. Both side stringers must be cut to exactly the same dimensions so that the stairs are square and sound.

6. Lay out the right and left hand stringers and check them for accuracy and length before doing any template or housing work. Figure 23-33 shows how to check the stringers for length and accuracy of layout.

7. Place the tread template with its straight edge exactly over a tread mark on the stringer. Adjust it so that the riser mark on the template is also exactly in line with the riser mark on the stringer. Make a point on the stringer by putting the scratch awl through the hole in the template

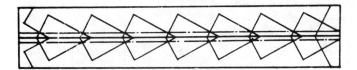

Checking stair layout
Figure 23-33

to locate the center of the nosing. Also mark along both sides of the template on the stringer. See B, Figure 23-30.

8. Bore a hole with a Forstner or center bit using the point marked through the tread template as a center. Bore the hole approximately 3/8 inch deep. Locate the center of a second hole so it will overlap the first and be within the top and bottom lines of the tread. See Figure 23-34.

9. Use the back of a 1½ inch butt chisel to chisel along the tread marks between the holes (Figure 23-34).

10. Cut 3/8 inch deep along the tread marks with a saw. Start with the tip of the saw at the holes and continue back the length of the tread mark.

Note: If using the back saw to cut the housed joints, tip the saw a trifle so that the edges of the joint will be undercut. This will allow a tighter joint between the top of the tread and the stringer cut and will help hold the wedge in between the bottom of the tread and the edge of the housing.

11. Chisel out the stock between the cuts. Take this stock out carefully and to a depth of about ¼ inch. Leave the remaining 1/8 inch to be taken out smoothly by the router plane.

12. Set the router plane blade to take a cut 3/8 inch deep and use it to bottom out the joint to an even depth.

13. Cut the other tread housings in the same manner. Naturally, a power router will speed and simplify steps 8 to 13 above and steps 15 and 16 below.

14. Mark the riser cuts by placing the riser template with its straight edge exactly over the riser mark on the stringer. Mark along the opposite edge of the template on the stringer.

15. Cut along these lines with the back saw. Chisel and rout out the stock the same as for the treads.

16. Finish cutting and routing for all the treads and risers including the nosings at the top of both stringers.

17. Make the top and bottom cuts of both stringers with a crosscut saw.

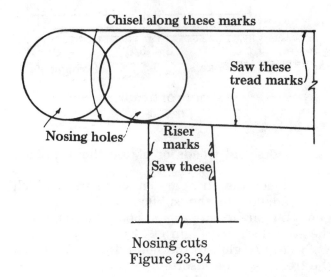

Nosing cuts
Figure 23-34

How To Assemble The Stairs

1. Select the tread stock. Use only clear stock free from imperfections and sanded to a finish on all surfaces that will show in the assembled stairs.

2. Square the pieces and cut them to length.

Note: Assume that the distance between the two walls of the stair well is 3 feet 6 inches.

3. Check the width of the stair well at the top and bottom and several intermediate points to see that no distance is less than 3 feet 6 inches so that the assembled stairs will easily fit between the walls.

4. Deduct from this assumed distance of 3 feet 6 inches twice the thickness of a stringer from the bottom of the housed joint to the outside face of the stringer (Figure 23-35). From this figure subtract 1 inch.

Note: Assuming that the stringers are 1-1/16 inch thick, the distance from the bottom of the housed joint to the outside face of the stringer would be 11/16 inch. Adding 11/16 inch for the other stringer would give 22/16 or 1-3/8 inch. 3 feet 6 inches minus 1-3/8 inches = 3 feet 4-5/8 inches. Subtracting 1 inch more would give 3 feet 3-5/8 inches, the length of the treads. The 1 inch is an allowance for fitting the assembled stairs in the well hole. This allowance is also made because the stairs have a tendency to spread when being assembled and the ends of all of the joints may not come up tight. The space between the stringer and wall will later be covered up with moulding.

5. Square and cut the treads and the nosed piece for the top step to length.

6. Rip the required number of wedges for

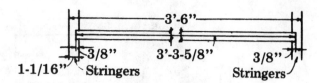

Length of treads
Figure 23-35

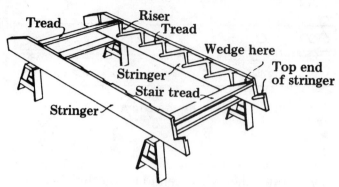

Partly assembled stairs
Figure 23-36

both risers and treads or obtain them already cut.

7. Place the stringers on saw horses which are toenailed to the subfloor and spaced far enough apart to properly support the length of the stringers. See Figure 23-36.

8. Apply glue to the housed joint in which the tread is to be inserted.

9. Insert a tread in the top housing of one stringer and tap it so the nosed section fits into the curved part of the housed joint.

10. Glue the wedge and drive it between the bottom side of the tread and the edge of the housing. Drive the tread and the wedge alternately until the tread nosing and the top surface of the tread fit perfectly against the edges of the housed joint that will show in the assembled stairs. The back edge of the tread must also be in line with the riser cut of the riser housing.

11. Drive an 8d common nail through the stringer into the nosing to pull the stringer up tightly against the tread end. Drive at least two more nails into the tread but use more if it is necessary to bring the tread up tightly against the bottom of the housed joint.

12. Insert the bottom tread in the bottom housing of the same stringer and fasten it to the stringer in the same manner.

13. Insert the opposite ends of these treads in the top and bottom housings of the opposite stringer and fasten them in the same manner.

14. Toenail the top edge of one stringer with 8d finishing nails to the tops of both saw horses. Be sure the stringer is straight. Place the steel square between the back edge of the top tread and the surface of the stringer. Bring the stairs into a square position at this point and toenail the loose stringer to the saw horses. Check the diagonally opposite corners of the stairs for squareness.

15. Insert the remaining treads and fasten them into the housings in the same manner.

Note: Be sure the back edge of each tread is perfectly flush with the front cut of each riser housing. If it does not reach this point, chisel off the riser cut until it is even with the back edge of the tread. If the tread projects beyond this point,

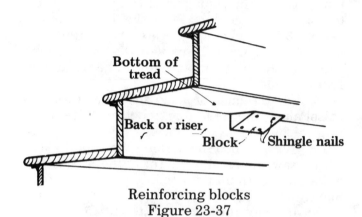

Reinforcing blocks
Figure 23-37

plane or chisel off the back edge of the tread very carefully to a straight line.

16. Cut the risers to the same length as the treads. The top riser will be about 1-3/8 inch longer than the rest as it must be face nailed to the stringers at the top cut.

17. Nail the top riser to the stringers and to the back edge of the top tread. Be sure the top of this riser is even with the bottom of the housed joint of the nosing.

18. Rip the bottom riser to width and insert it into the bottom riser housings in the same general manner as the treads.

19. Install the remaining risers and fasten them the same as the treads.

20. Nail the back of the risers to the back edges of the treads with 8d common nails. Space the nails about 8 inches apart.

21. Cut angle blocks from a 2 x 4 and glue and nail them in place with shingle nails. Put one block in the middle of the stair width at the intersection of the back surface of each riser and tread. See Figure 23-37.

22. Loosen the stringers from the horses, turn the stairs over and fit mouldings underneath the nosing. Nail them to both the riser and tread surfaces with 1¼ inch brads.

23. Nail through the top surfaces of the treads into the risers with 8d finishing nails.

Space the nails about 8 inches apart and set these nails.

Finally, place the stairs in the well hole.

1. Place the stairs in the well hole with the top riser against the header. Adjust the top edge of the housed joint for the nosing so it is level with the top of the finished floor. To do this, it may be necessary to shim the back of the top riser out from the face of the header on one side of the well.

2. Center the stairs between the two sides of the well and nail the riser securely to the header.

Note: If the finish floor has not been laid, be sure to use blocks of finish floor stock under the bottom ends of the stringers.

3. Locate the studs in the side walls and nail through the stringers into them with 10d or 12d finishing nails.

4. Insert the rabbeted nosed piece into the top nosing housed joints and fur it solidly to the top of the sub-floor or header over its entire length. Nail it to the header so that it will be forced tightly into the housed joints.

5. Set the nails and cover the stairs with building paper and wood cleats to protect the nosings and other surfaces.

6. Cut and fit mouldings on top of the stringers at the side walls of the well hole and under the top nosing of the stairs.

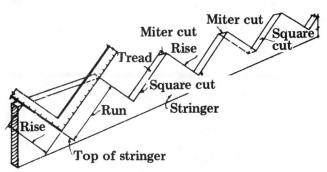

Layout of cut and mitered stringer
Figure 23-38

How To Build Housed And Open Stringers

Often housed stringer stairways are built along a single wall with an open stringer at the open side of the stairs.

1. Select the material for the open stringer. This is generally 1-1/16 inch x 11½ inches by the length of the stairs. Stock 25/32 inch thick is sometimes used but it is hardly strong enough, especially for stringers over 3 feet long.

2. Lay out the open stringer in the same manner as the notched stringer illustrated in Figure 23-27.

Note: Figure 23-38 shows how to lay out the

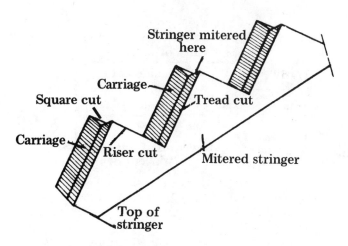

Assembly of carriage and stringer
Figure 23-39

stringer and how to miter the riser cuts.

3. Make the tread cuts of the stringer first. These cuts should be square with the face of the stringer.

4. Make the top and bottom cuts of the stringer. These cuts should also be square with the face.

5. Make the miter cuts along the riser marks. Use the riser marks as the long point of the miter.

6. Lay out and cut a carriage to fit behind the mitered stringer (23-39).

7. Adjust this carriage to the inside face of the stringer. Keep the tread cuts of the carriage even with the tread cuts of the stringer and the riser cuts of the carriage in line with the short ends of the miter cuts of the stringer. Nail the stringer and the carriage together temporarily.

8. Lay out and house the wall stringers as described previously for housed stringer stairways.

9. Determine the length of the risers.

Note: To find the length of the risers, lay out on the subfloor the position of the wall stringer and open stringer as shown in Figure 23-40. The outside stringer usually projects about 1 inch over the outside wall of the well hole. Measure the length of the riser as shown in Figure 23-40. This will give the length to the long point of the miter cut.

10. Cut a miter on one end of each riser and cut the opposite end square.

Note: Generally the top and bottom tread boards are assembled in the housings first and the stair frame is then fastened to the saw horses and squared as in assembling the housed stairs. The treads and risers may also be installed into the housed stringer first and then the open

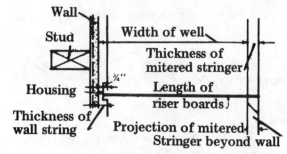

Length of mitered risers
Figure 23-40

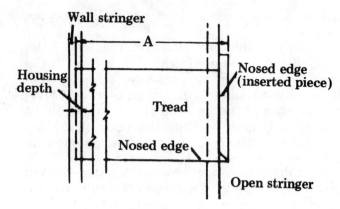

Returned tread
Figure 23-41

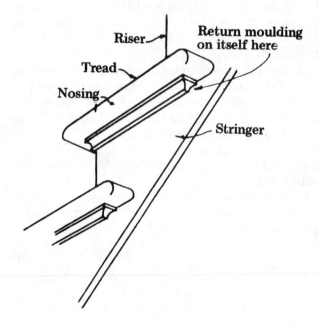

Moulding at returned treads
Figure 23-42

stringer assembled to the opposite ends of the treads and risers. This procedure sometimes makes it difficult to square the stairs and should be avoided, especially in long stairs. In short stairs it is satisfactory.

11. Place the square ends of the risers into the riser housings of the wall stringer. Nail, glue and wedge them as in the housed type of stairs.

12. Assemble the mitered ends of the risers to the mitered riser cuts of the open stringer. Keep the top edges of the riser boards even with the tread cuts of the open stringer. Nail the miters with finishing nails to the stair carriage riser cuts and to the miter of the open stringer.

13. Permanently nail the carriage to the open stringer.

14. Find the length of each tread by measuring from the bottom of the tread housing to the outside edge of the open stringer and adding on the amount the end of the tread will project over the open stringer. See A, Figure 23-41. This projection should be the same distance as that of the nosing over the riser on the front of the step.

15. Lay out and make the miter and straight cuts on the ends of the treads as shown in Figure 23-41.

16. Cut, glue and nail the mitered pieces to the ends of the treads.

Note: If balusters are to be tenoned into the treads, leave the end pieces loose until the hand rail has been erected.

17. Insert and fasten the treads the same as in the housed type. Face nail the treads to the tread cuts of the open stringer and to the risers.

18. Cut, fit and nail the moulding under the tread nosing at the face of the risers and under the return of the tread nosing at the stringer. Return the moulding on itself at the back edge of the tread. See Figure 23-42.

19. Erect the stairs against the header in the same way as in the housed stairs but be sure the outside stringer is parallel to the wall.

20. Nail the top of the stairs to the header and nail the wall stringer to the studs.

21. Cut and fit a moulding along the bottom edge of the open stringer where it meets the wall. Nail this moulding temporarily until the newel posts have been fitted to the stairs.

Chapter 24

Insulation

Most materials used in houses have some insulating value. Even air spaces between studs resist the passage of heat. However, when these stud spaces are filled or partially filled with a material high in resistance to heat transmission, namely thermal insulation, the stud space has many times the insulating value of the air alone.

The inflow of heat through outside walls and roofs in hot weather and its outflow during cold weather have important effects upon (a) the comfort of the occupants of a building and (b) the cost of providing either heating or cooling to maintain temperatures of acceptable limits for occupancy. During cold weather, high resistance to heat flow also means a saving in fuel. While the wood in the walls provides good insulation, commercial insulating materials are usually incorporated into exposed walls, ceilings, and floors to increase the resistance to heat passage. The use of insulation in warmer climates is justified with air conditioning, not only because of reduced operating costs but also because units of smaller capacity are required. Thus, whether from the standpoint of thermal insulation alone in cold climates or whether for the benefit of reducing cooling costs, adequate insulation should be a part of every building plan.

Insulating Materials

Commercial insulation is manufactured in a variety of forms and types, each with advantages for specific uses. Materials commonly used for insulation may be grouped in the following general classes: (1) flexible insulation (blanket and batt); (2) loose-fill insulation; (3) reflective insulation; (4) rigid insulation (structural); and (5) miscellaneous types.

The thermal properties of most building materials are known, and the rate of heat flow or coefficient of transmission for most combinations of construction can be calculated. This coefficient, or U-value, is a measure of heat transmission between air on the warm side and air on the cold side of the construction unit. The insulating value of the wall will vary with different types of construction, with materials used in construction, and with different types and thicknesses of insulation. Comparisons of U-values may be made and used to evaluate different combinations of materials and insulation based on overall heat loss, potential fuel savings, influence on comfort, and installation costs.

Air spaces add to the total resistance of a wall section to heat transmission, but an air space is not as effective as it would be if filled with an insulating material. Great importance is frequently given to dead-air spaces in speaking of a wall section. Actually the air is never dead in cells where there are differences in temperature on opposite sides of the space because the difference causes convection currents.

In some cases, the carpenter selects and installs insulation, while in others, the insulation contractor does the work. In either case, as a carpenter you must have at least a general knowledge of insulating methods. Insulation is now required by many building codes and should be installed in every home you work on. In most cases, the manufacturer of each kind of insulation furnishes instructions for installing that particular kind of insulation. These directions should be carefully followed.

Flexible Insulation

Flexible insulation is manufactured in two types, blanket and batt. Blanket insulation (Figure 24-1, A) is furnished in rolls or packages in widths suited to 16 and 24 inch stud and joist spacing. Usual thicknesses are 1½, 2, and 3 inches. The body of the blanket is made of felted mats of mineral or vegetable fibers, such as rock or glass wool, wood fiber, and cotton. Organic insulations are treated to make them resistant to fire, decay, insects, and vermin. Most blanket insulation is covered with paper or other sheet material with tabs on the sides for fastening to studs or joists. One covering sheet serves as a vapor barrier to resist movement of water vapor and should always face the warm side of the wall. Aluminum foil or asphalt or plastic laminated paper are commonly used as barrier materials.

Batt insulation (Figure 24-1, B) is also made of fibrous material preformed to thicknesses of 4 and 6 inches for 16 and 24 inch joist spacing. It is supplied with or without a vapor barrier. One

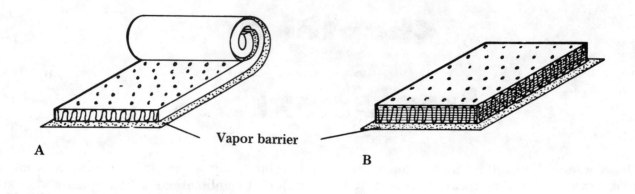

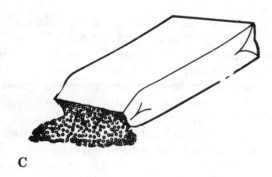

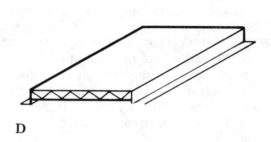

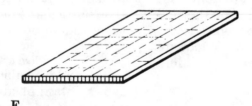

Types of insulation: *A*, Blanket; *B*, batt; *C*, fill;
D, reflective (one type), *E*, rigid
Figure 24-1

friction type of fibrous glass batt is supplied without a covering and is designed to remain in place without the normal fastening methods.

Loose Fill Insulation

Loose fill insulation (Figure 24-1, *C* and Figure 24-7, *C* is usually composed of materials used in bulk form, supplied in bags or bales, and placed by pouring, blowing or packing by hand. This includes rock or glass wool, wood fibers, shredded redwood bark, cork, wood pulp products, vermiculite, sawdust, and shavings.

Fill insulation is suited for use between first-floor ceiling joists in unheated attics. It is also used in sidewalls of existing houses that were not insulated during construction. Where no vapor barrier was installed during construction, suitable paint coatings, as described later in this chapter, should be used for vapor barriers when blown insulation is added to an existing house.

Reflective Insulation

Most materials reflect some radiant heat, and some materials have this property to a very high degree. Materials high in reflective properties include aluminum foil, sheet metal with tin coating, and paper products coated with a reflective oxide composition. Such materials can be used in enclosed stud spaces, in attics, and in similar locations to retard heat transfer by radiation. These reflective insulations are effective only when used where the reflective surface faces an air space at least ¾ inch or more deep. Where a reflective surface contacts another material, the reflective properties are lost and the material has little or no insulating value.

Reflective insulations are equally effective regardless of whether the reflective surface faces the warm or cold side. However, there is a decided difference in the equivalent conductance and the resistance to heat flow. The difference depends on (a) the orientation of the reflecting material and the dead air space, (b) the direction of heat flow (horizontal, up, or down), and (c) the mean summer or winter temperatures. Each possibility requires separate consideration. However, reflective insulation is perhaps more effective in preventing summer heat flow through ceilings and walls. It should likely be considered more for use in the southern portion of the United States than in the northern portion.

Reflective insulation of the foil type is sometimes applied to blankets and to the stud-surface side of gypsum lath. Metal foil suitably mounted on some supporting base makes an excellent vapor barrier. The type of reflective insulation shown in Figure 24-1, *D* includes reflective surfaces and air spaces between the outer sheets.

Rigid Insulation

Rigid insulation is usually a fiberboard material manufactured in sheet and other forms (Figure 24-1, *E*). However, rigid insulations are also made from such materials as inorganic fiber and glass fiber, though not commonly used in a house in this form. The most common types are made from processed wood, sugarcane, or other vegetable products. Structural insulating boards, in densities ranging from 15 to 31 pounds per cubic foot, are fabricated in such forms as building boards, roof decking, sheathing, and wallboard. While they have moderately good insulating properties, their primary purpose is structural.

Roof insulation is nonstructural and serves mainly to provide thermal resistance to heat flow in roofs. It is called "slab" or "block" insulation and is manufactured in rigid units ½ to 3 inches thick and usually 2 by 4 feet in size.

In house construction, perhaps the most

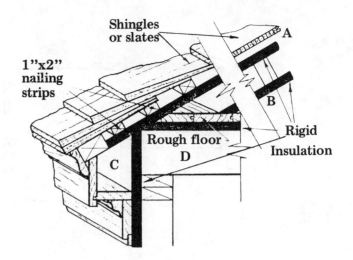

Rigid insulation at cornice
Figure 24-2

common forms of rigid insulation are sheathing and decorative coverings in sheets or in tile squares. Sheathing board is made in thicknesses of ½ and 25/32 inch. It is coated or impregnated with an asphalt compound to provide water resistance. Sheets are made in 2 by 8 foot size for horizontal application and 4 by 8 foot or longer for vertical application. Rigid insulation used as sheathing should be applied as described in Chapter 10. It may also be used to insulate at the cornice and roof lines of a building. Figure 24-2 shows how the material is installed underneath the roof boards (A), on the bottom edges of the rafters (B), behind the cornice members (C), and underneath the subflooring (D).

Miscellaneous Insulation

Some insulations do not fit in the classifications previously described, such as insulation blankets made up of multiple layers of corrugated paper. Other types, such as lightweight vermiculite and perlite aggregates, are sometimes used in plaster as a means of reducing heat transmission.

Other materials are foamed-in-place insulations, which include sprayed and plastic foam types. Sprayed insulation is usually inorganic fibrous material blown against a clean surface which has been primed with an adhesive coating. It is often left exposed for acoustical as well as insulating properties.

Expanded polystyrene and urethane plastic foams may be molded or foamed-in-place. Urethane insulation may also be applied by spraying. Polystyrene and urethane in board form can be obtained in thicknesses from ½ to 2 inches.

253

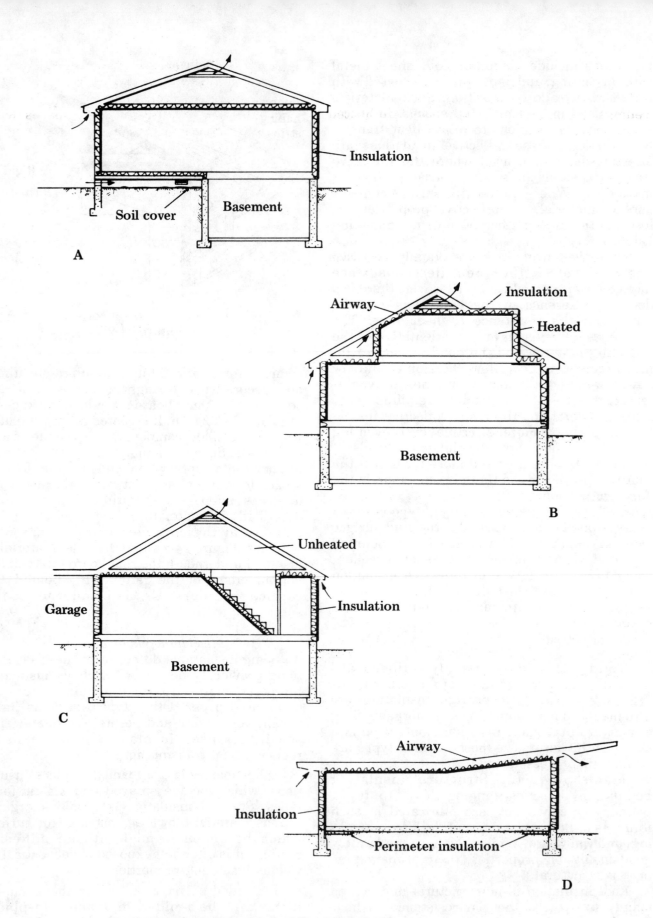

Placement of insulation: *A*, in walls, floor, and
ceiling; *B*, in 1½-story house; *C*, at attic
door; *D*, in flat roof
Figure 24-3

Where To Insulate

To reduce heat loss from the house during cold weather in most climates, all walls, ceilings, roofs, and floors that separate heated from unheated spaces should be insulated.

Insulation should be placed on all outside walls and in the ceiling (Figure 24-3, *A*). In houses involving unheated crawl spaces, it should be placed between the floor joists or around the wall perimeter. If a flexible type of insulation (blanket or batt) is used, it should be well supported between joists by slats and a galvanized wire mesh, or by a rigid board with the vapor barrier installed toward the subflooring. Press-fit or friction insulations fit tightly between joists and require only a small amount of support to hold them in place. Reflective insulation is often used for crawl spaces, but only one dead-air space should be assumed in calculating heat loss when the crawl space is ventilated. A ground cover of roll roofing or plastic film such as polyethylene should be placed on the soil of crawl spaces to decrease the moisture content of the space as well as of the wood members.

In 1½ story houses, insulation should be placed along all walls, floors, and ceilings that are adjacent to unheated areas (Figure 24-3, *B*). These include stairways, dwarf (knee) walls, and dormers. Provisions should be made for ventilation of the unheated areas.

Where attic space is unheated and a stairway is included, insulation should be used around the stairway as well as in the first floor ceiling (Figure 24-3, *C*). The door leading to the attic should be weatherstripped to prevent heat loss. Walls adjoining an unheated garage or porch should also be insulated.

In houses with flat or low pitched roofs (Figure 24-3, *D*), insulation should be used in the ceiling area with sufficient space allowed above for clear unobstructed ventilation between joists. Insulation should be used along the perimeter of houses built on slabs. A vapor barrier should be included under the slab.

In the summer, outside surfaces exposed to the direct rays of the sun may attain temperatures of 50 degrees fahrenheit or more above shade temperatures and, of course, tend to transfer this heat toward the inside of the house. Insulation in the walls and in attic areas retards the flow of heat and, consequently, less heat is transferred through such areas, resulting in improved summer comfort conditions.

Where air conditioning systems are used, insulation should be placed in all exposed ceilings and walls in the same manner as insulating against cold weather heat loss.

Shading of glass against direct rays of the sun and the use of insulated glass will aid in reducing the air conditioning load.

Ventilation of attic and roof spaces is an important adjunct to insulation. Without ventilation, an attic space may become very hot and hold the heat for many hours. Obviously, more heat will be transmitted through the ceiling when the attic temperature is 150 degrees fahrenheit than if it is 100 degrees to 120 degrees fahrenheit. Ventilation methods suggested for protection against cold weather condensation apply equally well to protection

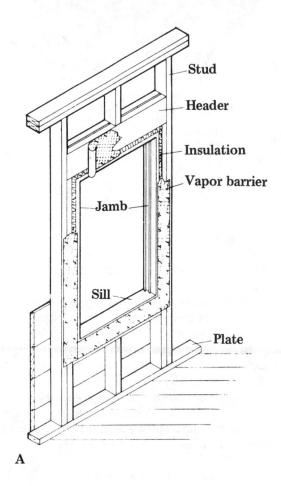

A

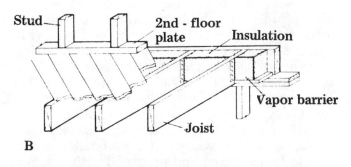

B

Precautions in insulating: *A*, Around openings; *B*, joist space in outside walls

Figure 24-4

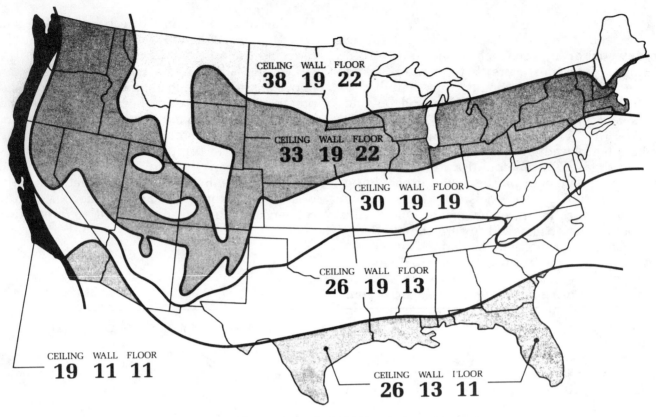

CEILING WALL FLOOR
38 19 22

CEILING WALL FLOOR
33 19 22

CEILING WALL FLOOR
30 19 19

CEILING WALL FLOOR
26 19 13

CEILING WALL FLOOR
19 11 11

CEILING WALL FLOOR
26 13 11

Recommended "R" values
Figure 24-5

against excessive hot weather roof temperatures.

The use of storm windows or insulated glass will greatly reduce heat loss. Almost twice as much heat loss occurs through a single glass as through a window glazed with insulated glass or protected by a storm sash. Furthermore, double glass will normally prevent surface condensation and frost from forming on inner glass surfaces in winter. When excessive condensation persists, paint failures or even decay of the sash rail or other parts can occur.

Precautions In Insulating

Areas over door and window frames and along side and head jambs also require insulation. Because these areas are filled with small sections of insulation, a vapor barrier must be used around the opening as well as over the header above the openings (Figure 24-4, *A*). Enveloping the entire wall eliminates the need for this type of vapor barrier installation.

In 1½ and 2 story houses and in basements, the area at the joist header at outside walls should be insulated and protected with a vapor barrier (Figure 24-4, *B*).

Insulation should be placed behind electrical outlet boxes and other utility connections in exposed walls to minimize condensation on cold surfaces.

The insulation industry has adopted "R" values as the measure of insulating performance. The higher the R rating, the greater the resistance to temperature transmission. It is always best to select insulation on the basis of cost per resistance unit rather than on cost per inch. R values are usually given on the insulation wrapper. Durability and resistance to flame spread and vermin should also be considered.

Energy savings are provided by resistance to heat flow and not thickness. Two different kinds of insulation may have the same thickness, but the one with the higher R value will perform better. For example, loose-fill mineral fiber (glass fiber or rock wool) insulation may have an R value as low as 2.2 per inch, while mineral fiber batts have an R value of about 3 per inch. Even if batts cost 30 to 40 percent more than loose fill per inch they might be a better investment.

Insulation batts are generally available in R-11 (about 4 inches thick), R-19 (about 6 inches), and R-22 (about 7 inches). If more than R-19 is recommended, batts should be combined. R-30

Ceilings, double layers of batts
R-38, Two layers of R-19 (6") mineral fiber
R-33, One layer of R-22 (6½") and one layer of R-11 (3½") mineral fiber
R-30, One layer of R-19 (6") and one layer of R-11 (3½") mineral fiber
R-26, Two layers of R-13 (3-5/8") mineral fiber

Ceilings, loose fill mineral wool and batts
R-38, R-19 (6") mineral fiber and 20 bags of wool per 1,000 S.F. (8¾")
R-33, R-22 (6½") mineral fiber and 11 bags of wool per 1,000 S.F. (5")
R-30, R-19 (6") mineral fiber and 11 bags of wool per 1,000 S.F. (5")
R-26, R-19 (6") mineral fiber and 8 bags of wool per 1,000 S.F. (3¼")

Walls, using 2"x6" framing
R-19, R-19 (6") mineral fiber batts

Walls, using 2"x4" framing
R-19, R-13 (3-5/8") mineral fiber batts and 1" polystyrene foam sheathing
R-13, R-13 (3-5/8") mineral fiber batts
R-11, R-11 (3½") mineral fiber batts

Floors
R-22, R-22 (6½") mineral fiber
R-19, R-19 (6") mineral fiber
R-13, R-13 (3-5/8") mineral fiber
R-11, R-11 (3½") mineral fiber

Insulation recommendations
Table 24-6

may consist of an R-11 and an R-19 batt. Note that R values are additive.

How Much Insulation Is Needed?

Figure 24-5 summarizes the optimum insulation values for residential heating and cooling as calculated for each climate zone. The figures are the recommendation of Owens-Corning Fiberglas and take into account national weather data, energy costs and projected increases, and insulation costs. Table 24-6 shows how much insulation is required to reach the recommended R values.

Figure 24-5 shows that most areas in the United States require R-19 wall insulation. Filling the cavity in a 2"x4" frame wall will produce no more than a R-14 rating. A nominal 6" cavity will carry R-19 insulation. Six inch studs spaced 24 inches on center will give the same strength as 4 inch studs 16 inches on center. The 6 inch stud wall uses less studs, less lumber and provides better insulation because the studs transmit heat better than the fully insulated spaces between the studs.

How To Install Insulation

Blanket insulation or batt insulation with a vapor barrier should be placed between framing to make up the desired resistance. For example,

members so that the tabs of the barrier lap the edge of the studs as well as the top and bottom plates. This method is not often popular with the contractor because it is more difficult to apply the dry wall or rock lath (plaster base)). However, it assures a minimum amount of vapor loss compared to the loss when tabs are stapled to the sides of the studs. To protect the head and soleplate as well as the headers over openings, it is good practice to use narrow strips of vapor barrier material along the top and bottom of the wall (Figure 24-7, *A*). Ordinarily, these areas are not covered too well by the barrier on the blanket or batt. A hand stapler is commonly used to fasten the insulation and the barriers in place.

For insulation without a barrier (press-fit or friction type), a plastic film vapor barrier such as 4 mil polyethylene is commonly used to envelop the entire exposed wall and ceiling (Figure 24-7, *B*). It covers the openings as well as window and door headers and edge studs. This system is one of the best from the standpoint of resistance to vapor movement. Furthermore, it does not have the installation inconveniences encountered when tabs of the insulation are stapled over the edges of the studs. After the dry wall is installed or plastering is completed, the film is trimmed around the window and door openings.

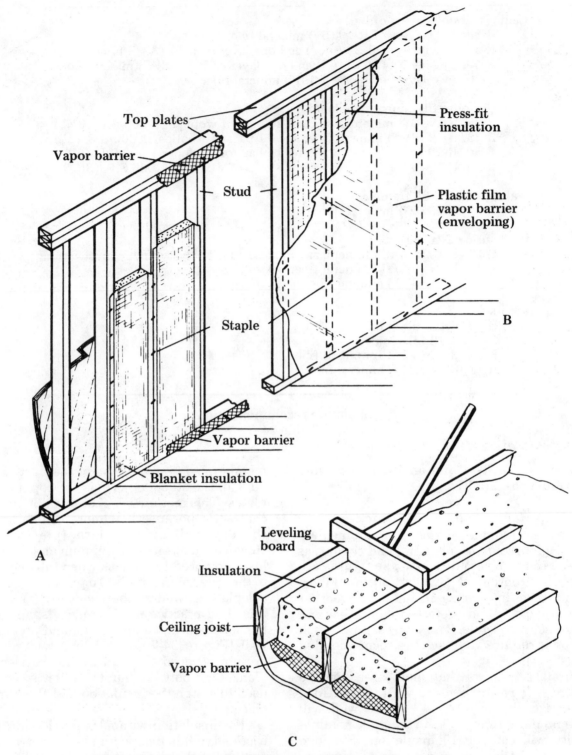

Application of insulation: *A*, Wall section with
blanket type; *B*, wall section with "press-fit"
insulation; *C*, ceiling with fill insulation
Figure 24-7

How To Install Blanket Insulation
Note: Be sure the right width blanket is used to fill the spaces between the joists, studs or rafters.
1. Place the blanket between the studs. The

moisture proof side of the blanket should face the inside of the room.

2. Nail or staple the flanges of the blankets to the studs at intervals specified by the maker of the material. See Figure 24-8. The blankets

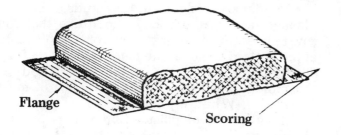
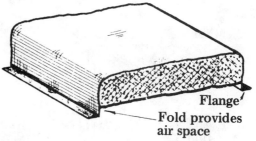

Blanket insulation with flanges
Figure 24-8

should be fastened between the studs so as to allow a space on both sides as shown in Figure 24-9. When necessary, cut the blankets with a hook knife. Short strokes with the knife are the most successful.

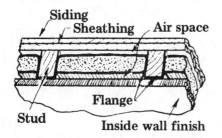

Blanket insulation installed in frame wall
Figure 24-9

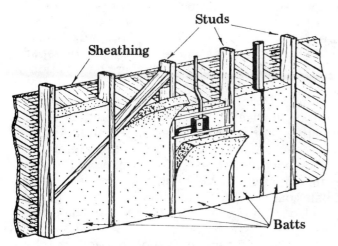

Batt insulation in frame wall
Figure 24-10

How To Apply Batt Insulation

1. Apply batts between the studs. In most cases, they merely need to be pushed into place. See Figure 24-10.

2. If it is necessary to fit the insulation around pipes, electrical fixtures or in odd shaped spaces, cut it to shape with a knife.

Note: Sometimes the batts are attached to waterproofed paper and have flanges

which are to be fastened to the studs or rafters as shown in Figure 24-11.

3. If moisture proof vapor barrier is not attached to the batts, staple sheets of this type of film to the inside surfaces of the studs, joists or rafters. This film should cover the entire inside surface of the insulated area.

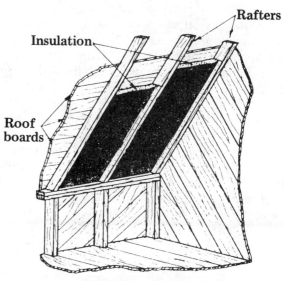

Insulation batts with flanges
Figure 24-11

How To Apply Reflective Insulation

Reflective insulation, in a single sheet form with two reflective surfaces, should be placed to divide the space formed by the framing members into two approximately equal spaces. Some reflective insulations include air spaces and are furnished with nailing tabs. This type is fastened to the studs to provide at least a ¾ inch space on each side of the reflective surfaces.

1. Use tin snips if necessary to cut metallic insulation between studs and around pipes, electrical equipment and odd shaped spaces. It is sometimes difficult to shape the metallic insulation to fit all the small openings. In these cases, the spaces could be filled with loose rock wool or glass insulation.

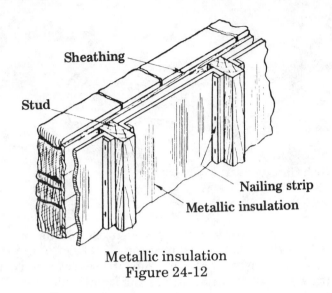

Metallic insulation
Figure 24-12

2. Nail the metal to the studs as shown in Figure 24-12.

Vapor Barriers

Some discussion of vapor barriers has been included previously because vapor barriers are usually a part of flexible insulation. However, further information is included in the following paragraphs.

Most building materials are permeable to water vapor. This presents problems because considerable water vapor is generated in a house from cooking, dishwashing, laundering, bathing, humidifiers, and other sources. In cold climates, this vapor may pass through wall and ceiling materials and condense in the wall or attic space; subsequently, in severe cases, it may damage the exterior paint and interior finish, or even result in decay in structural members. For protection, a material highly resistive to vapor transmission, called a vapor barrier, should be used on the warm side of a wall or below the insulation in an attic space.

Among the effective vapor barrier materials are asphalt laminated papers, aluminum foil, and plastic films. Most blanket and batt insulations are provided with a vapor barrier on one side, some of them with paper backed aluminum foil. Foil backed gypsum lath or gypsum boards are also available and serve as excellent vapor barriers.

The perm values of vapor barriers vary, but ordinarily it is good practice to use those which have values less than ¼ (0.25) perm. Although a value of ½ perm is considered adequate, aging reduces the effectiveness of some materials.

Some types of flexible blanket and batt insulations have a barrier material on one side. Such flexible insulations should be attached with the tabs at their sides fastened on the inside (narrow) edges of the studs, and the blanket should be cut long enough so that the cover sheet can lap over the face of the sole plate at the bottom and over the plate at the top of the stud space. However, such a method of attachment is not the common practice of most installers. When a positive seal is desired, wall-height rolls of plastic film vapor barriers should be applied over studs, plates, and window and door headers. This system, called "enveloping," is used over insulation having no vapor barrier or to insure excellent protection when used over any type of insulation. The barrier should be fitted tightly around outlet boxes and sealed if necessary. A ribbon of sealing compound around an outlet or switch box will minimize vapor loss at this area. Cold air returns in outside walls should consist of metal ducts to prevent vapor loss and subsequent paint problems.

Paint coatings on plaster may be very effective as vapor barriers if materials are properly chosen and applied. They do not however, offer protection during the period of construction, and moisture may cause paint blisters on exterior paint before the interior paint can be applied. This is most likely to happen in buildings that are constructed during periods when outdoor temperatures are 25 degrees fahrenheit or more below inside temperatures. Paint coatings cannot be considered a substitute for the membrane types of vapor barriers, but they do provide some protection for houses where other types of vapor barriers were not installed during construction.

Of the various types of paint, one coat of aluminum primer followed by two decorative coats of flat wall or lead and oil paint is quite effective. For rough plaster or for buildings in very cold climates, two coats of the aluminum primer may be necessary. A primer and sealer of the pigmented type, followed by decorative finish coats or two coats of rubber base paint, are also effective in retarding vapor transmission.

Because no type of vapor barrier can be considered 100 percent resistive, and some vapor leakage into the wall may be expected, the flow of vapor to the outside should not be impeded by materials of relatively high vapor resistance on the cold side of the vapor barrier. For example, sheathing paper should be of a type that is waterproof but not highly vapor resistant. This also applies to "permanent" outer coverings or siding. In such cases, the vapor barrier should have an equally low perm value. This will reduce the danger of condensation on cold surfaces within the wall.

Appendix A
Theory of Structures

A structure is built of parts, called members, intended to support and to transmit loads while remaining in equilibrium relative to each other. The points at which members are connected are called joints, and the methods of connection are determined principally by the kind of material being connected. All structures, whether bridges, buildings or towers, have certain factors in common which are considered in the design. As a builder whose tasks include the construction of various types of structures, an understanding of these basic factors is essential.

1. *Loads.* The principal loads present in every structure are classified as dead loads and live loads.

Dead load means the weight of the structure itself, which increases gradually as the structure is being built and remains constant, in most cases, once it has been completed. It must be remembered, however, that any modifications made to the structure must be considered as to their effect on the existing load-bearing members. In some cases, the addition of partitions or heating and air conditioning equipment, not compensated for in the original design, might require the relocation or addition of load-bearing members so as to safely handle the increased dead load. The weight of all structural members plus floors, walls, heating equipment, and all other nonmovable items in a building are also considered as dead load.

Live load means the weight of movable objects on the floor of the building or deck of a bridge or forces acting on their exteriors. This would mean people and furniture on the floor of a building, the traffic across a bridge, and other external forces such as wind, snow, ice, wave action, etc.

2. *Load distribution.* All structures are designed with the same basic theory of load distribution, which states that all live loads are supported by horizontal structural members. The loads are transmitted from the horizontal members to the supporting vertical members. Obviously, outside forces acting on the structure must be counteracted by members whose lengths are parallel to the external forces. Vertical members are supported by footings or foundation walls that rest on the ground. All structural loads are therefore dispersed into the ground supporting the structure. The ability of the ground to support loads is called soil bearing capacity.

3. *Soil bearing capacity.* The bearing capacity of the ground is measured in pounds per square foot (psf) and is determined by test; it varies for different soils. The area over which a footing extends is a factor in distributing the load received from a column in accordance with the bearing capacity of the soil. The size of the footing, therefore, is determined by the bearing capacity of the soil on which the structure is built.

4. *Uniformly distributed and concentrated loads.* The load of a structural member is classified as uniformly distributed or concentrated. A concentrated load is exerted at a particular point on the member as, for example, the load of a beam on a girder. A uniformly distributed load is one which is spread throughout a member, such as a slab on a beam.

5. *Eccentric loading.* Eccentric loading occurs when force on the member such as a column is not applied at the center of the column, but rather off center.

6. *Stress.* The loading of structural members has a tendency to deform them. The ability of a member to withstand certain kinds of deformation is called stress.

a. Tension. Tension is the longitudinal stress or pull that tends to lengthen a member. The ability of the member to resist this pull is the tensile strength of the member.

b. Compression. Compression is the force that tends to shorten or compress a member into a smaller area. Actually, compression is the opposite of tension. The stresses set up in the member are called compressive stresses.

c. Flexural or bending. The combinations of compressive and tensile stresses are called flexural or bending stresses. These stresses are greatest at the top and bottom of the member and zero at the neutral axis which is usually near the center of the depth, depending on the shape and composition of the member. This generally happens on a horizontal member, such as a beam. The material on the upper side of the beam is compressed or shortened, and that on the lower side is elongated. The stiffness of a member is its ability to resist bending.

d. Shear. Shear is the force that is applied to a horizontal surface tending to cut it in the vertical plane. The stresses set up in the member are called shearing stresses.

e. Torsion. Torsion is expressed as the ability of a member to withstand the stresses set up when the member remains stationary at one end while the other end is revolved or twisted.

f. Deflection. Deflection is caused by bending, by the shortening and lengthening of opposite sides of a member. This is a very important factor in the design of long beams and columns.

g. Strength. The strength of a structural member refers to that member's ability to withstand loads without failure. The design and use of a member determines strength in a particular instance. A structural member is never used in a case where its ultimate strength is needed to support design conditions. A safety factor must always be allowed for.

The above factors may all be considered in the design of a wood frame building, depending upon its type, purpose and cost. The type and use of a building play an important role in its design. It not only affects the building's architectural design but also affects its structural design. Based on a building's intended use, its physical and geographical location, the materials available, and the construction methods used, loads are calculated by which its structural members are designed.

From a design standpoint, it is impractical to consider the design factors under discussion individually. They must be considered collectively as each affects one or more of the others depending on the circumstances.

Design For Girders

In determining the number and location of girders, consideration must be given to the permissable length of joists, to the room arrangement, and to the location of bearing partitions.

In most houses one girder is enough. But if the joist span exceeds 14 or 15 feet, for which a 2 by 10 inch member is usually required, considerable increase in joist sizes becomes necessary in most species, thus making another girder advisable.

To illustrate this point, consider a building that is 17 feet 4 inches between inner faces of bearing walls. If no girders were used, with some species and grades of lumber, 2 by 14 inch or 3 by 10 inch joists would be required.

On the other hand, if a girder were used at the center of the building, the joist span would be only half of 17 feet 4 inches, or 8 feet 8 inches, for which, in most species of lumber, a 2 by 6 could be used. Furthermore, as joists over 14 or 16 feet long require two rows of bridging, it is well, as a means of economy, to keep joist lengths within 14 feet. If a building is more than 14 feet wide, it is usually better to use a girder.

As it is desirable to locate the girder directly under, or very close to, bearing partitions to avoid the necessity of additional or larger joists, the room arrangement will usually determine the location of girders.

If girders are far apart, joists must be larger and girders stronger, which will tend to increase the cost. For this and other reasons it is desirable to place the girders as close together as will not unduly increase the cost of foundation work. From 8 to 14 feet is usually an economical spacing for girders.

Dimensions And Strength

The girder should be strong enough for its load, but any size larger than needed is waste. There are three principal factors that must be understood before attempting to determine the size of a wood girder: (1) the effect of length on strength; (2) the effect of width on strength; and (3) the effect of depth on strength.

Length. If an 8 foot plank is supported at the ends on sawhorses, and loads are evenly distributed throughout its length, it will tend to bend. If the plank is 16 feet long, with the same load per foot of length, it will bend more and will be likely to break. It is natural to think that if the length is doubled, the safe load will be reduced one-half; but experience has shown, and the theory of mechanics bears it out, that instead of carrying safely half the load the 16 foot plank is good for only one-quarter.

This applies to planks or beams carrying a load distributed uniformly along their entire lengths and to all joists and girders. The reason is that for each foot of length added to the beam, another foot of load is also added. For a single concentrated load at the center, however, increasing the span does not increase the load;

and hence doubling the length decreases the safe load by only one-half.

Consequently, the greater the unsupported length or span, the stronger the girder must be. The strength can be increased in two ways, by using a stronger material or by using a larger beam. The beam may be enlarged by increasing the width or depth, or both.

Width. If the width is doubled, the strength is doubled, as is shown by many tests. This is made clearer by considering two beams, each of which will carry the same load. The fact that they are side by side will have no effect on what each will carry. It will be the same whether the beam is in two pieces or one piece. To double the width, therefore, doubles the capacity. Two inches in width added to a solid, dressed 6 by 10 inch girder, however, adds a little more than one third to its strength, since its actual width in the first place is only 5½ inches.

Depth. Doubling the depth of a girder much more than doubles the carrying capacity. Actually the increase is four-fold. In other words, a beam 3 inches wide and 12 inches deep will carry four times as much as one 3 inches wide and 6 inches deep. Therefore, as a general principle in the efficient use of material, to obtain greater carrying capacity, it is better to increase the depth of a beam than the width.

On the other hand, it is well to keep down the height of horizontal material in members such as girders, joists, and the like. Too much depth in a girder, especially if placed under the joists, decreases headroom in the basement. It is desirable, therefore, to adopt any of the following courses rather than increase the girder depth to much more than 10 inches, or 12 inches at the most.

1. Increase the width.
2. Put in additional supports, thus reducing the girder span and permitting a smaller girder.
3. Use a stronger material.

Thoroughly dry lumber should always be used, especially if the girder is under joists.

Determining The Size of Girders

To determine the size of a girder, seven steps are necessary:

1. Find the distance between girder supports.
2. Find the "half widths."
3. Find the "total floor load" per square foot carried by joists and bearing partitions to the girder.
4. Find the load per linear foot on the girder.
5. Find the total load on the girder.
6. Select the material for the girder.

7. Find the proper size of the girder in the material chosen.

Length of girder. Before it is possible to determine the girder size, the length of the girder between supports must be settled upon. This length will be determined by the spacing of supporting posts. These posts must be spaced according to some suitable division of the total length of the girder between walls. This may be determined at will, with due regard to the avoidance of excessive spans. Suppose you are construction a building 35 feet wide by 31 feet 8 inches deep. As posts carry a relatively large part of the weight, and it is well to use short spans as a means of reducing the size of girders, it will be better in this case to use three posts, thus fixing the span at about 8 feet 5 inches center to center of posts.

Having determined the location of supports, the next step is to find out what proportion of the joist load the girder must carry. This involves the determination of what in the tables is called the "half width."

General rule for half widths: A girder will carry the weight of the floor on each side to the midpoint of the joists which rest upon it. (See following discussion for special cases.)

This statement assumes that joists are butted or lapped over the girder. Butted or lapped over the girder, fully loaded joists tend to sag between supports, as shown in exaggerated fashion in Figure 4-11. Under such conditions there is no resistance to bending over the girder because of being merely butted, or at best, lapped and spiked. Suppose, however, the joists were continuous. Under load they would tend to assume the shape indicated in Figure 4-12. Being in one piece, they offer resistance to bending over the center support, and the girder is forced to carry a larger proportion of the load than if the joists were cut. The proportion actually taken can be found by formulas worked out by civil engineers. The method is complicated and involves higher mathematics, but the results may be stated as follows: If the girder is at the midpoint of continuous joists it will take five-eighths instead of one-half of the floor load.

If the girder is not located midway between adjacent supports, the determination of its load is further complicated. As the girder support moves to one side, the proportion of the floor load carried gradually decreases from five-eighths toward one-half. As it is seldom economical to put the girder far off center for continuous joists, all girders supporting continuous joists should be designed to support five-eighths of the total floor load of adjacent spans.

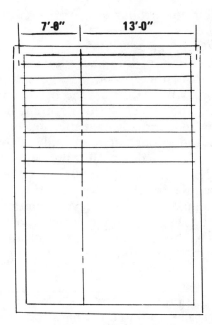

Example 1: Single girder
Figure A-1

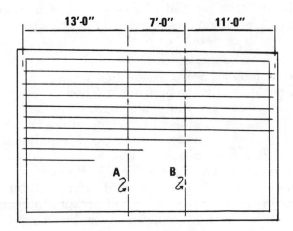

Example 2: Two girders
Figure A-2

The following may be used as a working rule for half widths:

To ascertain the half width for a girder, find the distance from the centerline of the girder to the nearest joist support on one side (which distance is generally known as the joist span). Add to this figure the corresponding span or distance from the girder to the nearest girder or wall on the other side. One-half of the total corresponds to the "half width," provided the joists are lapped or butted over the girder. If the joists are continuous, five-eighths of the total corresponds to the "half width."

To show the application of this rule, the following examples are given:

Example 1. This is the simplest case. (See Figure A-1.) There is a single girder running the length of the building. The heavy line shows the position of the girder and the light lines running at right angles thereto indicate the joists which rest upon it. The joist spans—that is, the distances to the nearest joint supports on either side of this girder—are 7 and 13 feet, a total of 20 feet. Half of this total, or 10 feet, is the "half width" for this girder, if the joists are cut. If the joists are continuous, the "half width" is five-eighths of 20 feet, or 12½ feet.

Example 2. Here the problem is complicated by the presence of two girders. (See Figure A-2.) In such cases, each girder must be taken separately. Applying the rule to girder A, the joist spans on either side are 13 and 7 feet, respectively, or a total of 20 feet. Half of this, or 10 feet for cut joists, is the half width for this girder.

In the case of girder B, the joist spans are 7 and 11 feet, or a total of 18 feet. The figure corresponding to the half width for girder B is 9 feet.

Figuring Floor Loads

The method for calculating the total floor load per square foot carried by joists and bearing partitions to the girder is illustrated in Figure A-3. The live loads specified also form the basis for determining the size of joists; and while larger than commonly occur, are used to provide adequate stiffness in the floor for all occasions. It may be possible that the maximum load provided for will never be applied, but consideration must be given to the possibility of crowded rooms and unusual loading situations, as when furniture is moved to the center of the room to permit painting or papering. When calculating loads for small house framing, you will find the information given in Figure A-3 valuable.

Assume that Figure A-3 represents an end view of a building. The girder being studied carries the first floor joist load. The bearing partition above, which rests directly upon the girder, will carry the second floor joist load also down to the girder. Hence, the girder will carry the weight of the second floor in addition to the first. Similarly, there is a bearing partition on the second floor which carries the second floor ceiling, attic, and roof loads. Thus the girder must carry not only the first floor, but the second floor, attic, and roof.

In some buildings the roof is framed so that the roof loads are carried entirely by the outside walls. In this case, no allowance would be made for roof loads when calculating total floor load to be carried by the girder. For this reason a plan should be carefully studied, and framing

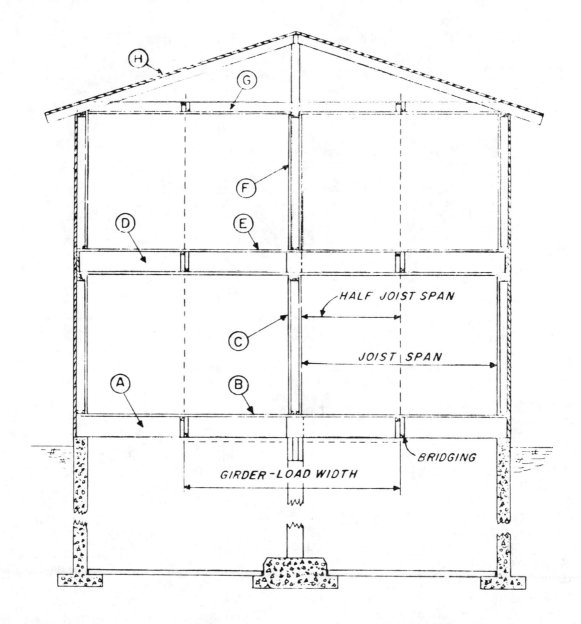

[A] Dead load of first floor = 10 pounds per sq. ft. Dead load of first floor with basement ceiling plastered = 20 pounds per sq. ft.

[B] Live load on first floor = 40 pounds per sq. ft. unless otherwise specified by local building code requirements.

[C] Dead load of partitions = 10 pounds per sq. ft. unless otherwise specified.

[D] Dead load on second floor = 20 pounds per sq. ft.

[E] Live load on second floor = 40 pounds per sq. ft. unless otherwise specified by local building code requirements.

[F] Dead load of partitions = 10 pounds per sq. ft.

unless otherwise specified.

[G] Live load on attic floor = 20 pounds per sq. ft. when used for storage only. Dead load of attic floor not floored = 10 pounds per sq. ft. Dead load of attic floor when floored = 20 pounds per sq. ft. when used for storage only.

[H] Roof of light construction, including both live and dead loads = 20 pounds per sq. ft. Roof of medium construction with light slate or asbestos roofing including both live and dead loads = 30 pounds per sq. ft. Roof of heavy construction with heavy slate or tile roofing, including both live and dead loads = 40 pounds per sq. ft. Note: Local building code requirements should be consulted as they may specify other minimum load values.

Calculating "total floor load" per square foot carried by joists and bearing partitions to the girder
Figure A-3

Total S.F. floor load	Total girder load per linear foot by "half widths"										
	5	6	7	8	9	10	12	14	16	18	20
10	50	60	70	80	90	100	120	140	160	180	200
20	100	120	140	160	180	200	240	280	320	360	400
30	150	180	210	240	270	300	360	420	480	540	600
40	200	240	280	320	360	400	480	560	640	720	800
50	250	300	350	400	450	500	600	700	800	900	1,000
60	300	360	420	480	540	600	720	840	960	1,080	1,200
70	350	420	490	560	630	700	840	980	1,120	1,260	1,400
80	400	480	560	640	720	800	960	1,120	1,280	1,440	1,600
90	450	540	630	720	810	900	1,080	1,260	1,440	1,620	1,800
100	500	600	700	800	900	1,000	1,200	1,400	1,600	1,800	2,000
110	550	660	770	880	990	1,100	1,320	1,540	1,760	1,980	2,200
120	600	720	840	960	1,080	1,200	1,440	1,680	1,920	2,160	2,400
130	650	780	910	1,040	1,170	1,300	1,560	1,820	2,080	2,340	2,600
140	700	840	980	1,120	1,260	1,400	1,680	1,960	2,240	2,520	2,800
150	750	900	1,050	1,200	1,350	1,500	1,800	2,100	2,400	2,700	3,000
160	800	960	1,120	1,280	1,440	1,600	1,920	2,240	2,560	2,880	3,200
170	850	1,020	1,190	1,360	1,530	1,700	2,040	2,380	2,720	3,060	3,400
180	900	1,080	1,260	1,440	1,620	1,800	2,160	2,520	2,880	3,240	3,600
190	950	1,140	1,330	1,520	1,710	1,900	2,280	2,660	3,040	3,420	3,800
200	1,000	1,200	1,400	1,600	1,800	2,000	2,400	2,800	3,200	3,600	4,000
210	1,050	1,260	1,470	1,680	1,890	2,100	2,520	2,940	3,360	3,780	4,200
220	1,100	1,320	1,540	1,760	1,980	2,200	2,640	3,080	3,520	3,960	4,400

Girder loads per linear foot: Total load per linear foot on the girder of various "half-widths" for various total floor loadings
Table A-4

methods analyzed, to determine just what loads are carried through the bearing partitions to the girder.

Using the live and dead load figures given, the total combined square foot floor load to be carried by the girder in Figure A-3 may be represented as follows:

(A)	Dead load of first floor	10
(B)	Live load on first floor	40
(C)	Dead load of partitions	10
(D)	Dead load of second floor	20
(E)	Live load on second floor	40
(F)	Dead load of partitions	10
(G)	Live load on attic floor	20
	Dead load of attic floor	10
(H)	Dead and live load of light roof	20
	Total	180

The minimum floor load to be allowed for the girder under discussion is 180 pounds per square foot. This is based on the assumption that there is no attic flooring, and that the ceiling of the basement is not plastered.

Load Per Linear Foot

To get the load per linear foot of girder, the total live and dead load per square foot for all floors supported, for the roof if it rests on a bearing partition, and for the bearing partitions, is multiplied by the half width, previously determined. Table A-4 gives loads per linear foot on girders for various half widths and types of buildings, as explained previously in this appendix. In the left-hand column are figures for various total loads per square foot. Each column is for a different half width. The table shows the total load per linear foot on the girder, the figures being the product of the total square foot load by the half width.

The table shows that for a half width of 15 feet (5 feet + 10 feet) and a total square foot floor load of 180 pounds, the load on the girder is 2,700 pounds per foot of length. In this case, the load represents the combined weights of strips of the first, second, and attic floors measured 1 foot along the girder and 15 feet across it, carried by each foot of the girder. It will be noted that the half widths are given in feet. In cases where half widths are not in even feet, the nearest whole foot should be used.

Total Load On The Girder

If the load upon the girder is 2,700 pounds for each foot of length, as determined in the usual way, the total load which it must carry is that figure multiplied by the distance in feet between girder supports. The total load per foot in this case is 2,700 pounds, which multiplied by the length of the girder, or 8 feet 5 inches (8.42 feet), if it is assumed that posts are equally spaced, is 22,750 pounds. This is the total load that the girder must carry. As this load does not occur at any one point, but is, for all practical purposes, spread equally over the whole length,

Solid dressed sizes	Span in feet														
	4,5,6	7	8	9	10	11	12	13	14	15	16	17	18	19	20
3 by 6	2,435	2,078	1,812	1,602	1,434	1,296	1,180	1,080	995	921	855	796	744	696	654
4 by 6	3,366	2,875	2,505	2,213	1,981	1,791	1,630	1,493	1,375	1,272	1,183	1,101	1,030	965	905
6 by 6	4,885	4,165	3,633	3,211	2,873	2,596	2,363	2,166	1,995	1,845	1,714	1,595	1,492	1,396	1,310
2 by 8	2,145	2,145	2,016	1,785	1,600	1,446	1,320	1,213	1,119	1,038	967	902	846	795	749
3 by 8	3,460	3,460	3,233	2,864	2,567	2,322	2,118	1,944	1,794	1,663	1,549	1,445	1,354	1,263	1,197
4 by 8	4,770	4,770	4,470	3,960	3,549	3,212	2,930	2,690	2,484	2,304	2,145	2,002	1,878	1,765	1,662
6 by 8	7,260	7,260	6,783	6,008	5,386	4,875	4,446	4,082	3,768	3,495	3,255	3,039	2,850	2,678	2,522
8 by 8	9,880	9,880	9,247	8,193	7,344	6,646	6,063	5,566	5,139	4,766	4,437	4,143	3,886	3,651	3,438
2 by 10	2,700	2,700	2,700	2,700	2,564	2,323	2,120	1,949	1,802	1,672	1,561	1,459	1,371	1,290	1,217
3 by 10	4,370	4,370	4,370	4,370	4,139	3,749	3,424	3,146	2,908	2,699	2,517	2,353	2,210	2,079	1,961
4 by 10	6,035	6,035	6,035	6,035	5,719	5,177	4,731	4,348	4,019	3,732	3,480	3,255	3,057	2,878	2,715
6 by 10	9,160	9,160	9,160	9,160	8,680	7,862	7,179	6,598	6,100	5,664	5,283	4,940	4,641	4,368	4,121
8 by 10	12,500	12,500	12,500	12,500	11,835	10,720	9,790	9,000	8,318	7,724	7,205	6,738	6,329	5,957	5,620
10 by 10	15,805	15,805	15,805	15,805	14,992	13,581	12,401	11,399	10,536	9,785	9,126	8,535	8,017	7,546	7,120
2 by 12	3,265	3,265	3,265	3,265	3,265	3,265	3,122	2,871	2,657	2,469	2,305	2,158	2,028	1,911	1,806
3 by 12	5,260	5,260	5,260	5,260	5,260	5,260	5,037	4,633	4,285	3,982	3,716	3,478	3,270	3,081	2,909
4 by 12	7,270	7,270	7,270	7,270	7,270	7,270	6,963	6,404	5,925	5,507	5,140	4,813	4,525	4,265	4,029
6 by 12	11,050	11,050	11,050	11,050	11,050	11,050	10,566	9,718	8,991	8,357	7,802	7,303	6,869	6,472	6,115
8 by 12	15,050	15,050	15,050	15,050	15,050	15,050	14,408	13,252	12,260	11,396	10,638	9,959	9,366	8,836	8,337
10 by 12	19,080	19,080	19,080	19,080	19,080	19,080	18,249	16,786	15,529	14,435	13,475	12,615	11,863	11,180	10,560
12 by 12	23,130	23,130	23,130	23,130	23,130	23,130	22,090	20,320	18,797	17,474	16,311	15,270	14,360	13,533	12,783
2 by 14	4,115	4,115	4,115	4,115	4,115	4,115	4,115	4,115	3,669	3,412	3,186	2,885	2,808	2,587	2,502
3 by 14	6,165	6,165	6,165	6,165	6,165	6,165	6,165	6,165	5,931	5,515	5,150	4,825	4,540	4,281	4,046
4 by 14	8,510	8,510	8,510	8,510	8,510	8,510	8,510	8,510	8,200	7,626	7,123	6,674	6,280	5,923	5,600
6 by 14	12,930	12,930	12,930	12,930	12,930	12,930	12,930	12,930	12,440	11,571	10,810	10,125	9,530	8,987	8,498
8 by 14	17,630	17,630	17,630	17,630	17,630	17,630	17,630	17,630	16,964	15,780	14,740	13,809	12,996	12,258	11,590
10 by 14	22,335	22,335	22,335	22,335	22,335	22,335	22,335	22,335	21,487	19,986	18,670	17,490	16,460	15,524	14,676
12 by 14	27,040	27,040	27,040	27,040	27,040	27,040	27,040	27,040	26,010	24,194	22,600	21,171	19,925	18,792	17,766
14 by 14	31,760	31,760	31,760	31,760	31,760	31,760	31,760	31,760	30,512	28,390	26,530	24,860	23,390	22,040	20,838

Built-up girders

Multiply above figures by 0.897 when 4-inch girder is made up of two 2-inch pieces.
0.887 when 6-inch girder is made up of three 2-inch pieces.
0.867 when 8-inch girder is made up of four 2-inch pieces.
0.856 when 10-inch girder is made up of five 2-inch pieces.

Note: Built-up girders of dressed lumber will carry somewhat smaller loads than solid girders; that is, two 2-inch dressed planks will equal only 3¼, whereas dressed 4-inch lumber will equal 3-5/8. It is, therefore, necessary to multiply by the above figures in order to compute the loads for built-up girders.

Solid wood girders: Allowable fiber stress 1,600 pounds per square inch; modules of elasticity, E-1,600,000
Table A-5

it is what is termed as "uniformly distributed" load.

Material For The Girder

At this point you must decide on the material you will use for the girder. If wood is selected, the kind or species must be decided; and it also must be decided whether to use a built-up or solid member.

Appendix B lists all common softwoods used for building construction, with allowable unit working stresses for each species and grade. Definite working stresses have been assigned to all these grades by the manufacturers.

Select from Appendix B the type of lumber and grade you would like to use. Find the "Fb" value in the "Normal Duration" column. Next, select from Tables A-5, A-6, A-7 or A-8 the table that applies to the working stress value you have selected. If the working stress you have selected is not 1,600, 1,400, 1,200 or 900 pounds per square inch, use the next lowest value.

Tables A-5 through A-8 give the total safe loads for different solid-wood girder sizes and spans, with working stresses of 1,600, 1,400, 1,200, and 900 pounds per square inch. These working stresses cover the range in values specified in most building codes. Where girders larger than those shown in tables are found necessary, design considerations will usually be such as to make it desirable to consult a competent civil engineer or architect. It is assumed when the depth of the girder is greater than one-twelfth of the span, loads will be limited by horizontal shear strength instead of bending strength.

For example, if No. 1 common southern yellow pine is to be used, the value given under "Extreme fiber in bending" is 1,200 pounds per square inch. Hence, use Table A-6 based on 1,200 pounds per square inch. Similarly if a structural grade of Douglas fir is to be used with bending stress of 1,800 pounds per square inch, Table A-5 based on 1,600 pounds per square inch should be used. If your girder spaced 8'-5'' center to center and carrying 22,750 pounds is Douglas fir with a bending stress of 1,600 pounds per square inch, a 12 x 12 member would be adequate.

The size needed will naturally vary with the species and grade used. The girder may be one solid timber or several pieces spiked together. If

Solid dressed sizes	Span in feet														
	4,5,6	7	8	9	10	11	12	13	14	15	16	17	18	19	20
2 by 6	1,318	1,124	979	865	774	699	636	582	536	495	459	427	399	372	349
3 by 6	2,127	1,816	1,581	1,397	1,249	1,128	1,026	938	863	698	740	688	641	599	561
4 by 6	2,938	2,507	2,184	1,930	1,726	1,559	1,418	1,297	1,194	1,102	1,023	952	888	831	777
6 by 6	4,263	3,638	3,168	2,800	2,504	2,260	2,055	1,881	1,731	1,599	1,483	1,379	1,286	1,202	1,126
2 by 8	1,865	1,865	1,760	1,558	1,395	1,260	1,150	1,055	973	902	839	783	733	687	6 646
3 by 8	3,020	3,020	2,824	2,500	2,238	2,024	1,845	1,692	1,560	1,444	1,343	1,253	1,172	1,100	1,034
4 by 8	4,165	4,165	3,904	3,456	3,906	2,800	2,552	2,342	2,160	2,002	1,862	1,737	1,626	1,528	1.435
6 by 8	6,330	6,330	5,924	5,244	4,698	4,250	3,873	3,553	3,277	3,037	2,825	2,636	2,468	2,315	2,178
8 by 8	8,630	8,630	8,078	7,151	6,406	5,793	5,281	4,845	4,469	4,141	3,851	3,595	3,365	3,157	2,969
2 by 10	2,360	2,360	2,360	2,360	2,237	2,026	1,848	1,699	1,569	1,455	1,356	1,268	1,190	1,118	1,054
3 by 10	3,810	3,810	3,810	3,810	3,612	3,271	2,984	2,740	2,531	2,348	2,267	2,045	1,917	1,803	1,698
4 by 10	5,265	5,265	5,265	5,265	4,992	4,520	4,125	3,788	3,500	3,248	3,026	2,830	2,653	2,496	2,352
6 by 10	7,990	7,990	7,990	7,990	7,576	6,860	6,261	5,751	5,312	4,929	4,593	4,294	4,029	3,787	3,571
8 by 10	10,920	10,920	10,920	10,920	10,330	9,351	8,537	7,841	7,244	6,922	6,264	5,857	5,493	5,165	4,868
10 by 10	13,825	13,825	13,825	13,825	13,085	11,849	10,813	9,933	9,176	8,815	7,934	7,419	6,958	6,543	6,168
2 by 12	2,845	2,845	2,845	2,845	2,845	2,845	2,724	2,503	2,315	2,150	2,006	1,878	1,763	1,660	1,567
3 by 12	4,590	4,590	4,590	4,590	4,590	4,590	4,394	4,039	3,734	3,468	3,234	3,026	2,842	2,675	2,524
4 by 12	6,350	6,350	6,350	6,350	6,350	6,350	6,075	5,585	5,165	4,797	4,474	4,189	3,933	3,705	3,496
6 by 12	9,640	9,640	9,640	9,640	9,640	9,640	9,220	8,475	7,837	7,280	6,791	6,357	5,970	5,622	5,307
8 by 12	13,160	13,160	13,160	13,160	13,160	13,160	12,570	11,556	10,685	9,926	9,260	8,669	8,141	7,666	7,237
10 by 12	16,670	16,670	16,670	16,670	16,670	16,670	15,923	14,638	13,536	12,573	11,728	10,980	10,311	9,710	9,166
12 by 12	20,170	20,170	20,170	20,170	20,170	20,170	19,274	17,709	16,384	15,220	14,197	13,291	12,482	11,753	11,096
2 by 14	3,595	3,595	3,595	3,595	3,595	3,595	3,595	3,595	3,199	2,973	2,776	2,490	2,443	2,301	2,173
3 by 14	5,365	5,365	5,365	5,365	5,365	5,365	5,365	5,365	5,171	4,805	4,485	4,202	3,949	3,721	3,514
4 by 14	7,420	7,420	7,420	7,420	7,420	7,420	7,420	7,420	7,151	6,646	6,225	5,814	5,465	5,150	4,876
6 by 14	11,290	11,290	11,290	11,290	11,290	11,290	11,290	11,290	10,849	10,086	9,415	8,821	8,292	7,814	7,384
8 by 14	15,360	15,360	15,360	15,360	15,360	15,360	15,360	15,360	14,796	13,754	12,840	12,030	11,307	10,658	10,071
10 by 14	19,465	19,465	19,465	19,465	19,465	19,465	19,465	19,465	18,740	17,420	16,261	15,236	14,321	13,498	12,755
12 by 14	23,590	23,590	23,590	23,590	23,590	23,590	23,590	23,590	22,685	21,088	19,685	18,445	17,236	16,340	15,440
14 by 14	27,690	27,690	27,690	27,690	27,690	27,690	27,690	27,690	26,630	24,754	23,108	21,652	20,350	20,142	18,125

Built-up girders

Multiply above figures by 0.897 when 4-inch girder is made up of two 2-inch pieces.
0.887 when 6-inch girder is made up of three 2-inch pieces.
0.867 when 8-inch girder is made up of four 2-inch pieces.
0.856 when 10-inch girder is made up of five 2-inch pieces.

Note: Built-up girders of dressed lumber will carry somewhat smaller loads than solid girders: that is, two 2-inch dressed planks will equal only 3¼, whereas dressed 4-inch lumber will equal 3-5/8. It is, therefore, necessary to multiply by the above figures in order to compute the loads for built-up girders.

Solid wood girders: Allowable fiber stress 1,400 pounds per square inch
Table A-6

a built-up girder is used, care should be taken to use the percentage figures at the bottom of Tables A-5 to A-8 applying to built-up girders. These allow for the lesser thickness of dressed material. A built-up girder is as satisfactory as a solid one. In fact, a built-up girder may be better, as it affords an opportunity to select and arrange the material for the best possible results.

The joist and girder tables show that it will usually be found possible to use either of two timbers, one of which is 2 inches deeper than the other. For example, Table A-7 shows that either a 10 by 10 or an 8 by 12 will safely support a load of 10,000 pounds over an 11 foot span. The deeper size will nearly always be preferable. In this example the 8 by 12 is both stronger and stiffer.

In the house discussed, suppose a No. 1 common grade of southern yellow pine is to be used. Refer to Table A-7, based on a 1,200 pound fiber stress per square inch. Referring to the allowable girder loads, it is seen that from a span of 8 feet 5 inches a solid beam of dressed material 14 by 14 inches will carry 23,640 pounds, or more than the 22,750 pound load to be carried on this girder. Although this beam is larger than is customarily seen in small house

work, it is, nevertheless, required to carry properly the loads imposed.

In Tables A-5, A-6 and A-7 are zigzag lines starting at the left at the top and ending at the right at the bottom of the tables. Loads given on the right of these lines will cause a deflection in timbers of corresponding size and span greater than 1/360 of the span. This is the limit usually set for deflection of girders and joists to avoid plaster cracks, sticking doors, and other difficulties resulting from sag under load. Loads given to the right of this line will result in serious deflection if certain types of lumber are used.

In most cases the lumber used for framing floors will be southern pine or Douglas fir. The zigzag lines apply for these two species only, but apply for all grades of these two species, even though the working stresses for lower grades are less than those given at the top of the tables. When other species of lower strength (working stress) are used, the larger sizes of timbers required to support these or lesser loads safely will give, in practically all cases, a stiffer girder or joist. Consequently, for lumber other than Douglas fir or southern pine, the zigzag lines can be ignored. If you are using Douglas fir or southern pine, refer to Table 4-15 for the girder

Solid dressed sizes	Span in feet														
	4,5,6	7	8	9	10	11	12	13	14	15	16	17	18	19	20
2 by 6	1,127	961	837	738	660	595	541	494	454	419	388	360	335	312	292
3 by 6	1,820	1,552	1,351	1,192	1,065	960	872	796	731	675	625	580	539	502	469
4 by 6	2,514	2,144	1,866	1,647	1,471	1,327	1,206	1,102	1,012	933	864	802	747	697	650
6 by 6	3,650	3,111	2,708	2,389	2,134	1,924	1,747	1,597	1,467	1,352	1,252	1,161	1,081	1,007	941
2 by 8	1,605	1,605	1,503	1,331	1,191	1,075	980	898	827	766	712	662	619	580	544
3 by 8	2,580	2,580	2,414	2,135	1,911	1,726	1,572	1,439	1,325	1,225	1,139	1,060	990	927	870
4 by 8	3,570	3,570	3,340	2,953	2,643	2,388	2,175	1,993	1,836	1,700	1,590	1,472	1,375	1,288	1,210
6 by 8	5,420	5,420	5,064	4,481	4,011	3,625	3,300	3,025	2,786	2,579	2,396	2,232	2,086	1,953	1,834
8 by 8	7,390	7,390	6,905	6,110	5,464	4,941	4,500	4,125	3,799	3,516	3,265	3,043	2,845	2,664	2,500
2 by 10	2,020	2,020	2,020	2,020	1,912	1,730	1,578	1,449	1,336	1,238	1,153	1,077	1,009	946	891
3 by 10	3,255	3,255	3,255	3,255	3,088	2,792	2,546	2,336	2,152	1,997	1,859	1,735	1,624	1,525	1,435
4 by 10	4,500	4,500	4,500	4,500	4,267	3,864	3,520	3,230	2,981	2,763	2,572	2,402	2,249	2,113	1,988
6 by 10	6,830	6,830	6,830	6,830	6,473	5,860	5,341	4,902	4,524	4,193	3,904	3,647	3,416	3,206	3,020
8 by 10	9,320	9,320	9,320	9,320	8,827	7,980	7,284	6,685	6,179	5,719	5,324	4,972	4,657	4,374	4,116
10 by 10	11,795	11,795	11,795	11,795	11,181	10,110	9,226	8,468	7,815	7,245	6,744	6,299	5,900	5,540	5,215
2 by 12	2,435	2,435	2,435	2,435	2,435	2,435	2,328	2,136	1,973	1,832	1,708	1,597	1,497	1,410	1,328
3 by 12	3,920	3,920	3,920	3,920	3,920	3,920	3,754	3,446	3,183	2,953	2,752	2,572	2,413	2,269	2,138
4 by 12	5,430	5,430	5,430	5,430	5,430	5,430	5,191	4,766	4,393	4,087	3,809	3,553	3,341	3,143	2,964
6 by 12	8,250	8,250	8,250	8,250	8,250	8,250	7,870	7,232	6,682	6,202	5,781	5,407	5,072	4,770	4,499
8 by 12	11,240	11,240	11,240	11,240	11,240	11,240	10,733	9,862	9,111	8,456	7,783	7,374	6,916	6,506	6,134
10 by 12	14,250	14,250	14,250	14,250	14,250	14,250	13,597	12,491	11,541	10,711	9,985	9,338	8,760	8,240	7,769
12 by 12	17,240	17,240	17,240	17,240	17,240	17,240	16,460	15,121	13,960	12,967	12,086	11,304	10,605	9,974	9,404
2 by 14	3,065	3,065	3,065	3,065	3,065	3,065	3,065	3,065	2,729	2,534	2,366	2,212	2,076	1,954	1,845
3 by 14	4,600	4,600	4,600	4,600	4,600	4,600	4,600	4,600	4,097	3,825	3,577	3,358	3,161	2,984	
4 by 14	6,340	6,340	6,340	6,340	6,340	6,340	6,340	6,340	6,102	5,668	5,288	4,950	4,648	4,377	4,132
6 by 14	9,630	9,630	9,630	9,630	9,630	9,630	9,630	9,630	9,258	8,601	8,020	7,511	7,054	6,642	6,270
8 by 14	13,140	13,140	13,140	13,140	13,140	13,140	13,140	13,140	12,628	11,730	10,942	10,244	9,619	9,058	8,552
10 by 14	16,640	16,640	16,640	16,640	16,640	16,640	16,640	16,640	16,002	14,854	13,860	12,975	12,183	11,472	10,830
12 by 14	20,140	20,140	20,140	20,140	20,140	20,140	20,140	20,140	19,167	17,982	16,780	15,706	14,748	13,887	13,110
14 by 14	23,640	23,640	23,640	23,640	23,640	23,640	23,640	23,640	22,742	21,120	19,690	18,440	17,310	16,370	15,388

Built-up girders

Multiply above figures by 0.897 when 4-inch girder is made up of two 2-inch pieces.
0.887 when 6-inch girder is made up of three 2-inch pieces.
0.867 when 8-inch girder is made up of four 2-inch pieces.
0.856 when 10-inch girder is made up of five 2-inch pieces.

Note: Built-up girders of dressed lumber will carry somewhat smaller loads than solid girders, that is, two 2-inch dressed planks will equal only 3¼, whereas dressed 4-inch lumber will equal 3-5/8. It is, therefore, necessary to multiply by the above figures in order to compute the loads for built-up girders.

Solid wood girders: Allowable fiber stress 1,200 pounds per square inch
Table A-7

size which will result in deflection less than 1/360 of the span.

Planning Column Loads

The supporting post or column in a house that is approximately square and large enough to require a girder, supported at the center, may carry one-fourth of the entire weight of the building. Therefore, the utmost care should be exercised in deciding upon the size, type, and material of the column, and in seeing that the column is properly seated on an adequate foundation.

The subject of spacing was fully explained in the section on girders but needs brief mention here. Great distances between posts should be avoided for two reasons:

1. To avoid great concentration of weight on one footing.

2. To avoid the necessity of large girders required by long spans.

In general, it is wise to limit spans to between 8 and 10 feet. For spans over 12 feet in a 2 story house, the girders required must be deep; and, consequently, will reduce basement headroom, will be heavier and harder to handle, will naturally increase the cost, and will concentrate heavy loads on individual post footings. If in addition to long girder spans, the joist spans are also long, the disadvantages are even more noticeable.

Before determining size, consider the manner in which a post acts. A post that is long in proportion to its sectional area will tend to bend. Bending occurs most readily across the narrowest measurement. Tests show that the smaller dimension in any relatively long post is the principal governing factor in its capacity for support. Tests further show that if a wood column is more than fifty times as long as its least dimension, it is quite unsafe and its use should be avoided. Thus a dressed 2 by 6 inch or 2 by 4 inch column longer than about 6 or 7 feet is unsafe, unless braced sideways. For this reason, all strength tables for columns show decreased bearing capacity with increase in length.

Computing The Load

The first step in determining the size of a column is to find the load it must carry. As the post in the usual dwelling supports the girder, it must carry the weight brought to it by the girder. The girder, so far as the post is concerned, is a simple beam. In fact, the general rule, giving the proportion of the joist load carried by the girder, may be reworded slightly

Solid dressed sizes	Span in feet														
	4,5,6	7	8	9	10	11	12	13	14	15	16	17	18	19	20
2 by 6	842	717	623	548	489	439	398	362	332	304	281	259	240	222	207
3 by 6	1,359	1,159	1,005	884	788	708	642	583	534	490	452	417	385	357	331
4 by 6	1,877	1,600	1,390	1,223	1,089	980	887	808	739	678	625	577	534	496	459
6 by 6	2,723	2,319	2,013	1,773	1,579	1,430	1,285	1,170	1,071	983	905	835	773	715	663
2 by 8	1,195	1,195	1,123	991	885	796	725	663	608	561	520	492	449	418	392
3 by 8	1,928	1,928	1,801	1,589	1,419	1,278	1,162	1,061	974	898	831	771	717	668	624
4 by 8	2,666	2,666	2,488	2,198	1,963	1,770	1,610	1,470	1,351	1,247	1,154	1,072	997	931	869
6 by 8	4,045	4,045	3,781	3,335	2,980	2,687	2,441	2,232	2,050	1,891	1,751	1,626	1,513	1,411	1,318
8 by 8	5,515	5,515	5,148	4,547	4,062	3,663	3,328	3,043	2,795	2,578	2,386	2,216	2,063	1,924	1,797
2 by 10	1,504	1,504	1,504	1,504	1,423	1,286	1,170	1,072	987	912	847	789	737	689	647
3 by 10	2,428	2,428	2,428	2,428	2,296	2,074	1,888	1,728	1,591	1,472	1,365	1,271	1,185	1,109	1,040
4 by 10	3,357	3,357	3,357	3,357	3,174	2,864	2,611	2,391	2,202	2,036	1,891	1,761	1,644	1,539	1,443
6 by 10	5,095	5,095	5,095	5,095	4,818	4,352	3,962	3,629	3,342	3,092	2,870	2,673	2,496	2,336	2,191
8 by 10	6,947	6,947	6,947	6,947	6,570	5,933	5,403	4,949	4,558	4,215	3,914	3,645	3,404	3,186	2,989
10 by 10	8,800	8,800	8,800	8,800	8,523	7,518	6,844	6,270	5,774	5,340	4,958	4,617	4,312	4,037	3,786
2 by 12	1,812	1,812	1,812	1,812	1,812	1,812	1,729	1,585	1,462	1,354	1,260	1,176	1,099	1,032	970
3 by 12	2,918	2,918	2,918	2,918	2,918	2,918	2,787	2,556	2,357	2,182	2,029	1,890	1,770	1,660	1,560
4 by 12	4,038	4,038	4,038	4,038	4,038	4,038	3,856	3,536	3,262	3,022	2,810	2,623	2,454	2,303	2,165
6 by 12	6,130	6,130	6,130	6,130	6,130	6,130	5,852	5,367	4,950	4,587	4,266	3,980	3,725	3,495	3,286
8 by 12	8,360	8,360	8,360	8,360	8,360	8,360	7,978	7,319	6,750	6,253	5,818	5,427	5,080	4,766	4,482
10 by 12	10,595	10,595	10,595	10,595	10,595	10,595	10,107	9,270	8,550	7,920	7,367	6,875	6,434	6,037	5,674
12 by 12	12,822	12,822	12,822	12,822	12,822	12,822	12,233	11,221	10,349	9,588	8,918	8,321	7,788	7,308	6,870
2 by 14	2,278	2,278	2,278	2,278	2,278	2,278	2,278	2,278	2,876	1,875	1,746	1,631	1,628	1,434	1,350
3 by 14	3,409	3,409	3,409	3,409	3,409	3,409	3,409	3,409	3,273	3,035	2,825	2,638	2,472	2,322	2,186
4 by 14	4,717	4,717	4,717	4,717	4,717	4,717	4,717	4,717	4,530	4,200	3,918	3,654	3,425	3,219	3,031
6 by 14	7,157	7,517	7,157	7,157	7,157	7,157	7,157	7,157	6,872	6,373	5,930	5,545	5,198	4,884	4,599
8 by 14	9,751	9,751	9,751	9,751	9,751	9,751	9,751	9,751	9,371	8,692	8,092	7,563	7,088	6,663	6,274
10 by 14	12,365	12,635	12,365	12,365	12,365	12,365	12,365	12,365	11,869	11,008	10,250	9,578	8,977	8,436	7,943
12 by 14	14,965	14,965	14,965	14,965	14,965	14,965	14,965	14,965	14,367	13,326	12,410	11,595	10,867	10,212	9,618
14 by 14	17,567	17,567	17,567	17,567	17,567	17,567	17,567	17,567	16,872	15,646	14,580	13,620	12,760	11,990	11,288

Built-up girders

Multiply above figures by 0.897 when 4-inch girder is made up of two 2-inch pieces.
0.887 when 6-inch girder is made up of three 2-inch pieces.
0.867 when 8-inch girder is made up of four 2-inch pieces.
0.856 when 10-inch girder is made up of five 2-inch pieces.

Note: Built-up girders of dressed lumber will carry somewhat smaller loads than solid girders, that is, two 2-inch dressed planks will equal only 3¼, whereas dressed 4-inch lumber will equal 3-5/8. It is, therefore, necessary to multiply by the above figures in order to compute the loads for built-up girders.

Solid wood girders: Allowable fiber stress 900 pounds per square inch
Table A-8

and used to show the proportion of the girder load supported by the posts: A post will carry the load on a girder to the midpoint of the span on both sides.

In computing post loads, follow these steps:

1. Find the span in feet from the center of the post to the nearest girder support on one side.

2. Multiply this span in feet by the girder load per linear foot, using the method suggested in computing the girder size for this span.

3. Find the span in feet from the center of the post to the nearest girder support on the other side.

4. Multiply this span in feet by the girder load per linear foot as in No. 2.

5. Find the load on the post by consideration of which type of construction is used. For this purpose, follow the procedure for one of the four cases outlined below. These four cases represent the types of construction commonly encountered.

Case 1. Girder cut over the post in question and also over the two nearest supports. (See Figure 5-2.) In this instance, post B takes one-half the total girder load on each side, so that the load on the post is one-half the girder weight itself and one-half the load carried by each length of girder.

Case 2. Girder continuous over the post in question but cut over the support on each nearest support. (See Figure 5-3, Page 78.) In this instance, post B carries approximately five-eighths of the load on the girder from A to C.

Case 3. Figure 5-4 gives a case of a continuous girder with two equidistant intermediate supports. It must be remembered that when more than one post is used, consideration must be given to each post separately. Although not exactly correct in all cases, the following rules are near enough for all practical purposes in considerating this type of construction. Post B can be assumed to take five-eighths of the girder load from A to B and one-half the girder load from B to C. In a similar manner post C may be considered to take five-eighths of the girder load from C to D and one-half of the girder load from C to B.

Case 4. In Figure 5-5 a single girder is supported at five points, the three center posts equidistant and dividing the girder into four

Nominal size, inches	3 by 4	4 by 4	4 by 6	6 by 6	6 by 8	8 by 8
Actual size, inches	2-5/8 by 3-5/8	3-5/8 by 3-5/8	3-5/8 by 5-5/8	5-1/2 by 5-1/2	5-1/2 by 7-1/2	7-1/2 by 7-1/2
Area in square inches	9.51	13.14	20.39	30.25	41.25	56.25
Height of column:						
4 feet	8,720	12,920	19,850	30,250	41,250	56,250
5 feet	7,430	12,400	19,200	30,050	41,000	56,250
6 feet	5,630	11,600	17,950	29,500	40,260	56,250
6 feet 6 inches	4,750	10,880	16,850	29,300	39,950	56,000
7 feet	4,130	10,040	15,550	29,000	39,600	55,650
7 feet 6 inches	--	9,300	14,400	28,800	39,000	55,300
8 feet	--	8,350	12,950	28,150	38,300	55,000
9 feet	--	6,500	10,100	26,850	36,600	54,340
10 feet	--	--	--	24,670	33,600	53,400
11 feet	--	--	--	22,280	30,380	52,100
12 feet	--	--	--	19,630	26,800	50,400
13 feet	--	--	--	16,920	23,070	47,850
14 feet	--	--	--	14,360	19,580	44,700

Maximum load allowance in pounds for lumber
columns of Douglas fir, southern pine, and North
Carolina pine, No. 1 common grade
Table A-9

Nominal size, inches	3 by 4	4 by 4	4 by 6	6 by 6	6 by 8	8 by 8
Actual size, inches	2-5/8 by 3-5/8	3-5/8 by 3-5/8	3-5/8 by 5-5/8	5-1/2 by 5-1/2	5-1/2 by 7-1/2	7-1/2 by 7-1/2
Area in square inches	9.51	13.14	20.39	30.25	41.25	56.25
Height of column:						
4 feet	4,950	7,280	11,300	16,940	23,100	31,500
5 feet	4,380	7,100	11,000	16,900	23,060	31,500
6 feet	3,460	6,650	10,300	16,700	22,850	31,500
6 feet 6 inches	2,960	6,320	9,800	16,600	22,700	31,400
7 feet	---	5,960	9,270	16,400	22,400	31,300
7 feet 6 inches	---	5,630	8,720	16,200	22,100	31,100
8 feet	---	5,160	7,930	15,950	21,800	31,000
9 feet	---	4,060	6,300	15,350	20,950	30,640
10 feet	---	---	---	14,400	19,600	30,240
11 feet	---	---	---	13,350	18,200	29,650
12 feet	---	---	---	12,200	16,600	28,800
13 feet	---	---	---	10,500	14,350	27,700
14 feet	---	---	---	8,950	12,200	26,300

Maximum load allowance in pounds for lumber
columns of eastern hemlock, western red cedar,
white fir, white pines, and spruces, No. 1
common grade
Table A-10

equal parts. As mentioned earlier, each post must be considered separately and a note made as to whether the girder is cut or continuous over the post in question. In addition, consideration must also be made as to whether the girder is cut or continuous over the nearest supports on both sides of the post in question.

Post C may be said to carry one-half of the load between C and B and C and D. Post B may be considered as carrying one-half the load between B and C and five-eighths of the load between B and A. In a similar manner post D may be said to carry one-half the girder load between D and C and five-eighths of the girder load between D and E.

Having determined the load on each post, the next step is to find the size required. This will depend on the material and type of post selected.

Selecting The Column Size

Other things being equal, round steel-pipe posts or square timber posts are preferred over other shapes because they give equal strength in two directions and provide greater strength for the same area than any other cross sectional area. In order to determine the size of timber post required, refer to Tables A-9 and A-10. To determine the size of steel-pipe post required, refer to Table A-11. From the column on the left

Nominal size, inches	6	5	4½	4	3½	3	2½	2	1½
External diameter, inches	6,625	5,563	5,000	4,500	4,000	3,500	2,875	2,375	1,900
Thickness, inches	.280	.258	.247	.237	.226	.216	.203	.154	.145
Effective length:									
5 feet	72.5	55.9	48.0	41.2	34.8	29.0	21.6	12.2	7.5
6 feet	72.5	55.9	48.0	41.2	34.8	28.6	19.4	10.6	6.0
7 feet	72.5	55.9	48.0	41.2	34.1	26.3	17.3	9.0	5.0
8 feet	72.5	55.9	48.0	40.1	31.7	24.0	15.1	7.4	4.2
9 feet	72.5	55.9	46.4	37.6	29.3	21.7	12.9	6.6	3.5
10 feet	72.5	54.2	43.8	35.1	26.9	19.4	11.4	5.8	2.7
11 feet	72.5	51.5	41.2	32.6	24.5	17.1	10.3	5.0	--
12 feet	70.2	48.7	38.5	30.0	22.1	15.2	9.2	4.1	--
13 feet	67.3	46.0	35.9	27.5	19.7	14.0	8.1	3.3	--
14 feet	64.3	43.2	33.3	25.0	18.0	12.9	7.0	--	--
Area in square inches	5.58	4.30	3.69	3.17	2.68	2.23	1.70	1.08	0.80
Weight per pound per foot	18.97	14.62	12.54	10.79	9.11	7.58	5.79	3.65	2.72

Allowable fiber stress per square inch, 13,000 pounds for lengths of 60 radii or under, reduced for length over 60 radii.

**Steel-pipe columns-standard pipe safe loads
in thousands of pounds
Table A-11**

side of the appropriate table, select the height or length required and follow the line across to get the size of post that will carry the assumed load. For example, determine the size column required for a house 17 feet wide with a load on the girder of 2,700 pounds per linear foot and one center post.

It will be recalled (Case 1) that where the girder is cut over the center post the post will carry one-half of the total girder load. The total load on the post will be:

$$\frac{8\ \text{ft. 5 in. by}}{2}\ \frac{2,700\ \text{lb.}}{2} + \frac{8\ \text{ft 5 in. by}}{2}\ \frac{2,700\ \text{lb.}}{2} = 22,750$$

Referring to the column tables for wood and pipe columns, suitable for a height of 6 feet 6 inches, the following sizes and materials will be found adequate:

No. 1 common Douglas fir (West Coast), 6 by 6 inches, dressed size, will carry 29,300 pounds.

No. 2 common eastern hemlock, 6 by 8 inches, dressed size, will carry 22,700 pounds.

Steel-pipe columns, 7 feet long and 3 inches in diameter, will carry 26,300 pounds.

Any of the foregoing posts could be used for this building. As to which would be best would depend upon preference, availability, and cost.

In small light frame structures it will often be found that a timber column smaller than 4 by 6 inches or 5 by 5 inches is satisfactory. However, no post smaller than 4 by 6 inches, and preferably 6 by 6 inches, should be used in the basement because of the possibility of damage from a severe blow.

Rafter And Ceiling Joists

Use Tables A-12 to A-21 to figure ceiling joist and rafter sizes. Assume a horizontal projection span of 13 feet, 0 inches, a live load of 20 p.s.f., dead load of 7 p.s.f., and rafters spaced 16 inches on centers. Table A-20 shows that a 2 x 6 having a bending stress (Fb) value of 1,200 p.s.i. and an E value of 940,000 p.s.i. would have an allowable span of 13 feet of horizontal projection. Convert horizontal to sloping distance with Table A-22. Finally, find the lumber grade and species you should use in Appendix B under the "Normal Duration" column.

Floor Openings

Whenever it is necessary to cut regular joists to provide an opening as, for example, at the stairwell, it is necessary to provide auxiliary joists (headers) at right angles to the regular joists, to carry the ends of the cut joists which are called tail beams. These headers, in turn, are supported by double or triple joists called trimmers. Whatever its strength requirements, a header cannot be of greater depth than the joists, except perhaps on the first floor where projection below the ceiling line of the basement is not objectionable. Custom has usually decreed the doubling of all headers and trimmers. As a matter of fact, size should be determined on the basis of loads to be carried. In many cases it is unnecessary to double the members. On the other hand, there are cases in which doubling of headers may be insufficient.

The header is similar to a girder in that it carries the ends of certain floor joists. Before the load on the trimmer can be found, it is necessary to find the load carried by the header. This load may be found by following the same methods as outlined for girders. As a matter of convenience, however, the general rule for girders may be modified for the header as follows: The header will carry the weight of the floor to the midpoint of the tail beams which rest upon it. Therefore, the load on the header is: Length of tail beam multiplied by length of header multiplied by floor load (both live and dead) in pounds per square foot divided by 2. An illustration of a case may be taken where a tail beam is 12 feet long, the length of the header 6 feet, and the floor load 50 pounds to the square foot. The figures for this example are as follows:

$$\frac{12 \times 6 \times 50}{2} = 1{,}800 \text{ pounds}$$

It was seen in the simple case of the joist resting on supports at either end that the load on the joist is divided equally between the supports. In like manner, the two trimmers will divide the total header load. Summarizing the foregoing: The header takes half the floor load on the tail beams and the trimmers each take half of the header load, so the trimmer carries one-fourth of the floor load carried by the tail beams. The following rule may therefore be formulated: Trimmer load = one-fourth length of header by length of tail beams by their total floor load, live plus dead, in pounds per square foot.

The figures in Table A-23 are arranged according to total floor load (live plus dead) and length of tail beams. These figures multiplied by the length of the header will give the total uniformly distributed load on the header. Half this amount gives the concentrated load on each trimmer corresponding to the foregoing formula.

The next step in determining the trimmer is to note the point of application of the header load and to ascertain its effect upon the trimmer. Nearly all tables for beams are based on uniformly distributed loads. It will be recalled that a concentrated load of 1,000 pounds applied at the center of a span produces the same effect as a uniformly distributed one of twice the amount, or 2,000 pounds. Simple reference to the girder tables will give the proper size of beams required for any such concentrated load, for any span or material, merely by doubling it and treating as a uniformly distributed load. If the relation can be found between the bending effect of a concentrated load at points other than the center, and a

corresponding uniformly distributed load of equal bending effect, the size of beam may be quickly found from the tables by using the equivalent uniformly distributed load. This relationship has been established and is shown in Table A-24. The concentrated load multiplied by the proper factor (determined by position of load) from Table A-24 gives the uniform load which can be used to select from the tables of maximum spans the necessary size of joist or trimmer to be used. The best way to summarize the discussion dealing with loadbearing and trimmers around floor openings is to work an example.

Example. Assume a hearth 6 feet long in a living room 14 feet wide; the header is to be 2 feet from one end of the trimmer (see Figure A-25); assume the usual live load of 40 pounds and a dead load allowance for the weight of the floor of 10 pounds with no ceiling cover below. The tail beams, therefore, are 12 feet long.

1. According to the tables, the joists, if of No. 1 common southern pine, would be 2 by 10's; hence, the header and trimmer depth would be limited to 10 inches.

2. The load on the header, according to Table A-23, would be 300 by 6, or 1,800 pounds. Girder tables show that a 2 by 10 of 6 foot span, for example, will carry 2,020 pounds; therefore, it is not necessary to double the header.

3. Load on trimmer is one-half header load, or 900 pounds.

4. The trimmer load is applied 2 feet from one end of the trimmer, so it is one-seventh of the span from one end. According to Table A-24, this is equivalent in bending effect to an equal load of 900 pounds distributed over the whole beam.

5. A 2 by 10 of common structural Douglas fir with a 14 foot span will carry 1,336 pounds. This is ample to carry the header.

6. If the regular joist is used as a trimmer also, it must support the header load in addition to its share of the floor load, which it is seldom able to do.

In such instances the joist must carry a strip of floor 16 inches wide with a combined live and dead load of 50 pounds. The total loads on the joist in the example above, therefore, would be: Joist load equals 14 feet (length) by 1⅓ feet (spacing) by 50 pounds load per square foot, or 932 pounds, plus the header load in terms of its equivalent uniformly distributed load, or 900 pounds. The trimmer joist to carry both loads must have a capacity of 1,832 pounds, but reference to girder tables show that a 2 by 10 of Douglas fir has a capacity of only 1,336 pounds as determined earlier. Therefore, if a single piece is to act as both joist and trimmer, its size will have to be increased to 3 by 10, which for a 1,200 pound fiber stress will carry 2,152 pounds.

JOIST SIZE (IN)	SPACING (IN)	Modulus of Elasticity, "E", 1,000,000 psi													
		0.8	0.9	1.0	1.1	1.2	1.3	1.4	1.5	1.6	1.7	1.8	1.9	2.0	2.2
2x4	12.0	7-10 / 900	8-1 / 970	8-5 / 1040	8-8 / 1110	8-11 / 1170	9-2 / 1240	9-5 / 1300	9-8 / 1360	9-10 / 1420	10-0 / 1480	10-3 / 1540	10-5 / 1600	10-7 / 1650	10-11 / 1760
	16.0	7-1 / 990	7-5 / 1070	7-8 / 1140	7-11 / 1220	8-1 / 1290	8-4 / 1360	8-7 / 1430	8-9 / 1500	8-11 / 1570	9-1 / 1630	9-4 / 1690	9-6 / 1760	9-8 / 1820	9-11 / 1940
	24.0	6-2 / 1130	6-5 / 1220	6-8 / 1310	6-11 / 1400	7-1 / 1480	7-3 / 1560	7-6 / 1640	7-8 / 1720	7-10 / 1790	8-0 / 1870	8-1 / 1940	8-3 / 2010	8-5 / 2080	8-8 / 2220
2x6	12.0	12-3 / 900	12-9 / 970	13-3 / 1040	13-8 / 1110	14-1 / 1170	14-5 / 1240	14-9 / 1300	15-2 / 1360	15-6 / 1420	15-9 / 1480	16-1 / 1540	16-4 / 1600	16-8 / 1650	17-2 / 1760
	16.0	11-2 / 990	11-7 / 1070	12-0 / 1140	12-5 / 1220	12-9 / 1290	13-1 / 1360	13-5 / 1430	13-9 / 1500	14-1 / 1570	14-4 / 1630	14-7 / 1690	14-11 / 1760	15-2 / 1820	15-7 / 1940
	24.0	9-9 / 1130	10-2 / 1220	10-6 / 1310	10-10 / 1400	11-2 / 1480	11-5 / 1560	11-9 / 1640	12-0 / 1720	12-3 / 1790	12-6 / 1870	12-9 / 1940	13-0 / 2010	13-3 / 2080	13-8 / 2220
2x8	12.0	16-2 / 900	16-10 / 970	17-5 / 1040	18-0 / 1110	18-6 / 1170	19-0 / 1240	19-6 / 1300	19-11 / 1360	20-5 / 1420	20-10 / 1480	21-2 / 1540	21-7 / 1600	21-11 / 1650	22-8 / 1760
	16.0	14-8 / 990	15-3 / 1070	15-10 / 1140	16-4 / 1220	16-10 / 1290	17-3 / 1360	17-9 / 1430	18-2 / 1500	18-6 / 1570	18-11 / 1630	19-3 / 1690	19-7 / 1760	19-11 / 1820	20-7 / 1940
	24.0	12-10 / 1130	13-4 / 1220	13-10 / 1310	14-3 / 1400	14-8 / 1480	15-1 / 1560	15-6 / 1640	15-10 / 1720	16-2 / 1790	16-6 / 1870	16-10 / 1940	17-2 / 2010	17-5 / 2080	18-0 / 2220
2x10	12.0	20-8 / 900	21-6 / 970	22-3 / 1040	22-11 / 1110	23-8 / 1170	24-3 / 1240	24-10 / 1300	25-5 / 1360	26-0 / 1420	26-6 / 1480	27-1 / 1540	27-6 / 1600	28-0 / 1650	28-11 / 1760
	16.0	18-9 / 990	19-6 / 1070	20-2 / 1140	20-10 / 1220	21-6 / 1290	22-1 / 1360	22-7 / 1430	23-2 / 1500	23-8 / 1570	24-1 / 1630	24-7 / 1690	25-0 / 1760	25-5 / 1820	26-3 / 1940
	24.0	16-5 / 1130	17-0 / 1220	17-8 / 1310	18-3 / 1400	18-9 / 1480	19-3 / 1560	19-9 / 1640	20-2 / 1720	20-8 / 1790	21-1 / 1870	21-6 / 1940	21-10 / 2010	22-3 / 2080	22-11 / 2220

Note: The required extreme fiber stress in bending, "F_b", in pounds per square inch is shown below each span.

CEILING JOISTS
20 Lbs. Per Sq. Ft. Live Load
(Limited attic storage where development of future rooms is not possible)
(Drywall Ceiling)

DESIGN CRITERIA:
Deflection - For 20 lbs. per sq. ft. live load.
Limited to span in inches divided by 240.

Strength - live load of 20 lbs. per sq. ft. plus dead load of 10 lbs. per sq. ft. determines required fiber stress value.

Table A-12

JOIST SIZE (IN)	SPACING (IN)	Modulus of Elasticity, "E", in 1,000,000 psi													
		0.8	0.9	1.0	1.1	1.2	1.3	1.4	1.5	1.6	1.7	1.8	1.9	2.0	2.2
2x4	12.0	9-10 / 710	10-3 / 770	10-7 / 830	10-11 / 880	11-3 / 930	11-7 / 980	11-10 / 1030	12-2 / 1080	12-5 / 1130	12-8 / 1180	12-11 / 1220	13-2 / 1270	13-4 / 1310	13-9 / 1400
	16.0	8-11 / 780	9-4 / 850	9-8 / 910	9-11 / 970	10-3 / 1030	10-6 / 1080	10-9 / 1140	11-0 / 1190	11-3 / 1240	11-6 / 1290	11-9 / 1340	11-11 / 1390	12-2 / 1440	12-6 / 1540
	24.0	7-10 / 900	8-1 / 970	8-5 / 1040	8-8 / 1110	8-11 / 1170	9-2 / 1240	9-5 / 1300	9-8 / 1360	9-10 / 1420	10-0 / 1480	10-3 / 1540	10-5 / 1600	10-7 / 1650	10-11 / 1760
2x6	12.0	15-6 / 710	16-1 / 770	16-8 / 830	17-2 / 880	17-8 / 930	18-2 / 980	18-8 / 1030	19-1 / 1080	19-6 / 1130	19-11 / 1180	20-3 / 1220	20-8 / 1270	21-0 / 1310	21-8 / 1400
	16.0	14-1 / 780	14-7 / 850	15-2 / 910	15-7 / 970	16-1 / 1030	16-6 / 1080	16-11 / 1140	17-4 / 1190	17-8 / 1240	18-1 / 1290	18-5 / 1340	18-9 / 1390	19-1 / 1440	19-8 / 1540
	24.0	12-3 / 900	12-9 / 970	13-3 / 1040	13-8 / 1110	14-1 / 1170	14-5 / 1240	14-9 / 1300	15-2 / 1360	15-6 / 1420	15-9 / 1480	16-1 / 1540	16-4 / 1600	16-8 / 1650	17-2 / 1760
2x8	12.0	20-5 / 710	21-2 / 770	21-11 / 830	22-8 / 880	23-4 / 930	24-0 / 980	24-7 / 1030	25-2 / 1080	25-8 / 1130	26-2 / 1180	26-9 / 1220	27-2 / 1270	27-8 / 1310	28-7 / 1400
	16.0	18-6 / 780	19-3 / 850	19-11 / 910	20-7 / 970	21-2 / 1030	21-9 / 1080	22-4 / 1140	22-10 / 1190	23-4 / 1240	23-10 / 1290	24-3 / 1340	24-8 / 1390	25-2 / 1440	25-11 / 1540
	24.0	16-2 / 900	16-10 / 970	17-5 / 1040	18-0 / 1110	18-6 / 1170	19-0 / 1240	19-6 / 1300	19-11 / 1360	20-5 / 1420	20-10 / 1480	21-2 / 1540	21-7 / 1600	21-11 / 1650	22-8 / 1760
2x10	12.0	26-0 / 710	27-1 / 770	28-0 / 830	28-11 / 880	29-9 / 930	30-7 / 980	31-4 / 1030	32-1 / 1080	32-9 / 1130	33-5 / 1180	34-1 / 1220	34-8 / 1270	35-4 / 1310	36-5 / 1400
	16.0	23-8 / 780	24-7 / 850	25-5 / 910	26-3 / 970	27-1 / 1030	27-9 / 1080	28-6 / 1140	29-2 / 1190	29-9 / 1240	30-5 / 1290	31-0 / 1340	31-6 / 1390	32-1 / 1440	33-1 / 1540
	24.0	20-8 / 900	21-6 / 970	22-3 / 1040	22-11 / 1110	23-8 / 1170	24-3 / 1240	24-10 / 1300	25-5 / 1360	26-0 / 1420	26-6 / 1480	27-1 / 1540	27-6 / 1600	28-0 / 1650	28-11 / 1760

Note: The required extreme fiber stress in bending, "F_b", in pounds per square inch is shown below each span.

CEILING JOISTS
10 Lbs. Per Sq. Ft. Live Load
(No attic storage and roof slope not steeper than 3 in 12)
(Drywall Ceiling)

DESIGN CRITERIA:
Deflection - For 10 lbs. per sq. ft. live load.
Limited to span in inches divided by 240.

Strength - live load of 10 lbs. per sq. ft. plus dead load of 5 lbs. per sq. ft. determines required fiber stress value.

Table A-13

Table A-14

RAFTER SIZE (IN)	SPACING (IN)	Allowable Extreme Fiber Stress in Bending, "F$_b$" (psi).														
		500	600	700	800	900	1000	1100	1200	1300	1400	1500	1600	1700	1800	1900
2x6	12.0	8-6 / 0.26	9-4 / 0.35	10-0 / 0.44	10-9 / 0.54	11-5 / 0.64	12-0 / 0.75	12-7 / 0.86	13-2 / 0.98	13-8 / 1.11	14-2 / 1.24	14-8 / 1.37	15-2 / 1.51	15-8 / 1.66	16-1 / 1.81	16-7 / 1.96
	16.0	7-4 / 0.23	8-1 / 0.30	8-8 / 0.38	9-4 / 0.46	9-10 / 0.55	10-5 / 0.65	10-11 / 0.75	11-5 / 0.85	11-10 / 0.97	12-4 / 1.07	12-9 / 1.19	13-2 / 1.31	13-7 / 1.44	13-11 / 1.56	14-4 / 1.70
	24.0	6-0 / 0.19	6-7 / 0.25	7-1 / 0.31	7-7 / 0.38	8-1 / 0.45	8-6 / 0.53	8-11 / 0.61	9-4 / 0.70	9-8 / 0.78	10-0 / 0.88	10-5 / 0.97	10-9 / 1.07	11-1 / 1.17	11-5 / 1.28	11-8 / 1.39
2x8	12.0	11-2 / 0.26	12-3 / 0.35	13-3 / 0.44	14-2 / 0.54	15-0 / 0.64	15-10 / 0.75	16-7 / 0.86	17-4 / 0.98	18-0 / 1.11	18-9 / 1.24	19-5 / 1.37	20-0 / 1.51	20-8 / 1.66	21-3 / 1.81	21-10 / 1.96
	16.0	9-8 / 0.23	10-7 / 0.30	11-6 / 0.38	12-3 / 0.46	13-0 / 0.55	13-8 / 0.65	14-4 / 0.75	15-0 / 0.85	15-7 / 0.96	16-3 / 1.07	16-9 / 1.19	17-4 / 1.31	17-10 / 1.44	18-5 / 1.56	18-11 / 1.70
	24.0	7-11 / 0.19	8-8 / 0.25	9-4 / 0.31	10-0 / 0.38	10-7 / 0.45	11-2 / 0.53	11-9 / 0.61	12-3 / 0.70	12-9 / 0.78	13-3 / 0.88	13-8 / 0.97	14-2 / 1.07	14-7 / 1.17	15-0 / 1.28	15-5 / 1.39
2x10	12.0	14-3 / 0.26	15-8 / 0.35	16-11 / 0.44	18-1 / 0.54	19-2 / 0.64	20-2 / 0.75	21-2 / 0.86	22-1 / 0.98	23-0 / 1.11	23-11 / 1.24	24-9 / 1.37	25-6 / 1.51	26-4 / 1.66	27-1 / 1.81	27-10 / 1.96
	16.0	12-4 / 0.23	13-6 / 0.30	14-8 / 0.38	15-8 / 0.46	16-7 / 0.55	17-6 / 0.65	18-4 / 0.75	19-2 / 0.85	19-11 / 0.96	20-8 / 1.07	21-5 / 1.19	22-1 / 1.31	22-10 / 1.44	23-5 / 1.56	24-1 / 1.70
	24.0	10-1 / 0.19	11-1 / 0.25	11-11 / 0.31	12-9 / 0.38	13-6 / 0.45	14-3 / 0.53	15-0 / 0.61	15-8 / 0.70	16-3 / 0.78	16-11 / 0.88	17-6 / 0.97	18-1 / 1.07	18-7 / 1.17	19-2 / 1.28	19-8 / 1.39
2x12	12.0	17-4 / 0.26	19-0 / 0.35	20-6 / 0.44	21-11 / 0.54	23-3 / 0.64	24-7 / 0.75	25-9 / 0.86	26-11 / 0.98	28-0 / 1.11	29-1 / 1.24	30-1 / 1.37	31-1 / 1.51	32-0 / 1.66	32-11 / 1.81	33-10 / 1.96
	16.0	15-0 / 0.23	16-6 / 0.30	17-9 / 0.38	19-0 / 0.46	20-2 / 0.55	21-3 / 0.65	22-4 / 0.75	23-3 / 0.85	24-3 / 0.96	25-2 / 1.07	26-0 / 1.19	26-11 / 1.31	27-9 / 1.44	28-6 / 1.56	29-4 / 1.70
	24.0	12-3 / 0.19	13-5 / 0.25	14-6 / 0.31	15-6 / 0.38	16-6 / 0.45	17-4 / 0.53	18-2 / 0.61	19-0 / 0.70	19-10 / 0.78	20-6 / 0.88	21-3 / 0.97	21-11 / 1.07	22-8 / 1.17	23-3 / 1.28	23-11 / 1.39

Note: The required modulus of elasticity, "E", in 1,000,000 pounds per square inch is shown below each span.

DESIGN CRITERIA:
Strength - 15 lbs. per sq. ft. dead load plus 20 lbs. per sq. ft. live load determines required fiber stress.
Deflection - For 20 lbs. per sq. ft. live load. Limited to span in inches divided by 240.

LOW OR HIGH SLOPE RAFTERS
20 Lbs. Per Sq. Ft. Live Load
(Supporting Drywall Ceiling)

RAFTERS: Spans are measured along the horizontal projection and loads are considered as applied on the horizontal projection.

Table A-14

Table A-15

RAFTER SIZE (IN)	SPACING (IN)	Allowable Extreme Fiber Stress in Bending, "F$_b$" (psi).														
		500	600	700	800	900	1000	1100	1200	1300	1400	1500	1600	1700	1800	1900
2x6	12.0	7-6 / 0.27	8-2 / 0.36	8-10 / 0.45	9-6 / 0.55	10-0 / 0.66	10-7 / 0.77	11-1 / 0.89	11-7 / 1.01	12-1 / 1.14	12-6 / 1.28	13-0 / 1.41	13-5 / 1.56	13-10 / 1.71	14-2 / 1.86	14-7 / 2.02
	16.0	6-6 / 0.24	7-1 / 0.31	7-8 / 0.39	8-2 / 0.48	8-8 / 0.57	9-2 / 0.67	9-7 / 0.77	10-0 / 0.88	10-5 / 0.99	10-10 / 1.10	11-3 / 1.22	11-7 / 1.35	11-11 / 1.48	12-4 / 1.61	12-8 / 1.75
	24.0	5-4 / 0.19	5-10 / 0.25	6-3 / 0.32	6-8 / 0.39	7-1 / 0.46	7-6 / 0.54	7-10 / 0.63	8-2 / 0.72	8-6 / 0.81	8-10 / 0.90	9-2 / 1.00	9-6 / 1.10	9-9 / 1.21	10-0 / 1.31	10-4 / 1.43
2x8	12.0	9-10 / 0.27	10-10 / 0.36	11-8 / 0.45	12-6 / 0.55	13-3 / 0.66	13-11 / 0.77	14-8 / 0.89	15-3 / 1.01	15-11 / 1.14	16-6 / 1.28	17-1 / 1.41	17-8 / 1.56	18-2 / 1.71	18-9 / 1.86	19-3 / 2.02
	16.0	8-7 / 0.24	9-4 / 0.31	10-1 / 0.39	10-10 / 0.48	11-6 / 0.57	12-1 / 0.67	12-8 / 0.77	13-3 / 0.88	13-9 / 0.99	14-4 / 1.10	14-10 / 1.22	15-3 / 1.35	15-9 / 1.48	16-3 / 1.61	16-8 / 1.75
	24.0	7-0 / 0.19	7-8 / 0.25	8-3 / 0.32	8-10 / 0.39	9-4 / 0.46	9-10 / 0.54	10-4 / 0.63	10-10 / 0.72	11-3 / 0.81	11-8 / 0.90	12-1 / 1.00	12-6 / 1.10	12-10 / 1.21	13-3 / 1.31	13-7 / 1.43
2x10	12.0	12-7 / 0.27	13-9 / 0.36	14-11 / 0.45	15-11 / 0.55	16-11 / 0.66	17-10 / 0.77	18-8 / 0.89	19-6 / 1.01	20-4 / 1.14	21-1 / 1.28	21-10 / 1.41	22-6 / 1.56	23-3 / 1.71	23-11 / 1.86	24-6 / 2.02
	16.0	10-11 / 0.24	11-11 / 0.31	12-11 / 0.39	13-9 / 0.48	14-8 / 0.57	15-5 / 0.67	16-2 / 0.77	16-11 / 0.88	17-7 / 0.99	18-3 / 1.10	18-11 / 1.22	19-6 / 1.35	20-1 / 1.48	20-8 / 1.61	21-3 / 1.75
	24.0	8-11 / 0.19	9-9 / 0.25	10-6 / 0.32	11-3 / 0.39	11-11 / 0.46	12-7 / 0.54	13-2 / 0.63	13-9 / 0.72	14-4 / 0.81	14-11 / 0.90	15-5 / 1.00	15-11 / 1.10	16-5 / 1.21	16-11 / 1.31	17-4 / 1.43
2x12	12.0	15-4 / 0.27	16-9 / 0.36	18-1 / 0.45	19-4 / 0.55	20-6 / 0.66	21-8 / 0.77	22-8 / 0.89	23-9 / 1.01	24-8 / 1.14	25-7 / 1.28	26-6 / 1.41	27-5 / 1.56	28-3 / 1.71	29-1 / 1.86	29-10 / 2.02
	16.0	13-3 / 0.24	14-6 / 0.31	15-8 / 0.39	16-9 / 0.48	17-9 / 0.57	18-9 / 0.67	19-8 / 0.77	20-6 / 0.88	21-5 / 0.99	22-2 / 1.10	23-0 / 1.22	23-9 / 1.35	24-5 / 1.48	25-2 / 1.61	25-10 / 1.75
	24.0	10-10 / 0.19	11-10 / 0.25	12-10 / 0.32	13-8 / 0.39	14-6 / 0.46	15-4 / 0.54	16-1 / 0.63	16-9 / 0.72	17-5 / 0.81	18-1 / 0.90	18-9 / 1.00	19-4 / 1.10	20-0 / 1.21	20-6 / 1.31	21-1 / 1.43

Note: The required modulus of elasticity, "E", in 1,000,000 pounds per square inch is shown below each span.

DESIGN CRITERIA:
Strength - 15 lbs. per sq. ft. dead load plus 30 lbs. per sq. ft. live load determines required fiber stress.
Deflection - For 30 lbs. per sq. ft. live load. Limited to span in inches divided by 240.

LOW OR HIGH SLOPE RAFTERS
30 Lbs. Per Sq. Ft. Live Load
(Supporting Drywall Ceiling)

RAFTERS: Spans are measured along the horizontal projection and loads are considered as applied on the horizontal projection.

Table A-15

Table A-16

RAFTER SIZE (IN)	SPACING (IN)	Allowable Extreme Fiber Stress in Bending, "F_b" (psi).														
		500	600	700	800	900	1000	1100	1200	1300	1400	1500	1600	1700	1800	1900
2x6	12.0	9-2 / 0.33	10-0 / 0.44	10-10 / 0.55	11-7 / 0.67	12-4 / 0.80	13-0 / 0.94	13-7 / 1.09	14-2 / 1.24	14-9 / 1.40	15-4 / 1.56	15-11 / 1.73	16-5 / 1.91	16-11 / 2.09	17-5 / 2.28	17-10 / 2.47
	16.0	7-11 / 0.29	8-8 / 0.38	9-5 / 0.48	10-0 / 0.58	10-8 / 0.70	11-3 / 0.82	11-9 / 0.94	12-4 / 1.07	12-10 / 1.21	13-3 / 1.35	13-9 / 1.50	14-2 / 1.65	14-8 / 1.81	15-1 / 1.97	15-6 / 2.14
	24.0	6-6 / 0.24	7-1 / 0.31	7-8 / 0.39	8-2 / 0.48	8-8 / 0.57	9-2 / 0.67	9-7 / 0.77	10-0 / 0.88	10-5 / 0.99	10-10 / 1.10	11-3 / 1.22	11-7 / 1.35	11-11 / 1.48	12-4 / 1.61	12-8 / 1.75
2x8	12.0	12-1 / 0.33	13-3 / 0.44	14-4 / 0.55	15-3 / 0.67	16-3 / 0.80	17-1 / 0.94	17-11 / 1.09	18-9 / 1.24	19-6 / 1.40	20-3 / 1.56	20-11 / 1.73	21-7 / 1.91	22-3 / 2.09	22-11 / 2.28	23-7 / 2.47
	16.0	10-6 / 0.29	11-6 / 0.38	12-5 / 0.48	13-3 / 0.58	14-0 / 0.70	14-10 / 0.82	15-6 / 0.94	16-3 / 1.07	16-10 / 1.21	17-6 / 1.35	18-2 / 1.50	18-9 / 1.65	19-4 / 1.81	19-10 / 1.97	20-5 / 2.14
	24.0	8-7 / 0.24	9-4 / 0.31	10-1 / 0.39	10-10 / 0.48	11-6 / 0.57	12-1 / 0.67	12-8 / 0.77	13-3 / 0.88	13-9 / 0.99	14-4 / 1.10	14-10 / 1.22	15-3 / 1.35	15-9 / 1.48	16-3 / 1.61	16-8 / 1.75
2x10	12.0	15-5 / 0.33	16-11 / 0.44	18-3 / 0.55	19-6 / 0.67	20-8 / 0.80	21-10 / 0.94	22-10 / 1.09	23-11 / 1.24	24-10 / 1.40	25-10 / 1.56	26-8 / 1.73	27-7 / 1.91	28-5 / 2.09	29-3 / 2.28	30-1 / 2.47
	16.0	13-4 / 0.29	14-8 / 0.38	15-10 / 0.48	16-11 / 0.58	17-11 / 0.70	18-11 / 0.82	19-10 / 0.94	20-8 / 1.07	21-6 / 1.21	22-4 / 1.35	23-2 / 1.50	23-11 / 1.65	24-7 / 1.81	25-4 / 1.97	26-0 / 2.14
	24.0	10-11 / 0.24	11-11 / 0.31	12-11 / 0.39	13-9 / 0.48	14-8 / 0.57	15-5 / 0.67	16-2 / 0.77	16-11 / 0.88	17-7 / 0.99	18-3 / 1.10	18-11 / 1.22	19-6 / 1.35	20-1 / 1.48	20-8 / 1.61	21-3 / 1.75
2x12	12.0	18-9 / 0.33	20-6 / 0.44	22-2 / 0.55	23-9 / 0.67	25-2 / 0.80	26-6 / 0.94	27-10 / 1.09	29-1 / 1.24	30-3 / 1.40	31-4 / 1.56	32-6 / 1.73	33-6 / 1.91	34-7 / 2.09	35-7 / 2.28	36-7 / 2.47
	16.0	16-3 / 0.29	17-9 / 0.38	19-3 / 0.48	20-6 / 0.58	21-9 / 0.70	23-0 / 0.82	24-1 / 0.94	25-2 / 1.07	26-2 / 1.21	27-2 / 1.35	28-2 / 1.50	29-1 / 1.65	29-11 / 1.81	30-10 / 1.97	31-8 / 2.14
	24.0	13-3 / 0.24	14-6 / 0.31	15-8 / 0.39	16-9 / 0.48	17-9 / 0.57	18-9 / 0.67	19-8 / 0.77	20-6 / 0.88	21-5 / 0.99	22-2 / 1.10	23-0 / 1.22	23-9 / 1.35	24-5 / 1.48	25-2 / 1.61	25-10 / 1.75

Note: The required modulus of elasticity, "E", in 1,000,000 pounds per square inch is shown below each span.

DESIGN CRITERIA:
Strength - 10 lbs. per sq. ft. dead load plus 20 lbs. per sq. ft. live load determines required fiber stress.
Deflection - For 20 lbs. per sq. ft. live load. Limited to span in inches divided by 240.

LOW SLOPE RAFTERS
Slope 3 in 12 or less - 20 Lbs. Per Sq. Ft. Live Load
(No Finished Ceiling)

RAFTERS: Spans are measured along the horizontal projection and loads are considered as applied on the horizontal projection.

Table A-16

Table A-17

RAFTER SIZE (IN)	SPACING (IN)	Allowable Extreme Fiber Stress In Bending, "F_b" (psi).														
		500	600	700	800	900	1000	1100	1200	1300	1400	1500	1600	1700	1800	1900
2x6	12.0	7-11 / 0.32	8-8 / 0.43	9-5 / 0.54	10-0 / 0.66	10-8 / 0.78	11-3 / 0.92	11-9 / 1.06	12-4 / 1.21	12-10 / 1.36	13-3 / 1.52	13-9 / 1.69	14-2 / 1.86	14-8 / 2.04	15-1 / 2.22	15-6 / 2.41
	16.0	6-11 / 0.28	7-6 / 0.37	8-2 / 0.47	8-8 / 0.57	9-3 / 0.68	9-9 / 0.80	10-2 / 0.92	10-8 / 1.05	11-1 / 1.18	11-6 / 1.32	11-11 / 1.46	12-4 / 1.61	12-8 / 1.76	13-1 / 1.92	13-5 / 2.08
	24.0	5-7 / 0.23	6-2 / 0.30	6-8 / 0.38	7-1 / 0.46	7-6 / 0.55	7-11 / 0.65	8-4 / 0.75	8-8 / 0.85	9-1 / 0.96	9-5 / 1.08	9-9 / 1.19	10-0 / 1.31	10-4 / 1.44	10-8 / 1.57	10-11 / 1.70
2x8	12.0	10-6 / 0.32	11-6 / 0.43	12-5 / 0.54	13-3 / 0.66	14-0 / 0.78	14-10 / 0.92	15-6 / 1.06	16-3 / 1.21	16-10 / 1.36	17-6 / 1.52	18-2 / 1.69	18-9 / 1.86	19-4 / 2.04	19-10 / 2.22	20-5 / 2.41
	16.0	9-1 / 0.28	9-11 / 0.37	10-9 / 0.47	11-6 / 0.57	12-2 / 0.68	12-10 / 0.80	13-5 / 0.92	14-0 / 1.05	14-7 / 1.18	15-2 / 1.32	15-8 / 1.46	16-3 / 1.61	16-9 / 1.76	17-2 / 1.92	17-8 / 2.08
	24.0	7-5 / 0.23	8-1 / 0.30	8-9 / 0.38	9-4 / 0.46	9-11 / 0.55	10-6 / 0.65	11-0 / 0.75	11-6 / 0.85	11-11 / 0.96	12-5 / 1.08	12-10 / 1.19	13-3 / 1.31	13-8 / 1.44	14-0 / 1.57	14-5 / 1.70
2x10	12.0	13-4 / 0.32	14-8 / 0.43	15-10 / 0.54	16-11 / 0.66	17-11 / 0.78	18-11 / 0.92	19-10 / 1.06	20-8 / 1.21	21-6 / 1.36	22-4 / 1.52	23-2 / 1.69	23-11 / 1.86	24-7 / 2.04	25-4 / 2.22	26-0 / 2.41
	16.0	11-7 / 0.28	12-8 / 0.37	13-8 / 0.47	14-8 / 0.57	15-6 / 0.68	16-4 / 0.80	17-2 / 0.92	17-11 / 1.05	18-8 / 1.18	19-4 / 1.32	20-0 / 1.46	20-8 / 1.61	21-4 / 1.76	21-11 / 1.92	22-6 / 2.08
	24.0	9-5 / 0.23	10-4 / 0.30	11-2 / 0.38	11-11 / 0.46	12-8 / 0.55	13-4 / 0.65	14-0 / 0.75	14-8 / 0.85	15-3 / 0.96	15-10 / 1.08	16-4 / 1.19	16-11 / 1.31	17-5 / 1.44	17-11 / 1.57	18-5 / 1.70
2x12	12.0	16-3 / 0.32	17-9 / 0.43	19-3 / 0.54	20-6 / 0.66	21-9 / 0.78	23-0 / 0.92	24-1 / 1.06	25-2 / 1.21	26-2 / 1.36	27-2 / 1.52	28-2 / 1.69	29-1 / 1.86	29-11 / 2.04	30-10 / 2.22	31-8 / 2.41
	16.0	14-1 / 0.28	15-5 / 0.37	16-8 / 0.47	17-9 / 0.57	18-10 / 0.68	19-11 / 0.80	20-10 / 0.92	21-9 / 1.05	22-8 / 1.18	23-6 / 1.32	24-4 / 1.46	25-2 / 1.61	25-11 / 1.76	26-8 / 1.92	27-5 / 2.08
	24.0	11-6 / 0.23	12-7 / 0.30	13-7 / 0.38	14-6 / 0.46	15-5 / 0.55	16-3 / 0.65	17-0 / 0.75	17-9 / 0.85	18-6 / 0.96	19-3 / 1.08	19-11 / 1.19	20-6 / 1.31	21-2 / 1.44	21-9 / 1.57	22-5 / 1.70

Note: The required modulus of elasticity, "E", in 1,000,000 pounds per square inch is shown below each span.

DESIGN CRITERIA:
Strength - 10 lbs. per sq. ft. dead load plus 30 lbs. per sq. ft. live load determines required fiber stress.
Deflection - For 30 lbs. per sq. ft. live load. Limited to span in inches divided by 240.

LOW SLOPE RAFTERS
Slope 3 in 12 or less - 30 Lbs. Per Sq. Ft. Live Load
(No Finished Ceiling)

RAFTERS: Spans are measured along the horizontal projection and loads are considered as applied on the horizontal projection.

Table A-17

RAFTER SIZE (IN)	SPACING (IN)	Allowable Extreme Fiber Stress in Bending, "F_b" (psi).														
		500	600	700	800	900	1000	1100	1200	1300	1400	1500	1600	1700	1800	1900
2x4	12.0	6-2 0.29	6-9 0.38	7-3 0.49	7-9 0.59	8-3 0.71	8-8 0.83	9-1 0.96	9-6 1.09	9-11 1.23	10-3 1.37	10-8 1.52	11-0 1.68	11-4 1.84	11-8 2.00	12-0 2.17
	16.0	5-4 0.25	5-10 0.33	6-4 0.42	6-9 0.51	7-2 0.61	7-6 0.72	7-11 0.83	8-3 0.94	8-7 1.06	8-11 1.19	9-3 1.32	9-6 1.45	9-10 1.59	10-1 1.73	10-5 1.88
	24.0	4-4 0.21	4-9 0.27	5-2 0.34	5-6 0.42	5-10 0.50	6-2 0.59	6-5 0.68	6-9 0.77	7-0 0.87	7-3 0.97	7-6 1.08	7-9 1.19	8-0 1.30	8-3 1.41	8-6 1.53
2x6	12.0	9-8 0.29	10-7 0.38	11-5 0.49	12-3 0.59	13-0 0.71	13-8 0.83	14-4 0.96	15-0 1.09	15-7 1.23	16-2 1.37	16-9 1.52	17-3 1.68	17-10 1.84	18-4 2.00	18-10 2.17
	16.0	8-4 0.25	9-2 0.33	9-11 0.42	10-7 0.51	11-3 0.61	11-10 0.72	12-5 0.83	13-0 0.94	13-6 1.06	14-0 1.19	14-6 1.32	15-0 1.45	15-5 1.59	15-11 1.73	16-4 1.88
	24.0	6-10 0.21	7-6 0.27	8-1 0.34	8-8 0.42	9-2 0.50	9-8 0.59	10-2 0.68	10-7 0.77	11-0 0.87	11-5 0.97	11-10 1.08	12-3 1.19	12-7 1.30	13-0 1.41	13-4 1.53
2x8	12.0	12-9 0.29	13-11 0.38	15-1 0.49	16-1 0.59	17-1 0.71	18-0 0.83	18-11 0.96	19-9 1.09	20-6 1.23	21-4 1.37	22-1 1.52	22-9 1.68	23-6 1.84	24-2 2.00	24-10 2.17
	16.0	11-0 0.25	12-1 0.33	13-1 0.42	13-11 0.51	14-10 0.61	15-7 0.72	16-4 0.83	17-1 0.94	17-9 1.06	18-5 1.19	19-1 1.32	19-9 1.45	20-4 1.59	20-11 1.73	21-6 1.88
	24.0	9-0 0.21	9-10 0.27	10-8 0.34	11-5 0.42	12-1 0.50	12-9 0.59	13-4 0.68	13-11 0.77	14-6 0.87	15-1 0.97	15-7 1.08	16-1 1.19	16-7 1.30	17-1 1.41	17-7 1.53
2x10	12.0	16-3 0.29	17-10 0.38	19-3 0.49	20-7 0.59	21-10 0.71	23-0 0.83	24-1 0.96	25-2 1.09	26-2 1.23	27-2 1.37	28-2 1.52	29-1 1.68	30-0 1.84	30-10 2.00	31-8 2.17
	16.0	14-1 0.25	15-5 0.33	16-8 0.42	17-10 0.51	18-11 0.61	19-11 0.72	20-10 0.83	21-10 0.94	22-8 1.06	23-7 1.19	24-5 1.32	25-2 1.45	25-11 1.59	26-8 1.73	27-5 1.88
	24.0	11-6 0.21	12-7 0.27	13-7 0.34	14-6 0.42	15-5 0.50	16-3 0.59	17-1 0.68	17-10 0.77	18-6 0.87	19-3 0.97	19-11 1.08	20-7 1.19	21-2 1.30	21-10 1.41	22-5 1.53

Note: The required modulus of elasticity, "E", in 1,000,000 pounds per square inch is shown below each span.

DESIGN CRITERIA:
Strength - 7 lbs. per sq. ft. dead load plus 20 lbs. per sq. ft. live load determines required fiber stress.
Deflection - For 20 lbs. per sq. ft. live load. Limited to span in inches divided by 180.

HIGH SLOPE RAFTERS
Slope over 3 in 12 - 20 Lbs. Per Sq. Ft. Live Load
(Light Roof Covering)

RAFTERS: Spans are measured along the horizontal projection and loads are considered as applied on the horizontal projection.

Table A-18

RAFTER SIZE (IN)	SPACING (IN)	Allowable Extreme Fiber Stress in Bending, "F_b" (psi).														
		500	600	700	800	900	1000	1100	1200	1300	1400	1500	1600	1700	1800	1900
2x4	12.0	5-3 0.27	5-9 0.36	6-3 0.45	6-8 0.55	7-1 0.66	7-5 0.77	7-9 0.89	8-2 1.02	8-6 1.15	8-9 1.28	9-1 1.42	9-5 1.57	9-8 1.72	10-0 1.87	10-3 2.03
	16.0	4-7 0.24	5-0 0.31	5-5 0.39	5-9 0.48	6-1 0.57	6-5 0.67	6-9 0.77	7-1 0.88	7-4 0.99	7-7 1.11	7-11 1.23	8-2 1.36	8-5 1.49	8-8 1.62	8-10 1.76
	24.0	3-9 0.19	4-1 0.25	4-5 0.32	4-8 0.39	5-0 0.47	5-3 0.55	5-6 0.63	5-9 0.72	6-0 0.81	6-3 0.91	6-5 1.01	6-8 1.11	6-10 1.21	7-1 1.32	7-3 1.43
2x6	12.0	8-3 0.27	9-1 0.36	9-9 0.45	10-5 0.55	11-1 0.66	11-8 0.77	12-3 0.89	12-9 1.02	13-4 1.15	13-10 1.28	14-4 1.42	14-9 1.57	15-3 1.72	15-8 1.87	16-1 2.03
	16.0	7-2 0.24	7-10 0.31	8-5 0.39	9-1 0.48	9-7 0.57	10-1 0.67	10-7 0.77	11-1 0.88	11-6 0.99	12-0 1.11	12-5 1.23	12-9 1.36	13-2 1.49	13-7 1.62	13-11 1.76
	24.0	5-10 0.19	6-5 0.25	6-11 0.32	7-5 0.39	7-10 0.47	8-3 0.55	8-8 0.63	9-1 0.72	9-5 0.81	9-9 0.91	10-1 1.01	10-5 1.11	10-9 1.21	11-1 1.32	11-5 1.43
2x8	12.0	10-11 0.27	11-11 0.36	12-10 0.45	13-9 0.55	14-7 0.66	15-5 0.77	16-2 0.89	16-10 1.02	17-7 1.15	18-2 1.28	18-10 1.42	19-6 1.57	20-1 1.72	20-8 1.87	21-3 2.03
	16.0	9-5 0.24	10-4 0.31	11-2 0.39	11-11 0.48	12-8 0.57	13-4 0.67	14-0 0.77	14-7 0.88	15-2 0.99	15-9 1.11	16-4 1.23	16-10 1.36	17-4 1.49	17-11 1.62	18-4 1.76
	24.0	7-8 0.19	8-5 0.25	9-1 0.32	9-9 0.39	10-4 0.47	10-11 0.55	11-5 0.63	11-11 0.72	12-5 0.81	12-10 0.91	13-4 1.01	13-9 1.11	14-2 1.21	14-7 1.32	15-0 1.43
2x10	12.0	13-11 0.27	15-2 0.36	16-5 0.45	17-7 0.55	18-7 0.66	19-8 0.77	20-7 0.89	21-6 1.02	22-5 1.15	23-3 1.28	24-1 1.42	24-10 1.57	25-7 1.72	26-4 1.87	27-1 2.03
	16.0	12-0 0.26	13-2 0.34	14-3 0.43	15-2 0.53	16-2 0.63	17-0 0.74	17-10 0.85	18-7 0.97	19-5 1.09	20-1 1.22	20-10 1.35	21-6 1.49	22-2 1.63	22-10 1.78	23-5 1.93
	24.0	9-10 0.19	10-9 0.25	11-7 0.32	12-5 0.39	13-2 0.47	13-11 0.55	14-7 0.63	15-2 0.72	15-10 0.81	16-5 0.91	17-0 1.01	17-7 1.11	18-1 1.21	18-7 1.32	19-2 1.43

Note: The required modulus of elasticity, "E", in 1,000,000 pounds per square inch is shown below each span.

DESIGN CRITERIA:
Strength - 7 lbs. per sq. ft. dead load plus 30 lbs. per sq. ft. live load determines required fiber stress.
Deflection - For 30 lbs. per sq. ft. live load. Limited to span in inches divided by 180.

HIGH SLOPE RAFTERS
Slope over 3 in 12 - 30 Lbs. Per Sq. Ft. Live Load
(Light Roof Covering)

RAFTERS: Spans are measured along the horizontal projection and loads are considered as applied on the horizontal projection.

Table A-19

Table A-20

RAFTER SIZE (IN)	SPACING (IN)	Allowable Extreme Fiber Stress in Bending, "F_b" (psi).														
		500	600	700	800	900	1000	1100	1200	1300	1400	1500	1600	1700	1800	1900
2x4	12.0	5-5 / 0.20	5-11 / 0.26	6-5 / 0.33	6-10 / 0.40	7-3 / 0.48	7-8 / 0.56	8-0 / 0.65	8-4 / 0.74	8-8 / 0.83	9-0 / 0.93	9-4 / 1.03	9-8 / 1.14	9-11 / 1.24	10-3 / 1.36	10-6 / 1.47
	16.0	4-8 / 0.17	5-1 / 0.23	5-6 / 0.28	5-11 / 0.35	6-3 / 0.41	6-7 / 0.49	6-11 / 0.56	7-3 / 0.64	7-6 / 0.72	7-10 / 0.80	8-1 / 0.89	8-4 / 0.98	8-7 / 1.08	8-10 / 1.17	9-1 / 1.27
	24.0	3-10 / 0.14	4-2 / 0.18	4-6 / 0.23	4-10 / 0.28	5-1 / 0.34	5-5 / 0.40	5-8 / 0.46	5-11 / 0.52	6-2 / 0.59	6-5 / 0.66	6-7 / 0.73	6-10 / 0.80	7-0 / 0.88	7-3 / 0.96	7-5 / 1.04
2x6	12.0	8-6 / 0.20	9-4 / 0.26	10-0 / 0.33	10-9 / 0.40	11-5 / 0.48	12-0 / 0.56	12-7 / 0.65	13-2 / 0.74	13-8 / 0.83	14-2 / 0.93	14-8 / 1.03	15-2 / 1.14	15-8 / 1.24	16-1 / 1.36	16-7 / 1.47
	16.0	7-4 / 0.17	8-1 / 0.23	8-8 / 0.28	9-4 / 0.35	9-10 / 0.41	10-5 / 0.49	10-11 / 0.56	11-5 / 0.64	11-10 / 0.72	12-4 / 0.80	12-9 / 0.89	13-2 / 0.98	13-7 / 1.08	13-11 / 1.17	14-4 / 1.27
	24.0	6-0 / 0.14	6-7 / 0.18	7-1 / 0.23	7-7 / 0.28	8-1 / 0.34	8-6 / 0.40	8-11 / 0.46	9-4 / 0.52	9-8 / 0.59	10-0 / 0.66	10-5 / 0.73	10-9 / 0.80	11-1 / 0.88	11-5 / 0.96	11-8 / 1.04
2x8	12.0	11-2 / 0.20	12-3 / 0.26	13-3 / 0.33	14-2 / 0.40	15-0 / 0.48	15-10 / 0.56	16-7 / 0.65	17-4 / 0.74	18-0 / 0.83	18-9 / 0.93	19-5 / 1.03	20-0 / 1.14	20-8 / 1.24	21-3 / 1.36	21-10 / 1.47
	16.0	9-8 / 0.17	10-7 / 0.23	11-6 / 0.28	12-3 / 0.35	13-0 / 0.41	13-8 / 0.49	14-4 / 0.56	15-0 / 0.64	15-7 / 0.72	16-3 / 0.80	16-9 / 0.89	17-4 / 0.98	17-10 / 1.08	18-5 / 1.17	18-11 / 1.27
	24.0	7-11 / 0.14	8-8 / 0.18	9-4 / 0.23	10-0 / 0.28	10-7 / 0.34	11-2 / 0.40	11-9 / 0.46	12-3 / 0.52	12-9 / 0.59	13-3 / 0.66	13-8 / 0.73	14-2 / 0.80	14-7 / 0.88	15-0 / 0.96	15-5 / 1.04
2x10	12.0	14-3 / 0.20	15-8 / 0.26	16-11 / 0.33	18-1 / 0.40	19-2 / 0.48	20-2 / 0.56	21-2 / 0.65	22-1 / 0.74	23-0 / 0.83	23-11 / 0.93	24-9 / 1.03	25-6 / 1.14	26-4 / 1.24	27-1 / 1.36	27-10 / 1.47
	16.0	12-4 / 0.17	13-6 / 0.23	14-8 / 0.28	15-8 / 0.35	16-7 / 0.41	17-6 / 0.49	18-4 / 0.56	19-2 / 0.64	19-11 / 0.72	20-8 / 0.80	21-5 / 0.89	22-1 / 0.98	22-10 / 1.08	23-5 / 1.17	24-1 / 1.27
	24.0	10-1 / 0.14	11-1 / 0.18	11-11 / 0.23	12-9 / 0.28	13-6 / 0.34	14-3 / 0.40	15-0 / 0.46	15-8 / 0.52	16-3 / 0.59	16-11 / 0.66	17-6 / 0.73	18-1 / 0.80	18-7 / 0.88	19-2 / 0.96	19-8 / 1.04

Note: The required modulus of elasticity, "E", in 1,000,000 pounds per square inch is shown below each span.

DESIGN CRITERIA:
Strength - 15 lbs. per sq. ft. dead load plus 20 lbs. per sq. ft. live load determines required fiber stress.
Deflection - For 20 lbs. per sq. ft. live load. Limited to span in inches divided by 180.

HIGH SLOPE RAFTERS
Slope over 3 in 12 - 20 Lbs. Per Sq. Ft. Live Load
(Heavy Roof Covering)

RAFTERS: Spans are measured along the horizontal projection and loads are considered as applied on the horizontal projection.

Table A-20

Table A-21

RAFTER SIZE (IN)	SPACING (IN)	Allowable Extreme Fiber Stress in Bending, "F_b" (psi).														
		500	600	700	800	900	1000	1100	1200	1300	1400	1500	1600	1700	1800	1900
2x4	12.0	4-9 / 0.20	5-3 / 0.27	5-8 / 0.34	6-0 / 0.41	6-5 / 0.49	6-9 / 0.58	7-1 / 0.67	7-5 / 0.76	7-8 / 0.86	8-0 / 0.96	8-3 / 1.06	8-6 / 1.17	8-9 / 1.28	9-0 / 1.39	9-3 / 1.51
	16.0	4-1 / 0.18	4-6 / 0.23	4-11 / 0.29	5-3 / 0.36	5-6 / 0.43	5-10 / 0.50	6-1 / 0.58	6-5 / 0.66	6-8 / 0.74	6-11 / 0.83	7-2 / 0.92	7-5 / 1.01	7-7 / 1.11	7-10 / 1.21	8-0 / 1.31
	24.0	3-4 / 0.14	3-8 / 0.19	4-0 / 0.24	4-3 / 0.29	4-6 / 0.35	4-9 / 0.41	5-0 / 0.47	5-3 / 0.54	5-5 / 0.61	5-8 / 0.68	5-10 / 0.75	6-0 / 0.83	6-3 / 0.90	6-5 / 0.99	6-7 / 1.07
2x6	12.0	7-6 / 0.20	8-2 / 0.27	8-10 / 0.34	9-6 / 0.41	10-0 / 0.49	10-7 / 0.58	11-1 / 0.67	11-7 / 0.76	12-1 / 0.86	12-6 / 0.96	13-0 / 1.06	13-5 / 1.17	13-10 / 1.28	14-2 / 1.39	14-7 / 1.51
	16.0	6-6 / 0.18	7-1 / 0.23	7-8 / 0.29	8-2 / 0.36	8-8 / 0.43	9-2 / 0.50	9-7 / 0.58	10-0 / 0.66	10-5 / 0.74	10-10 / 0.83	11-3 / 0.92	11-7 / 1.01	11-11 / 1.11	12-4 / 1.21	12-8 / 1.31
	24.0	5-4 / 0.14	5-10 / 0.19	6-3 / 0.24	6-8 / 0.29	7-1 / 0.35	7-6 / 0.41	7-10 / 0.47	8-2 / 0.54	8-6 / 0.61	8-10 / 0.68	9-2 / 0.75	9-6 / 0.83	9-9 / 0.90	10-0 / 0.99	10-4 / 1.07
2x8	12.0	9-10 / 0.20	10-10 / 0.27	11-8 / 0.34	12-6 / 0.41	13-3 / 0.49	13-11 / 0.58	14-8 / 0.67	15-3 / 0.76	15-11 / 0.86	16-6 / 0.96	17-1 / 1.06	17-8 / 1.17	18-2 / 1.28	18-9 / 1.39	19-3 / 1.51
	16.0	8-7 / 0.18	9-4 / 0.23	10-1 / 0.29	10-10 / 0.36	11-6 / 0.43	12-1 / 0.50	12-8 / 0.58	13-3 / 0.66	13-9 / 0.74	14-4 / 0.83	14-10 / 0.92	15-3 / 1.01	15-9 / 1.11	16-3 / 1.21	16-8 / 1.31
	24.0	7-0 / 0.14	7-8 / 0.19	8-3 / 0.24	8-10 / 0.29	9-4 / 0.35	9-10 / 0.41	10-4 / 0.47	10-10 / 0.54	11-3 / 0.61	11-8 / 0.68	12-1 / 0.75	12-6 / 0.83	12-10 / 0.90	13-3 / 0.99	13-7 / 1.07
2x10	12.0	12-7 / 0.20	13-9 / 0.27	14-11 / 0.34	15-11 / 0.41	16-11 / 0.49	17-10 / 0.58	18-8 / 0.67	19-6 / 0.76	20-4 / 0.86	21-1 / 0.96	21-10 / 1.06	22-6 / 1.17	23-3 / 1.28	23-11 / 1.39	24-6 / 1.51
	16.0	10-11 / 0.18	11-11 / 0.23	12-11 / 0.29	13-9 / 0.36	14-8 / 0.43	15-5 / 0.50	16-2 / 0.58	16-11 / 0.66	17-7 / 0.74	18-3 / 0.83	18-11 / 0.92	19-6 / 1.01	20-1 / 1.11	20-8 / 1.21	21-3 / 1.31
	24.0	8-11 / 0.14	9-9 / 0.19	10-6 / 0.24	11-3 / 0.29	11-11 / 0.35	12-7 / 0.41	13-2 / 0.47	13-9 / 0.54	14-4 / 0.61	14-11 / 0.68	15-5 / 0.75	15-11 / 0.83	16-5 / 0.90	16-11 / 0.99	17-4 / 1.07

Note: The required modulus of elasticity, "E", in 1,000,000 pounds per square inch is shown below each span.

DESIGN CRITERIA:
Strength - 15 lbs. per sq. ft. dead load plus 30 lbs. per sq. ft. live load determines required fiber stress.
Deflection - For 30 lbs. per sq. ft. live load. Limited to span in inches divided by 180.

HIGH SLOPE RAFTERS
Slope over 3 in 12 - 30 Lbs. Per Sq. Ft. Live Load
(Heavy Roof Covering)

RAFTERS: Spans are measured along the horizontal projection and loads are considered as applied on the horizontal projection.

Table A-21

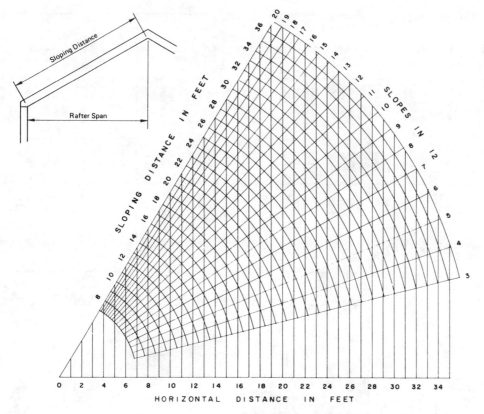

To use the diagram select the known horizontal distance and follow the vertical line to its intersection with the radial line of the specified slope, then proceed along the arc to read the sloping distance. In some cases it may be desirable to interpolate between the one foot separations. The diagram also may be used to find the horizontal distance corresponding to a given sloping distance or to find the slope when the horizontal and sloping distances are known.

Example: With a roof slope of 8 in 12 and a horizontal distance of 20 feet the sloping distance may be read as 24 feet.

Conversion diagram for rafters
Figure A-22

Length of tail beams	4	5	6	7	8	9	10	11	12	13	14	15	16	17	18	19	20
Total live and dead loads per square foot: 10	20	25	30	35	40	45	50	55	60	65	70	75	80	85	90	95	100
30	60	75	90	105	120	135	150	165	180	195	210	225	240	255	270	285	300
40	80	100	120	140	160	180	200	220	240	260	280	300	320	340	360	380	400
50	100	125	150	175	200	225	250	275	300	325	350	375	400	425	450	475	500
60	120	150	180	210	240	270	300	330	360	390	420	450	480	510	540	570	600
70	140	175	210	245	280	315	350	385	420	455	490	525	560	595	630	665	700
80	160	200	240	280	320	360	400	440	480	520	560	600	640	680	720	760	800
90	180	225	270	315	360	405	450	495	540	585	630	675	720	765	810	855	900
100	200	250	300	350	400	450	500	550	600	650	700	750	800	850	900	950	1,000

Table for use in figuring header and trimmer loads
Table A-23

Load	Position of load	Factor
Concentrated	Applied at center of span	Multiply by 2.
Do	Applied at one-third of span	Multiply by 1¾.[1]
Do	Applied at one-fourth of span	Multiply by 1½.
Do	Applied at one-fifth of span	Multiply by 1¼.[1]
Do	Applied at one-seventh of span	Multiply by 1.
Do	Applied at one-eighth of span	Multiply by 7/8.
Do	Applied at one-tenth of span	Multiply by ¾.[1]

[1] These factors are to the nearest simple fraction, but are close enough for practical purposes.

Relationship between bending effect of concentrated and uniformly distributed loads
Table A-24

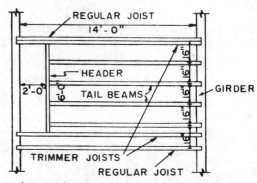

Hearth opening requiring calculations to determine header and trimmer sizes
Figure A-25

These "F_b" values are for use where repetitive members are spaced not more than 24 inches. For wider spacing, the "F_b" values should be reduced 13 percent.
Values for surfaced dry or surfaced green lumber apply at 19 percent maximum moisture content in use.

Species and Grade	Size	Allowable Unit Stress in Bending "F_b"			Modulus of Elasticity "E"	Grading Rules Agency
		Normal Duration	Snow Loading	7-Day Loading		
CALIFORNIA REDWOOD (Surfaced dry or surfaced green)						
Clear Heart Structural	2x4	2650	3050	3310	1,400,000	Redwood Inspection Service (See note 1)
Clear Structural		2650	3050	3310	1,400,000	
Select Structural		2350	2700	2940	1,400,000	
Select Structural, Open grain		1850	2130	2310	1,100,000	
No. 1		1950	2240	2440	1,400,000	
No. 1, Open grain		1550	1780	1940	1,100,000	
No. 2	2x4	1600	1840	2000	1,250,000	
No. 2, Open grain		1250	1440	1560	1,000,000	
No. 3		900	1030	1120	1,100,000	
No. 3, Open grain		725	830	910	900,000	
Stud		725	830	910	900,000	
Construction	2x4	950	1090	1190	900,000	
Standard		525	600	660	900,000	
Utility		250	290	310	900,000	
Clear Heart Structural	**2x6 & wider**	**2650**	**3050**	**3310**	**1,400,000**	
Clear Structural		**2650**	**3050**	**3310**	**1,400,000**	
Select Structural	**2x6 and wider**	**2000**	**2300**	**2500**	**1,400,000**	
Select Structural, Open grain		**1600**	**1840**	**2000**	**1,400,000**	
No. 1		**1700**	**1960**	**2120**	**1,400,000**	
No. 1, Open grain		**1350**	**1550**	**1690**	**1,100,000**	
No. 2	**2x6 and wider**	**1400**	**1610**	**1750**	**1,250,000**	
No. 2, Open grain		**1100**	**1260**	**1370**	**1,000,000**	
No. 3		**800**	**920**	**1000**	**1,100,000**	
No. 3, Open grain		**650**	**750**	**810**	**900,000**	
COAST SITKA SPRUCE (Surfaced dry or surfaced green)						
Select Structural	2x4	1700	1960	2130	1,700,000	Nat'l. Lumber Grades Auth. (A Canadian Agency—See notes 1 and 2)
No. 1 & Appearance		1450	1670	1810	1,700,000	
No. 2		1200	1380	1500	1,500,000	
No. 3		675	780	840	1,300,000	
Stud		675	780	840	1,300,000	
Construction	2x4	875	1010	1090	1,300,000	
Standard		500	580	630	1,300,000	
Utility		225	260	280	1,300,000	
Select Structural	**2x6 and wider**	**1500**	**1730**	**1880**	**1,700,000**	
No. 1 & Appearance		**1250**	**1440**	**1560**	**1,700,000**	
No. 2		**1050**	**1210**	**1310**	**1,500,000**	
No. 3		**600**	**690**	**750**	**1,300,000**	
COAST SPECIES (Surfaced dry or surfaced green)						
Select Structural	2x4	1700	1950	2120	1,500,000	National Lumber Grades Authority (A Canadian Agency—See notes 1 and 2)
No. 1 & Appearance		1450	1670	1810	1,500,000	
No. 2		1200	1380	1500	1,400,000	
No. 3		675	780	840	1,200,000	
Stud		675	780	840	1,200,000	
Construction	2x4	875	1010	1090	1,200,000	
Standard		500	570	620	1,200,000	
Utility		225	260	280	1,200,000	
Select Structural	**2x6 and wider**	**1500**	**1730**	**1870**	**1,500,000**	
No. 1 & Appearance		**1250**	**1440**	**1560**	**1,500,000**	
No. 2		**1050**	**1210**	**1310**	**1,400,000**	
No. 3		**600**	**690**	**750**	**1,200,000**	

Species and Grade	Size	Allowable Unit Stress in Bending "F_b"			Modulus of Elasticity "E"	Grading Rules Agency
		Normal Duration	Snow Loading	7-Day Loading		
ASPEN (Surfaced dry or surfaced green)						
Select Structural	2x4	1500	1720	1880	1,100,000	Northern Hardwood and Pine Manufacturers Association
No. 1 & Appearance		1300	1500	1620	1,100,000	
No. 2		1050	1210	1310	1,000,000	
No. 3		575	660	720	900,000	
Stud		575	660	720	900,000	
Construction	2x4	750	860	940	900,000	
Standard		425	490	530	900,000	
Utility		200	230	250	900,000	
Select Structural	**2x6 and wider**	**1300**	**1500**	**1620**	**1,100,000**	Western Wood Products Association (See note 1)
No. 1 & Appearance		**1100**	**1280**	**1380**	**1,100,000**	
No. 2		**900**	**1040**	**1120**	**1,000,000**	
No. 3		**525**	**600**	**660**	**900,000**	
BALSAM FIR (Surfaced dry or surfaced green)						
Select Structural	2x4	1550	1780	1940	1,200,000	Northeastern Lumber Manufacturers Association
No. 1		1300	1500	1620	1,200,000	
No. 2		1100	1260	1380	1,100,000	
No. 3		600	690	750	1,200,000	Northern Hardwood & Pine Manufacturers Association
Appearance		1150	1320	1440	1,200,000	
Stud		600	690	750	900,000	
Construction	2x4	800	920	1000	900,000	
Standard		450	520	560	900,000	
Utility		200	230	250	900,000	
Select Structural	**2x6 and wider**	**1350**	**1550**	**1680**	**1,200,000**	(See note 1)
No. 1 & Appearance		**1150**	**1320**	**1440**	**1,200,000**	
No. 2		**960**	**1090**	**1190**	**1,100,000**	
No. 3		**560**	**630**	**690**	**900,000**	
BLACK COTTONWOOD (Surfaced dry or surfaced green)						
Select Structural	2x4	1200	1380	1500	1,200,000	National Lumber Grades Authority (A Canadian Agency—See notes 1 and 2)
No. 1 & Appearance		1000	1150	1250	1,200,000	
No. 2		825	950	1030	1,100,000	
No. 3		450	520	560	900,000	
Stud		450	520	560	900,000	
Construction	2x4	600	690	750	900,000	
Standard		325	370	410	900,000	
Utility		150	170	190	900,000	
Select Structural	**2x6 and wider**	**1000**	**1150**	**1250**	**1,200,000**	
No. 1 & Appearance		**875**	**1010**	**1080**	**1,200,000**	
No. 2		**700**	**800**	**875**	**1,100,000**	
No. 3		**425**	**490**	**530**	**900,000**	

Footnotes for Appendix B

1. When 2" lumber is manufactured at a maximum moisture content of 15 percent (grade-marked MC-15) and used in a condition where the moisture content does not exceed 15 percent the design values shown in Table W-1 for "surfaced dry or surfaced green" lumber may be increased eight percent (8%) for allowable unit stress in bending "F_b", and five percent (5%) for Modulus of Elasticity "E"

2. National Lumber Grades Authority is the Canadian rules writing agency responsible for preparation, maintenance and dissemination of a uniform softwood lumber grading rule for all Canadian species.

Top Table

Species and Grade	Size	Normal Duration	Snow Loading	7-Day Loading	Modulus of Elasticity "E"	Grading Rules Agency
EASTERN HEMLOCK–TAMARACK (Surfaced dry or surfaced green)						
Select Structural	2x4	2050	2360	2560	1,300,000	Northeastern Lumber Manufacturers Association
No. 1		1750	2010	2190	1,300,000	
No. 2		1450	1670	1810	1,100,000	Northern Hardwood & Pine Manufacturers Association
No. 3		800	920	1000	1,100,000	
Appearance		1500	1720	1880	1,300,000	(See note 1)
Stud		800	920	1000	1,000,000	
Construction	2x4	1050	1210	1310	1,000,000	
Standard		575	660	720	1,000,000	
Utility		275	320	340	1,000,000	
Select Structural	**2x6 and wider**	**1750**	**2010**	**2190**	**1,300,000**	
No. 1 & Appearance		**1500**	**1720**	**1880**	**1,300,000**	
No. 2		**1200**	**1380**	**1500**	**1,100,000**	
No. 3		**725**	**830**	**910**	**1,000,000**	
EASTERN HEMLOCK–TAMARACK (NORTH) (Surfaced dry or surfaced green)						
Select Structural	2x4	2050	2360	2560	1,300,000	Nat'l Lumber Grades Auth. (A Canadian Agency—See notes 1 and 2)
No. 1		1750	2010	2190	1,300,000	
No. 2		1450	1670	1810	1,100,000	
No. 3		800	920	1000	1,100,000	
Appearance		1500	1720	1880	1,300,000	
Stud		800	920	1000	1,000,000	
Construction	2x4	1050	1210	1310	1,000,000	
Standard		575	660	720	1,000,000	
Utility		275	320	340	1,000,000	
Select Structural	**2x6 and wider**	**1750**	**2010**	**2190**	**1,300,000**	
No. 1 & Appearance		**1500**	**1720**	**1880**	**1,300,000**	
No. 2		**1200**	**1380**	**1500**	**1,100,000**	
No. 3		**725**	**830**	**910**	**1,000,000**	
EASTERN SPRUCE (Surfaced dry or surfaced green)						
Select Structural	2x4	1750	2010	2190	1,400,000	Northeastern Lumber Manufacturers Association
No. 1		1500	1720	1880	1,400,000	
No. 2		1250	1440	1560	1,200,000	Northern Hardwood & Pine Manufacturers Association
No. 3		675	780	840	1,100,000	
Stud		675	780	840	1,100,000	(See note 1)
Construction	2x4	875	1010	1090	1,100,000	
Standard		500	580	620	1,100,000	
Utility		225	260	280	1,100,000	
Select Structural	**2x6 and wider**	**1500**	**1720**	**1880**	**1,400,000**	
No. 1 & Appearance		**1250**	**1440**	**1560**	**1,400,000**	
No. 2		**1000**	**1150**	**1250**	**1,200,000**	
No. 3		**600**	**690**	**750**	**1,100,000**	
EASTERN WHITE PINE (Surfaced dry or surfaced green)						
Select Structural	2x4	1550	1780	1940	1,200,000	Northeastern Lumber Manufacturers Association
No. 1		1350	1550	1690	1,200,000	
No. 2		1100	1260	1370	1,100,000	NeLMA and NHPMA
No. 3		600	690	750	1,000,000	(See note 1)
Construction	2x4	800	920	1000	1,000,000	
Standard		450	520	560	1,000,000	
Utility		200	230	250	1,000,000	
Stud		600	690	750	1,000,000	
Select Structural	**2x6 and wider**	**1350**	**1550**	**1690**	**1,200,000**	
No. 1 & Appearance		**1150**	**1320**	**1440**	**1,200,000**	
No. 2		**950**	**1090**	**1190**	**1,100,000**	
No. 3		**550**	**630**	**690**	**1,000,000**	

Column group heading: **Allowable Unit Stress in Bending "Fb"**

Bottom Table

Species and Grade	Size	Normal Duration	Snow Loading	7-Day Loading	Modulus of Elasticity "E"	Grading Rules Agency
DOUGLAS FIR–LARCH (Surfaced dry or surfaced green)						
Dense Select Structural		2800	3220	3500	1,900,000	Western Wood Products Association (See note 1)
Select Structural		2400	2760	3000	1,800,000	
Dense No. 1		2400	2760	3000	1,900,000	
No. 1 & Appearance		2050	2360	2560	1,800,000	West Coast Lumber Inspection Bureau
Dense No. 2	2x4	1950	2240	2440	1,700,000	
No. 2		1650	1900	2060	1,700,000	
No. 3		925	1060	1160	1,500,000	
Stud		925	1060	1160	1,500,000	
Construction	2x4	1200	1380	1500	1,500,000	
Standard		675	780	840	1,500,000	
Utility		325	370	410	1,500,000	
Dense Select Structural		**2400**	**2760**	**3000**	**1,900,000**	
Select Structural		**2050**	**2360**	**2560**	**1,800,000**	
Dense No. 1		**2050**	**2360**	**2560**	**1,900,000**	
No. 1 & Appearance		**1750**	**2010**	**2190**	**1,800,000**	
Dense No. 2	**2x6 and wider**	**1700**	**1960**	**2120**	**1,700,000**	
No. 2		**1450**	**1670**	**1810**	**1,700,000**	
No. 3		**850**	**980**	**1060**	**1,500,000**	
DOUGLAS FIR–LARCH (NORTH) (Surfaced dry or surfaced green)						
Dense Select Structural		2800	3220	3500	1,900,000	Nat'l Lumber Grades Auth. (A Canadian Agency—See notes 1 and 2)
Select Structural		2400	2760	3000	1,800,000	
Dense No. 1		2400	2760	3000	1,900,000	
No. 1 & Appearance		2050	2360	2560	1,800,000	
Dense No. 2	2x4	1950	2240	2440	1,700,000	
No. 2		1650	1900	2060	1,700,000	
No. 3		925	1060	1160	1,500,000	
Stud		925	1060	1160	1,500,000	
Construction	2x4	1200	1380	1500	1,500,000	
Standard		675	780	840	1,500,000	
Utility		325	370	410	1,500,000	
Dense Select Structural		**2400**	**2760**	**3000**	**1,900,000**	
Select Structural		**2050**	**2360**	**2560**	**1,800,000**	
Dense No. 1		**2050**	**2360**	**2560**	**1,900,000**	
No. 1 & Appearance		**1750**	**2010**	**2190**	**1,800,000**	
Dense No. 2	**2x6 and wider**	**1700**	**1960**	**2120**	**1,700,000**	
No. 2		**1450**	**1670**	**1810**	**1,700,000**	
No. 3		**850**	**980**	**1060**	**1,500,000**	
DOUGLAS FIR SOUTH (Surfaced dry or surfaced green)						
Select Structural		2300	2640	2880	1,400,000	Western Wood Products Association (See note 1)
No. 1 & Appearance		1950	2240	2440	1,400,000	
No. 2	2x4	1600	1840	2000	1,300,000	
No. 3		875	1010	1090	1,100,000	
Stud		875	1010	1090	1,100,000	
Construction	2x4	1150	1320	1440	1,100,000	
Standard		650	750	810	1,100,000	
Utility		300	340	380	1,100,000	
Select Structural		**1950**	**2240**	**2440**	**1,400,000**	
No. 1 & Appearance		**1650**	**1900**	**2060**	**1,400,000**	
No. 2	**2x6 and wider**	**1350**	**1550**	**1690**	**1,300,000**	
No. 3		**800**	**920**	**1000**	**1,100,000**	

Column group heading: **Allowable Unit Stress in Bending "Fb"**

Top Table

Species and Grade	Size	Normal Duration	Snow Loading	7-Day Loading	Modulus of Elasticity "E"	Grading Rules Agency
HEM–FIR (NORTH) (Surfaced dry or surfaced green)						
Select Structural		1900	2180	2380	1,500,000	Nat'l. Lumber Grades Auth. (A Canadian Agency— See notes 1 and 2)
No. 1 & Appearance		1600	1840	2000	1,500,000	
No. 2	2x4	1300	1500	1620	1,400,000	
No. 3		725	830	910	1,200,000	
Stud		725	830	910	1,200,000	
Construction		950	1090	1190	1,200,000	
Standard	2x4	525	600	660	1,200,000	
Utility		250	290	310	1,200,000	
Select Structural		1650	1900	2060	1,500,000	
No. 1 & Appearance	2x6 and wider	1400	1610	1750	1,500,000	
No. 2		1150	1320	1440	1,400,000	
No. 3		675	780	840	1,200,000	
IDAHO WHITE PINE (Surfaced dry or surfaced green)						
Select Structural		1550	1780	1940	1,400,000	Western Wood Products Association (See note 1)
No. 1 & Appearance		1300	1500	1620	1,400,000	
No. 2	2x4	1050	1210	1310	1,300,000	
No. 3		600	690	750	1,200,000	
Stud		600	690	750	1,200,000	
Construction		775	890	970	1,200,000	
Standard	2x4	425	490	530	1,200,000	
Utility		200	230	250	1,200,000	
Select Structural		1300	1500	1620	1,400,000	
No. 1 & Appearance	2x6 and wider	1100	1260	1380	1,400,000	
No. 2		925	1060	1160	1,300,000	
No. 3		550	630	690	1,200,000	
LODGEPOLE PINE (Surfaced dry or surfaced green)						
Select Structural		1750	2010	2190	1,300,000	Western Wood Products Association (See note 1)
No. 1 & Appearance		1500	1720	1880	1,300,000	
No. 2	2x4	1200	1380	1500	1,200,000	
No. 3		675	780	840	1,000,000	
Stud		675	780	840	1,000,000	
Construction		875	1010	1090	1,000,000	
Standard	2x4	500	580	620	1,000,000	
Utility		225	260	280	1,000,000	
Select Structural		1500	1720	1880	1,300,000	
No. 1 & Appearance	2x6 and wider	1300	1500	1620	1,300,000	
No. 2		1050	1210	1310	1,200,000	
No. 3		625	720	780	1,000,000	
MOUNTAIN HEMLOCK (Surfaced dry or surfaced green)						
Select Structural		2000	2300	2500	1,300,000	Western Wood Products Association (See note 1)
No. 1 & Appearance		1700	1960	2120	1,300,000	
No. 2	2x4	1400	1610	1750	1,100,000	
No. 3		775	890	970	1,000,000	
Stud		775	890	970	1,000,000	
Construction		1000	1150	1250	1,000,000	
Standard	2x4	575	660	720	1,000,000	
Utility		275	320	340	1,000,000	
Select Structural		1700	1960	2120	1,300,000	West Coast Lumber Inspection Bureau
No. 1 & Appearance	2x6 and wider	1450	1670	1810	1,300,000	
No. 2		1200	1380	1500	1,100,000	
No. 3		700	800	880	1,000,000	

Bottom Table

Species and Grade	Size	Normal Duration	Snow Loading	7-Day Loading	Modulus of Elasticity "E"	Grading Rules Agency
EASTERN WHITE PINE (NORTH) (Surfaced dry or surfaced green)						
Select Structural		1550		1940	1,200,000	Nat'l. Lumber Grades Auth. (A Canadian Agency— See notes 1 and 2)
No. 1 & Appearance	2x4	1350		1690	1,200,000	
No. 2		1100		1380	1,100,000	
No. 3		600		750	1,000,000	
Stud						
Construction						
Standard	2x4					
Utility						
Select Structural		1350	1550	1690	1,200,000	
No. 1 & Appearance	2x6 and wider	1150	1320	1440	1,200,000	
No. 2		950	1090	1190	1,100,000	
No. 3		550	630	690	1,000,000	
EASTERN WOODS (Surfaced dry or surfaced green)						
Select Structural		1500	1720	1880	1,100,000	Northeastern Lumber Manufacturers Association; Northern Hardwood & Pine Manufacturers Association (See note 1)
No. 1		1300	1500	1620	1,100,000	
No. 2	2x4	1050	1210	1310	1,000,000	
No. 3		575	660	720	900,000	
Stud		575	660	720	900,000	
Construction		750	860	940	900,000	
Standard	2x4	425	490	530	900,000	
Utility		200	230	250	900,000	
Appearance	2x4	1300	1500	1620	1,100,000	Northern Hardwood & Pine Manufacturers Association (See note 1)
Select Structural		1300	1500	1620	1,100,000	
No. 1 & Appearance	2x6 and wider	1100	1260	1380	1,100,000	
No. 2		900	1040	1120	1,000,000	
No. 3		525	600	660	900,000	
ENGELMANN SPRUCE (ENGELMANN SPRUCE–LODGEPOLE PINE) (Surfaced dry or surfaced green)						
Select Structural		1550	1780	1940	1,300,000	Western Wood Products Association (See note 1)
No. 1 & Appearance		1350	1550	1690	1,300,000	
No. 2	2x4	1100	1260	1380	1,100,000	
No. 3		600	690	750	1,000,000	
Stud		600	690	750	1,000,000	
Construction		800	920	1000	1,000,000	
Standard	2x4	450	520	560	1,000,000	
Utility		200	230	250	1,000,000	
Select Structural		1350	1560	1690	1,300,000	
No. 1 & Appearance	2x6 and wider	1150	1320	1440	1,300,000	
No. 2		950	1090	1190	1,100,000	
No. 3		550	630	690	1,000,000	
HEM–FIR (Surfaced dry or surfaced green)						
Select Structural		1900	2180	2380	1,500,000	Western Wood Products Association (See note 1)
No. 1 & Appearance		1600	1840	2000	1,500,000	
No. 2	2x4	1350	1550	1690	1,400,000	
No. 3		725	830	910	1,200,000	
Stud		725	830	910	1,200,000	
Construction		975	1120	1220	1,200,000	
Standard	2x4	550	630	690	1,200,000	
Utility		250	290	310	1,200,000	
Select Structural		1650	1900	2060	1,500,000	West Coast Lumber Inspection Bureau
No. 1 & Appearance	2x6 and wider	1450	1610	1750	1,500,000	
No. 2		1150	1320	1440	1,400,000	
No. 3		675	780	840	1,200,000	

Top Table

Species and Grade	Size	Allowable Unit Stress in Bending "Fb"			Modulus of Elasticity "E"	Grading Rules Agency
		Normal Duration	Snow Loading	7-Day Loading		
NORTHERN WHITE CEDAR (Surfaced dry or surfaced green)						
Select Structural		1350	1550	1690	800,000	Northeastern Lumber Manufacturers Association (See note 1)
No. 1	2x4	1150	1320	1440	800,000	
No. 2		950	1090	1190	700,000	
No. 3		525	600	660	600,000	
Appearance		1000	1150	1250	800,000	
Stud		525	600	660	600,000	
Construction	2x4	675	780	840	600,000	
Standard		375	430	470	600,000	
Utility		175	200	220	600,000	
Select Structural	2x6 and wider	1150	1320	1440	800,000	
No. 1 & Appearance		1000	1150	1250	800,000	
No. 2		825	950	1030	700,000	
No. 3		475	550	590	600,000	
PONDEROSA PINE (Surfaced dry or surfaced green)						
Select Structural		1650	1900	2060	1,200,000	Nat'l. Lumber Grades Auth. (A Canadian Agency—See notes 1 and 2)
No. 1 & Appearance	2x4	1400	1610	1750	1,200,000	
No. 2		1150	1320	1440	1,100,000	
No. 3		625	720	780	1,000,000	
Stud		625	720	780	1,000,000	
Construction	2x4	825	950	1030	1,000,000	
Standard		450	520	560	1,000,000	
Utility		225	260	280	1,000,000	
Select Structural	2x6 and wider	1400	1610	1750	1,200,000	
No. 1 & Appearance		1200	1380	1500	1,200,000	
No. 2		975	1120	1220	1,100,000	
No. 3		575	660	720	1,000,000	
PONDEROSA PINE–SUGAR PINE (PONDEROSA PINE–LODGEPOLE PINE) (Surfaced dry or surfaced green)						
Select Structural		1650	1900	2060	1,200,000	Western Wood Products Association (See note 1)
No. 1 & Appearance	2x4	1400	1610	1750	1,200,000	
No. 2		1150	1320	1440	1,100,000	
No. 3		625	720	780	1,000,000	
Stud		625	720	780	1,000,000	
Construction	2x4	825	950	1030	1,000,000	
Standard		450	520	560	1,000,000	
Utility		225	260	280	1,000,000	
Select Structural	2x6 and wider	1400	1610	1750	1,200,000	
No. 1 & Appearance		1200	1380	1500	1,200,000	
No. 2		975	1120	1220	1,100,000	
No. 3		575	660	720	1,000,000	
RED PINE (Surfaced dry or surfaced green)						
Select Structural		1600	1840	2000	1,300,000	Nat'l. Lumber Grades Auth. (A Canadian Agency—See notes 1 and 2)
No. 1 & Appearance	2x4	1350	1550	1690	1,300,000	
No. 2		1100	1270	1380	1,200,000	
No. 3		625	720	780	1,000,000	
Stud		625	720	780	1,000,000	
Construction	2x4	800	920	1000	1,000,000	
Standard		450	520	560	1,000,000	
Utility		225	260	280	1,000,000	
Select Structural	2x6 and wider	1350	1550	1690	1,300,000	
No. 1 & Appearance		1150	1320	1440	1,300,000	
No. 2		950	1090	1190	1,200,000	
No. 3		550	630	690	1,000,000	

Bottom Table

Species and Grade	Size	Allowable Unit Stress in Bending "Fb"			Modulus of Elasticity "E"	Grading Rules Agency
		Normal Duration	Snow Loading	7-Day Loading		
MOUNTAIN HEMLOCK – HEM–FIR (Surfaced dry or surfaced green)						
Select Structural		1900	2180	2380	1,300,000	Western Wood Products Association (See note 1)
No. 1 & Appearance	2x4	1600	1840	2000	1,300,000	
No. 2		1350	1550	1690	1,100,000	
No. 3		725	830	910	1,000,000	
Stud		725	830	910	1,000,000	
Construction	2x4	975	1120	1220	1,000,000	
Standard		550	630	690	1,000,000	
Utility		250	290	310	1,000,000	
Select Structural	2x6 and wider	1650	1900	2060	1,300,000	
No. 1 & Appearance		1400	1610	1750	1,300,000	
No. 2		1150	1320	1440	1,100,000	
No. 3		675	780	840	1,000,000	
NORTHERN ASPEN (Surfaced dry or surfaced green)						
Select Structural		1500	1720	1870	1,400,000	National Lumber Grades Authority (A Canadian Agency - See notes 1 and 2)
No. 1 & Appearance	2x4	1250	1440	1560	1,400,000	
No. 2		1050	1210	1310	1,200,000	
No. 3		575	660	720	1,100,000	
Stud		575	660	720	1,100,000	
Construction	2x4	750	860	940	1,100,000	
Standard		425	490	530	1,100,000	
Utility		200	230	250	1,100,000	
Select Structural	2x6 and wider	1250	1440	1560	1,400,000	
No. 1 & Appearance		1100	1260	1370	1,400,000	
No. 2		900	1030	1120	1,200,000	
No. 3		525	600	660	1,100,000	
NORTHERN PINE (Surfaced dry or surfaced green)						
Select Structural		1850	2130	2310	1,400,000	Northeastern Lumber Manufacturers Association / Northern Hardwood & Pine Manufacturers Association (See note 1)
No. 1	2x4	1600	1840	2000	1,400,000	
No. 2		1300	1500	1620	1,300,000	
No. 3		725	830	910	1,100,000	
Appearance		1400	1610	1750	1,400,000	
Stud		725	830	910	1,100,000	
Construction	2x4	950	1090	1190	1,100,000	
Standard		525	600	660	1,100,000	
Utility		250	290	310	1,100,000	
Select Structural	2x6 and wider	1600	1840	2000	1,400,000	
No. 1 & Appearance		1400	1610	1750	1,400,000	
No. 2		1100	1260	1380	1,300,000	
No. 3		650	750	810	1,100,000	
NORTHERN SPECIES (Surfaced dry or surfaced green)						
Select Structural		1550	1780	1940	1,100,000	National Lumber Grades Authority (A Canadian Agency - See notes 1 and 2)
No. 1 & Appearance	2x4	1300	1490	1620	1,100,000	
No. 2		1050	1210	1310	1,000,000	
No. 3		600	690	750	900,000	
Stud		600	690	750	900,000	
Construction	2x4	775	890	970	900,000	
Standard		425	490	530	900,000	
Utility		200	230	250	900,000	
Select Structural	2x6 and wider	1300	1490	1620	1,100,000	
No. 1 & Appearance		1150	1320	1440	1,100,000	
No. 2		925	1060	1160	1,000,000	
No. 3		550	630	690	900,000	

Top Table

Species and Grade	Size	Allowable Unit Stress in Bending "Fb"			Modulus of Elasticity "E"	Grading Rules Agency
		Normal Duration	Snow Loading	7-Day Loading		
SPRUCE–PINE–FIR (Surfaced dry or surfaced green)						
Select Structural	2x4	1650	1900	2060	1,500,000	Nat'l. Lumber Grades Auth. (A Canadian Agency—See notes 1 and 2)
No. 1 & Appearance		1400	1610	1750	1,500,000	
No. 2		1150	1320	1440	1,300,000	
No. 3		650	750	810	1,200,000	
Stud		650	750	810	1,200,000	
Construction	2x4	850	980	1060	1,200,000	
Standard		475	550	590	1,200,000	
Utility		225	260	280	1,200,000	
Select Structural	2x6 and wider	1450	1670	1810	1,500,000	
No. 1 & Appearance		1200	1380	1500	1,500,000	
No. 2		1000	1150	1250	1,300,000	
No. 3		575	660	720	1,200,000	
WESTERN CEDARS (Surfaced dry or surfaced green)						
Select Structural	2x4	1750	2010	2190	1,100,000	Western Wood Products Association (See note 1) / West Coast Lumber Inspection Bureau
No. 1 & Appearance		1500	1720	1880	1,100,000	
No. 2		1200	1380	1500	1,000,000	
No. 3		675	780	840	900,000	
Stud		675	780	840	900,000	
Construction	2x4	875	1010	1090	900,000	
Standard		500	580	620	900,000	
Utility		225	260	280	900,000	
Select Structural	2x6 and wider	1500	1720	1880	1,100,000	
No. 1 & Appearance		1300	1500	1620	1,000,000	
No. 2		1050	1210	1310	1,000,000	
No. 3		625	720	780	900,000	
WESTERN CEDARS (NORTH) (Surfaced dry or surfaced green)						
Select Structural	2x4	1700	1960	2120	1,100,000	Nat'l. Lumber Grades Auth. (A Canadian Agency—See notes 1 and 2)
No. 1 & Appearance		1450	1670	1810	1,100,000	
No. 2		1200	1380	1500	1,000,000	
No. 3		650	750	810	900,000	
Stud		650	750	810	900,000	
Construction	2x4	850	980	1060	900,000	
Standard		475	550	590	900,000	
Utility		225	260	280	900,000	
Select Structural	2x6 and wider	1450	1670	1810	1,100,000	
No. 1 & Appearance		1250	1440	1560	1,100,000	
No. 2		1000	1150	1250	1,000,000	
No. 3		600	690	750	900,000	
WESTERN HEMLOCK (Surfaced dry or surfaced green)						
Select Structural	2x4	2100	2410	2620	1,600,000	Western Wood Products Association (See note 1) / West Coast Lumber Inspection Bureau
No. 1 & Appearance		1800	2070	2250	1,600,000	
No. 2		1450	1670	1810	1,400,000	
No. 3		800	920	1000	1,300,000	
Stud		800	920	1000	1,300,000	
Construction	2x4	1050	1210	1310	1,300,000	
Standard		600	690	750	1,300,000	
Utility		275	320	340	1,300,000	
Select Structural	2x6 and wider	1800	2070	2250	1,600,000	
No. 1 & Appearance		1550	1780	1940	1,600,000	
No. 2		1250	1440	1560	1,400,000	
No. 3		750	860	940	1,400,000	

Bottom Table

Species and Grade	Size	Allowable Unit Stress in Bending "Fb"			Modulus of Elasticity "E"	Grading Rules Agency
		Normal Duration	Snow Loading	7-Day Loading		
SITKA SPRUCE (Surfaced dry or surfaced green)						
Select Structural	2x4	1800	2070	2250	1,500,000	West Coast Lumber Inspection Bureau
No. 1 & Appearance		1550	1780	1940	1,500,000	
No. 2		1250	1440	1560	1,300,000	
No. 3		700	800	880	1,200,000	
Stud		700	800	880	1,200,000	
Construction	2x4	925	1060	1160	1,200,000	
Standard		500	580	620	1,200,000	
Utility		250	290	310	1,200,000	
Select Structural	2x6 and wider	1550	1780	1940	1,500,000	
No. 1 & Appearance		1300	1500	1620	1,500,000	
No. 2		1050	1210	1310	1,300,000	
No. 3		600	690	750	1,200,000	
SOUTHERN PINE (Surfaced dry)						
Select Structural	2x4	2400	2760	3000	1,800,000	Southern Pine Inspection Bureau
Dense Select Structural		2800	3220	3500	1,900,000	
No. 1		2000	2300	2500	1,800,000	
No. 1 Dense		2350	2700	2940	1,900,000	
No. 2	2x4	1450	1670	1810	1,400,000	
No. 2 MG		1650	1900	2060	1,600,000	
No. 2 Dense		1950	2240	2440	1,700,000	
No. 3		950	1090	1190	1,400,000	
No. 3 Dense		1100	1260	1380	1,500,000	
Stud		950	1090	1190	1,400,000	
Construction	2x4	1200	1380	1500	1,400,000	
Standard		700	800	880	1,400,000	
Utility		325	370	410	1,400,000	
Select Structural	2x6 and wider	2050	2360	2560	1,800,000	
Dense Select Structural		2400	2760	3000	1,900,000	
No. 1		1750	2010	2190	1,800,000	
No. 1 Dense		2050	2360	2560	1,900,000	
No. 2	2x6 and wider	1200	1380	1500	1,400,000	
No. 2 MG		1450	1670	1810	1,600,000	
No. 2 Dense		1650	1900	2060	1,700,000	
No. 3		825	950	1030	1,400,000	
No. 3 Dense		975	1120	1220	1,500,000	
SOUTHERN PINE (Surfaced at 15 percent moisture content—KD)						
Select Structural	2x4	2600	2990	3250	1,900,000	Southern Pine Inspection Bureau
Dense Select Structural		3050	3510	3810	2,000,000	
No. 1		2200	2530	2750	1,900,000	
No. 1 Dense		2600	2990	3250	2,000,000	
No. 2	2x4	1550	1780	1940	1,500,000	
No. 2 MG		1800	2070	2250	1,700,000	
No. 2 Dense		2150	2470	2690	1,800,000	
No. 3		1000	1150	1250	1,500,000	
No. 3 Dense		1200	1380	1500	1,600,000	
Stud		1000	1150	1250	1,500,000	
Construction	2x4	1300	1500	1620	1,500,000	
Standard		750	860	940	1,500,000	
Utility		350	400	440	1,500,000	
Select Structural	2x6 and wider	2250	2590	2810	1,900,000	
Dense Select Structural		2600	2990	3250	2,000,000	
No. 1		1900	2180	2380	1,900,000	
No. 1 Dense		2200	2530	2750	2,000,000	
No. 2	2x6 and wider	1300	1500	1620	1,500,000	
No. 2 MG		1550	1780	1940	1,700,000	
No. 2 Dense		1800	2070	2250	1,800,000	
No. 3		900	1040	1120	1,500,000	
No. 3 Dense		1050	1210	1310	1,600,000	

Working Stresses For Joists And Rafters, Machine Stress Rated

These "F_b" values are for use where repetitive members are spaced not more than 24 inches. For wider spacing, the "F_b" values should be reduced 13 percent. Values apply at 19 percent maximum moisture content in use.

Grade Designation	Size Classification	Normal Duration	Snow Loading	7-Day Loading	Modulus of Elasticity "E"	Species
WESTERN WOOD PRODUCTS ASSOCIATION						
2700f-2.2E	Machine stress rated lumber 2x4 and wider	3100	3560	3880	2,200,000	
2400f-2.0E		2750	3160	3440	2,000,000	
2100f-1.8E		2400	2760	3000	1,800,000	
1800f-1.6E		2050	2360	2560	1,600,000	
1650f-1.5E		1900	2180	2380	1,500,000	
1500f-1.4E		1750	2010	2190	1,400,000	
1200f-1.2E		1400	1610	1750	1,200,000	All Western Species
1800f-2.1E	Machine stress rated joists 2x4 and wider	2050	2360	2560	2,100,000	
1350f-1.8E		1550	1780	1940	1,800,000	
1200f-1.5E		1400	1610	1750	1,500,000	
900f-1.2E		1050	1210	1310	1,200,000	
900f-1.0E		1050	1210	1310	1,000,000	
WEST COAST LUMBER INSPECTION BUREAU						
2700f-2.2E	Machine stress rated lumber 2x4 and wider	3100	3560	3880	2,200,000	
2400f-2.0E		2750	3160	3440	2,000,000	
2100f-1.8E		2400	2760	3000	1,800,000	
1800f-1.6E		2050	2360	2560	1,600,000	
1650f-1.5E		1900	2180	2380	1,500,000	
1500f-1.4E		1750	2010	2190	1,400,000	
1450f-1.3E		1650	1900	2060	1,300,000	All West Coast Species
1200f-1.2E		1400	1610	1750	1,200,000	
900f-1.0E		1050	1210	1310	1,000,000	
1800f-2.1E	Machine stress rated joists 2x6 and wider	2050	2360	2560	2,100,000	
1500f-1.8E		1750	2010	2190	1,800,000	
1200f-1.5E		1400	1610	1750	1,500,000	
900f-1.2E		1050	1210	1310	1,200,000	
900f-1.0E		1050	1210	1310	1,000,000	
SOUTHERN PINE INSPECTION BUREAU						
3300f-2.6E	Machine stress rated lumber 2x4 and wider	3800	4370	4750	2,600,000	
3000f-2.4E		3450	3970	4310	2,400,000	
2700f-2.2E		3100	3560	3880	2,200,000	
2400f-2.0E		2750	3160	3440	2,000,000	
2100f-1.8E		2400	2760	3000	1,800,000	
1800f-1.6E		2050	2360	2560	1,600,000	
1650f-1.5E		1900	2180	2380	1,500,000	
1500f-1.4E		1750	2010	2190	1,400,000	Southern Pine
1200f-1.2E		1400	1610	1750	1,200,000	
1800f-2.1E	Machine stress rated joists 2x4 and wider	2050	2360	2560	2,100,000	
1350f-1.8E		1550	1780	1940	1,800,000	
1200f-1.5E		1400	1610	1750	1,500,000	
900f-1.2E		1050	1210	1310	1,200,000	
900f-1.0E		1050	1210	1310	1,000,000	

Species and Grade	Size	Normal Duration	Snow Loading	7-Day Loading	Modulus of Elasticity "E"	Grading Rules Agency
WESTERN WHITE PINE (Surfaced dry or surfaced green)						
Select Structural	2x4	1550	1780	1940	1,400,000	
No. 1 & Appearance		1300	1500	1630	1,400,000	
No. 2		1050	1210	1310	1,300,000	
No. 3		600	690	750	1,200,000	
Stud		600	690	750	1,200,000	Nat'l. Lumber Grades Auth. (A Canadian Agency— See notes 1 and 2)
Construction	2x4	775	890	970	1,200,000	
Standard		425	490	530	1,200,000	
Utility		200	230	250	1,200,000	
Select Structural	2x6 and wider	1300	1500	1630	1,400,000	
No. 1 & Appearance		1150	1320	1440	1,400,000	
No. 2		925	1060	1160	1,300,000	
No. 3		550	630	690	1,200,000	
WHITE WOODS (WESTERN WOODS) (Surfaced dry or surfaced green)						
Select Structural	2x4	1550	1780	1940	1,100,000	
No. 1 & Appearance		1300	1500	1620	1,100,000	
No. 2		1050	1210	1310	900,000	
No. 3		600	690	750	900,"00	
Stud		600	690	750	900,000	Western Wood Products Association (See note 1)
Construction	2x4	775	890	970	900,000	
Standard		425	490	530	900,000	
Utility		200	230	250	900,000	
Select Structural	2x6 and wider	1300	1500	1620	1,100,000	
No. 1 & Appearance		1100	1260	1380	1,100,000	
No. 2		925	1060	1160	1,000,000	
No. 3		550	630	690	900,000	

Species of timber	Grade	Extreme fiber in bending — Joist and plank sizes; 4 inches and less in thickness	Extreme fiber in bending — Beam and stringer sizes; 5 inches and thicker	Modulus of elasticity
Douglas fir, coast region	Dense superstructural	2,000	2,000	1,600,000
	Superstructural and dense structural	1,800	1,800	1,600,000
	Structural	1,600	1,600	1,600,000
Douglas fir, inland empire	Dense superstructural*	2,000	2,000	1,600,000
	Dense structural*	1,800	1,800	1,600,000
	No. 1 common dimension and timbers	1,135	1,135	1,500,000
Larch, western	No. 1 common dimension and timbers	1,135	1,135	1,300,000
Pine, southern yellow	Extra dense select structural	2,300	2,300	1,600,000
	Select structural	2,000	2,000	1,600,000
	Extra dense select heart	2,000	2,000	1,600,000
	Dense heart	1,800	1,800	1,600,000
	Structural square edge and sound	1,600	1,600	1,600,000
	Dense No. 1 common	1,200	1,200	1,600,000
Redwood	Superstructural	2,133	1,707	1,200,000
	Prime structural	1,707	1,494	1,200,000
	Select structural	1,280	1,322	1,200,000
	Heart structural	1,024	1,150	1,200,000

*When graded the same as corresponding grade of coat region Douglas fir.

Allowable unit stresses for structural timber

Index

Practical References for Builders

National Construction Estimator

Current building costs for residential, commercial, and industrial construction. Estimated prices for every common building material. Manhours, recommended crew, and labor cost for installation. Includes *Estimate Writer*, an electronic version of the book on computer disk, with a stand-alone estimating program — free on 5¼" high density (1.2Mb) disk. The National Construction Estimator and *Estimate Writer* on 1.2Mb disk cost $26.50. (Add $10 if you want *Estimate Writer* on 5¼" double density 360K disks or 3½" 720K disks.) **576 pages, 8½ x 11, $26.50. Revised annually**

Finish Carpentry

The time-saving methods and proven shortcuts you need to do first class finish work on any job: cornices and rakes, gutters and downspouts, wood shingle roofing, asphalt, asbestos and built-up roofing, prefabricated windows, door bucks and frames, door trim, siding, wallboard, lath and plaster, stairs and railings, cabinets, joinery, and wood flooring. **192 pages, 8½ x 11, $15.25**

Profits in Buying & Renovating Homes

Step-by-step instructions for selecting, repairing, improving, and selling highly profitable "fixer-uppers." Shows which price ranges offer the highest profit-to-investment ratios, which neighborhoods offer the best return, practical directions for repairs, and tips on dealing with buyers, sellers, and real estate agents. Shows you how to determine your profit before you buy, what "bargains" to avoid, and how to make simple, profitable, inexpensive upgrades. **304 pages, 8½ x 11, $19.75**

Wood-Frame House Construction

Step-by-step construction details, from the layout of the outer walls, excavation and formwork, to finish carpentry and painting, with clear illustrations and explanations. Everything you need to know about framing, roofing, siding, insulation and vapor barrier, interior finishing, floor coverings, and stairs — complete step-by-step "how to" information on building a frame house. **240 pages, 8½ x 11, $14.25. Revised edition**

Carpentry for Residential Construction

How to do professional quality carpentry work in residences. Illustrated instructions on everything from setting batterboards to framing floors and walls, installing floor, wall, and roof sheathing, and applying roofing. Covers finish carpentry: installing each type of cornice, frieze, lookout, ledger, fascia, and soffit; hanging windows and doors; installing siding, drywall, and trim. Each job description includes tools and materials needed, estimated manhours required, and a step-by-step guide to each part of the task. **400 pages, 5½ x 8½, $19.75**

Carpentry in Commercial Construction

Covers forming, framing, exteriors, interior finish, and cabinet installation in commercial buildings: how to design and build concrete forms, select lumber dimensions, what grades and species to use for a design load, how to select and install materials based on their fire rating or sound-transmission characteristics, and plan and organize a job efficiently. Loaded with illustrations, tables, charts, and diagrams. **272 pages, 5½ x 8½, $19.00**

Handbook of Construction Contracting

Volume 1: Everything you need to know to start and run your construction business; the pros and cons of each type of contracting, the records you'll need to keep, and how to read and understand house plans and specs so you find any problems before the actual work begins. All aspects of construction are covered in detail, including all-weather wood foundations, practical math for the job site, and elementary surveying. **416 pages, 8½ x 11, $24.75**
Volume 2: Everything you need to know to keep your construction business profitable; different methods of estimating, keeping and controlling costs, estimating excavation, concrete, masonry, rough carpentry, roof covering, insulation, doors and windows, exterior finishes, specialty finishes, scheduling work flow, managing workers, advertising and sales, spec building and land development, and selecting the best legal structure for your business. **320 pages, 8½ x 11, $24.75**

National Repair & Remodeling Estimator

The complete pricing guide for dwelling reconstruction costs. Reliable, specific data you can apply on every repair and remodeling job. Up-to-date material costs and labor figures based on thousands of jobs across the country. Provides

recommended crew sizes; average production rates; exact material, equipment, and labor costs; a total unit cost and a total price including overhead and profit. Separate listings for high- and low-volume builders, so prices shown are accurate for any size business. Estimating tips specific to repair and remodeling work to make your bids complete, realistic, and profitable. *New this year! The complete book on a disk with a built-in estimating program for an IBM-compatible hard-drive computer. FREE on a 5-1/4" high-density (1.2 Mb) disk when you buy the book.* (Add $10 for *Repair & Remodeling Estimate Writer* on extra 5¼" double density 360K disks or 3½" 720K disks.) **288 pages, 11 x 8½, $29.50. Revised annually.**

Building Layout

Shows how to use a transit to locate a building correctly on the lot, plan proper grades with minimum excavation, find utility lines and easements, establish correct elevations, lay out accurate foundations, and set correct floor heights. Explains how to plan sewer connections, level a foundation that's out of level, use a story pole and batterboards, work on steep sites, and minimize excavation costs. **240 pages, 5½ x 8½, $11.75**

10 Day Money Back Guarantee

Craftsman Book Company / 6058 Corte del Cedro / P.O. Box 6500 / Carlsbad, CA 92018

☎

In a hurry?
We accept phone orders charged to your MasterCard, Visa or American Express
Call 1-800-829-8123
FAX (619) 438-0398

Include a check with your order and we pay shipping

Total enclosed_____ (In Calif. add 7.25% tax)

Use your ☐ Visa ☐ MasterCard or ☐ American Express

Card # _____

Exp. date_____Initials_____

Call 1-800-829-8123 for a FREE Full Color Catalog with over 100 Titles

Name _____

Company _____

Address _____

City/State/Zip _____

☐ 95.00 Audiotape: Construction Field Supervision
☐ 65.00 Audiotape: Estimating Electrical Work
☐ 65.00 Audiotape: Estimating Remodeling
☐ 19.95 Audiotape: Plumbers Examination
☐ 22.00 Basic Plumbing with Illustrations
☐ 30.00 Berger Building Cost File
☐ 11.25 Blueprint Reading for Building Trades
☐ 19.75 Bookkeeping for Builders
☐ 24.95 Builder's & Estimator's Formbook
☐ 24.95 Builder's Comprehensive Dictionary
☐ 20.00 Builder's Guide to Accounting, Revised
☐ 15.25 Builder's Guide to Construction Financing
☐ 15.50 Builder's Office Manual Revised
☐ 16.50 Building Cost Manual
☐ 11.75 Building Layout
☐ 22.00 Cabinetmaking: Design to Finish
☐ 25.50 Carpentry Estimating
☐ 19.75 Carpentry for Residential Construction
☐ 19.00 Carpentry in Commercial Construction
☐ 16.25 Carpentry Layout
☐ 17.75 Computers: The Builder's New Tool
☐ 14.50 Concrete and Formwork
☐ 20.50 Concrete Construction & Estimating
☐ 26.00 Construction Estimating Reference Data
☐ 22.00 Construction Superintending
☐ 19.25 Construction Surveying & Layout
☐ 19.00 Contractor's Growth & Profit Guide
☐ 24.25 Contractor's Guide to Building Code
☐ 16.75 Contractor's Survival Manual
☐ 16.50 Contractor's Year-Round Tax Guide

☐ 15.75 Cost Records for Construction Estimating
☐ 9.50 Dial-A-Length Rafterule
☐ 18.25 Drywall Contracting
☐ 18.00 Electrical Blueprint Reading Revised
☐ 28.50 Electrical Construction Estimator with
   free *Electrical Estimate Writer* on
   5¼" (1.2Mb) disk. *Add $10 for extra*
   ☐ 5¼" (360K) or ☐ 3½" (720K) disks.
☐ 23.00 Electrician's Exam Preparation Guide
☐ 19.00 Estimating Electrical Construction
☐ 34.95 Estimating Framing Quantities
☐ 17.00 Estimating Home Building Costs
☐ 28.00 Estimating Painting Costs
☐ 22.50 Estimating Plumbing Costs
☐ 21.50 Estimating Tables for Home Building
☐ 22.75 Excavation & Grading Handbook Rev.
☐ 9.25 E-Z Square
☐ 23.25 Fences and Retaining Walls
☐ 15.25 Finish Carpentry
☐ 24.75 Handbook of Construction Contract. Vol. 1
☐ 24.75 Handbook of Construction Contract. Vol. 2
☐ 15.00 Home Wiring: Improve, Extension, Repairs
☐ 17.50 How to Sell Remodeling
☐ 19.50 How to Succeed w/Own Const. Bus.
☐ 24.50 HVAC Contracting
☐ 24.00 Illustrated Guide to Nt'l Electrical Code
☐ 20.25 Manual of Electrical Contracting
☐ 19.75 Manual of Professional Remodeling
☐ 17.25 Masonry & Concrete Construction
☐ 26.50 Masonry Estimating

☐ 26.50 National Construction Estimator with
   free *Estimate Writer* on 5¼" (1.2Mb) disk.
   Add $10 for extra *Estimate Writer on either*
   ☐ 5¼" (360K) or ☐ 3½" (720K) disks.
☐ 29.50 National Repair & Remodel Estimator w/
   free *Repair & Remodeling Estimate Writer*
   on 5¼" (1.2Mb) disk. *Add $10 for extra*
   ☐ 5¼" (360K) or ☐ 3½" (720K) disks.
☐ 19.25 Paint Contractor's Manual
☐ 21.25 Painter's Handbook
☐ 27.50 Painting Cost Guide
☐ 23.50 Pipe & Excavation Contracting
☐ 19.25 Planning Drain, Waste & Vent Systems
☐ 21.00 Plumber's Exam Preparation Guide
☐ 18.00 Plumber's Handbook Revised
☐ 19.75 Profits in Buying & Renovating Homes
☐ 14.25 Rafter Length Manual
☐ 27.00 Remodeler's Handbook
☐ 18.25 Remodeling Contractor's Handbook
☐ 26.25 Remodeling Kitchens & Baths
☐ 11.50 Residential Electrical Design
☐ 16.75 Residential Electrician's Handbook
☐ 18.25 Residential Wiring
☐ 22.00 Roof Framing
☐ 14.00 Roofers Handbook
☐ 17.00 Rough Carpentry
☐ 21.00 Running Your Remodeling Business
☐ 27.00 Spec Builder's Guide
☐ 15.50 Stair Builder's Handbook
☐ 14.25 Wood-Frame House Construction

Mail This Card Today For A FREE Full Color Catalog

Craftsman Book Company / 6058 Corte del Cedro / P.O. Box 6500 / Carlsbad, CA 92018

Over 100 books, videos, and audios at your fingertips with information that can save you time and money. Here you'll find information on carpentry, contracting, estimating, remodeling, electrical work, and plumbing.

All items come with an unconditional 10-day money-back guarantee. If they don't save you money, mail them back for a full refund.

Name _____

Company _____

Address _____

City/State/Zip _____

BUSINESS REPLY MAIL
FIRST CLASS PERMIT NO. 271 CARLSBAD, CA

POSTAGE WILL BE PAID BY ADDRESSEE

Craftsman Book Company
6058 Corte Del Cedro
P.O. Box 6500
Carlsbad, CA 92018-9974

BUSINESS REPLY MAIL
FIRST CLASS PERMIT NO. 271 CARLSBAD, CA

POSTAGE WILL BE PAID BY ADDRESSEE

Craftsman Book Company
6058 Corte Del Cedro
P.O. Box 6500
Carlsbad, CA 92018-9974